Polymer Melt Fracture

Polymer Melt Fracture

Rudy Koopmans
Jaap den Doelder
Jaap Molenaar

CRC Press
Taylor & Francis Group
Boca Raton London New York

CRC Press is an imprint of the
Taylor & Francis Group, an **informa** business

CRC Press
Taylor & Francis Group
6000 Broken Sound Parkway NW, Suite 300
Boca Raton, FL 33487-2742

First issued in paperback 2018

© 2011 by Taylor and Francis Group, LLC
CRC Press is an imprint of Taylor & Francis Group, an Informa business

No claim to original U.S. Government works

ISBN 13: 978-1-138-03405-1 (pbk)
ISBN 13: 978-1-57444-780-4 (hbk)

Library of Congress Cataloging-in-Publication Data

Koopmans, Rudy.
    Polymer melt fracture / authors Rudy Koopmans, Jaap den Doelder, and Jaap Molenaar.
        p. cm.
    "A CRC title."
    Includes bibliographical references and index.
    ISBN 978-1-57444-780-4 (alk. paper)
    1. Polymers--Fracture. 2. Deformations (Mechanics) 3. Plastics--Extrusion. 4. Polymer melting. I. Doelder, Jaap den. II. Molenaar, Jaap. III. Title.

TA455.P58K66 2010
668.4'13--dc22
                                                                        2010005494

Visit the Taylor & Francis Web site at
http://www.taylorandfrancis.com

and the CRC Press Web site at
http://www.crcpress.com

To our families and friends

There is nothing stable in this world; uproar's your only music.

John Keats
English poet (1795–1821)

Remember that there is nothing stable in human affairs; therefore avoid undue elation in prosperity, or undue depression in adversity.

Socrates
Greek philosopher in Athens (469 BC–399 BC)

Contents

Foreword

Extrusion defects present true limits to the processing of synthetic polymers. Productivity maximization of the processing equipment is restricted when extrusion defects occur in such polymer processing operations as extrusion blown film, blow molding, cable coating, and injection molding, and tube, sheet, and profile extrusions.

Different extrusion defects may be encountered depending on the polymer considered and the processing conditions. "Polymer melt fracture" thus refers to a wide variety of extrusion defects often identified with a descriptive terminology such as "sharkskin," "orange peel," "bamboo," "spurt," "helical," and "gross melt fracture." Typically, these extrusion defects are each associable to different polymer melt flow features. The existing nomenclature often induces confusion as the terms do not relate to the perceived origin of the extrusion defect. It makes understanding and pinpointing possible solutions for suppressing the extrusion defect more difficult.

In the second half of the twentieth century, at the beginning of the "plastics age," polymer producers emphasized the development of new polymers and compounds. The main focus was to target differentiated physicochemical properties while the polymer processing issues were believed to be of minor importance and would be solved eventually. Indeed polymer converters invested hugely, often through "trial-and-error" approaches, to optimize polymer processing and suppress, or at least postpone, the extrusion defects. Progress toward higher yield extrusion was being made by mechanical engineering feats, modifying extrusion die geometries, or by introducing "magic" additives. The issue now, however, was that for each polymer type or class, a similar optimization exercise was required.

Eventually, in the early 1990s, it was possible to claim that extrusion defects in polymer processing had been mastered. What apparently remained was an academic research topic aimed at defining an elegant theory explaining the underlying physics of the phenomena.

However, new developments in polymer chemistry allowed for innovative and industrially relevant polymer architectures to be presented to the market (e.g., metallocene catalyzed polyethylene). Polymer producers and converters alike discovered that the existing empirical approach to solving extrusion defect issues failed. The topic of polymer melt fracture was back on the agenda of many researchers. It clearly indicated the need to really understand the underlying physical mechanisms of these phenomena.

Between 2000 and 2004, a Fifth European Framework Research project entitled 3PI (Postpone Polymer Processing Instabilities—G5RD-CT2000-00238) aimed at resolving precisely this issue. A consortium of polymer and polymer additives suppliers, polymer converters, and some of the most prominent universities and research centers joined forces to take a very close look at the polymer architecture, the associated rheology, the processing behavior, and the occurrence of extrudate defects. This book by Rudy Koopmans, Jaap den Doelder, and Jaap Molenaar presents a number of the more important findings of this major collaborative work. However, the book aims to be much more than just a research report.

The first merit of this book is that it describes precisely and in detail the defects that may be encountered for various polymers during processing. This is depicted by a series of amazing illustrations, which, for many polymer converters, may be considered as a "museum of horror"; however, for scientists, these are things of beauty, reflecting shapes of polymer organizations that are still to be understood and predicted.

The second merit of this book is the identification of the thermomechanical conditions, i.e., pressure, stress, shear and elongational rate, and temperature for which extrusion defects appear. The processing operational maps that may be deduced from the reported experimental findings will be of direct use for polymer converters.

The third merit of this book is that it proposes a uniform nomenclature for polymer melt fracture and plausible mechanisms for the occurrence and the development of extrusion defects. Surface distortions, often referred to as "sharkskin" and visible as a slight periodic deformed or sometimes fractured extrudate skin at the die exit, are related to the elongational properties of the polymer melt stream. Volume distortions often appearing as "helical" extrudate deformations are equally defined by the elongational character of the polymer melt and the flow instability it generates upstream before the die entrance. The combination of polymer compressibility before the die entrance and a periodic transition between weak and strong slip at the die wall results in a "spurt" defect. The latter again is inspired by the elongational characteristics of the polymer melt. It can be mathematically described by relaxation oscillation functions that capture the pressure–flow rate behavior in a sawtooth pattern and relate the extrudate defect to a repetitive sequence of smooth, surface-, and volume-distorted features.

These insights represent a real step forward for polymer converters and polymer producers alike in order to find remedies to suppress the "sharkskin" defect, for example. It will be futile to modify the upstream die geometry but a reduction of the elongational stress at the die exit will instead postpone the occurrence of surface distortions. On the contrary, when faced with volume distortions, it is of no use to introduce polymer additives or to modify the die geometry outlet because the origin of the phenomenon is to be found before the die entry.

The book also indicates that the polymer melt fracture story has not come to an end. Further experiments and theoretical developments are needed. For all these reasons, this book is a major contribution to the world of polymers and polymer processing. It provides a baseline for the state of the art and is a source for marvelous challenges for theoreticians and mathematicians interested in developing advanced stability theories for nonlinear viscoelastic fluids in strong flow conditions.

Professor Dr. Ir. Jean-François Agassant
Department of Material Science and Engineering
Mines-Paristech
Sophia-Antipolis, France

Preface

The extrusion of polymer melts through dies at high rates often yields distorted extrudates. A 1945 paper by H.K. Nason published in the *Journal of Applied Physics*, volume 6, pp. 338–343 and entitled "A high temperature, high pressure rheometer for plastics" was the first to (briefly) mention the phenomenon of extrudate distortion for polymer melts. More than 65 years later, polymer extrudate distortion remains an important topic for research and development activities in industry and academia. The considerable interest in this topic, commonly referred to as polymer melt fracture, is mainly driven by economic (a polymer processing limit) and scientific interests (why does it occur?).

In the second half of the twentieth century, synthetic polymers (plastics) evolved from a research curiosity to a market of more than 250 million tons. Compared to steel and aluminum, the global volume of plastics continues to grow on a global basis at a steady, averaged rate of about 5% per year. The steady growth has, and is stimulating continued improvements in the area of polymer architecture and polymer extrusion techniques. Significant progress has been made in each area but polymer melt fracture always remained a challenge.

The linguist Noam Chomsky once suggested that our ignorance can be divided into problems and mysteries. When we face a problem, we may not know its solution, but we have insight, increasing knowledge, and an inkling of what we are looking for. When we face a mystery, however, we can only stare in wonder and bewilderment, not knowing what an explanation would even look like. Polymer melt fracture is not a mystery anymore but still remains a problem. Key to the problem solution are answers to the questions: "What triggers polymer melt fracture?," "How does polymer melt fracture relate to the polymer architecture?," and "Is polymer melt fracture predictable in terms of macroscopically defined material properties and processing conditions?" The efforts to find these answers resulted in a significant body of literature. The publications range from simple experimental observations to advanced theoretical considerations and conjectures. Several review papers are available discussing different aspects of the topic. The book edited by S.G. Hatzikiriakos and K.B. Migler—*Polymer Processing Instabilities*, Marcel Dekker, New York, 2005—focuses on specific issues of polymer melt fracture (as will be defined in this book) and general fluid flow instabilities such as draw resonance in fiber spinning and film blowing. However, bringing together industrial polymer melt fracture features with academic understanding has not been done in an extensive way. This would require a systematic in-depth overview of the available knowledge, covering a comprehensive body of literature and bringing together various aspects from experiment to modeling on polymer melt fracture. In addition, that knowledge should lead to insights on "how to" either prevent the phenomenon from occurring or remediate when it presents itself in practice. Therefore, in this book the aim is to provide both up-to-date understanding and practical guidance for identification and remediation. For all those faced with observations of "strange" extrudate behavior, this book should provide a basis for what options one has to address polymer melt fracture phenomena.

The route followed to bring all these aspects together has been challenging and at times frustrating. On the one hand, the subject matter is complex as it involves many different aspects of science and technology. It requires the collaboration of open-minded people who are specialists in different disciplines. On the other hand, confusing descriptions of polymer melt fracture, contradictory experimental results, and opposing theoretical explanations have tested our patience to present a final answer to the key questions. Moreover, scientific enquiry and economic striving are not always compatible. It has led many times to restricting the advancement of science and technology and limiting the progress of knowledge. Consequently, the writing and publishing process took longer than anticipated. Irrespective, we believe this book provides insight and guidance. We do, however, realize that no final, "one-size-fits-theory" has been formulated as yet. It implies that there are different remediative steps to be taken for different polymer melt fracture phenomena. Therefore, the book is divided into 10 chapters, and sufficient reference material and insight are provided for readers to develop either a pragmatic understanding of polymer melt fracture or a basis for a more mathematical and theoretically inclined approach.

Each chapter covers a different aspect of science and technology in relation to polymer melt fracture. A book on this topic is not complete without pictures. In Chapter 1, we collected a set of optical and scanning electron microscopy pictures of distorted capillary die extrudates for a number of commercially available polymers. These pictures illustrate the variety and "beauty" of polymer melt fracture irrespective of the specific nature of the polymers used. Throughout the book, more application-oriented polymer melt fracture pictures are shown. Chapters 2 and 3 present a brief introduction to the basic science and technology of polymers. They cover subjects that should be considered essential knowledge to study polymer melt fracture. Chapter 2 reviews what polymers are, how they are made, and how they can be characterized. Chapter 3 discusses some aspects of polymer rheology. The principles of continuum mechanics are reviewed and linear viscoelastic material functions are defined.

Next, it is appropriate to present how polymer melt fracture is experienced in the polymer processing industry. Therefore, Chapter 4 focuses on the various ways polymer melt fracture may appear during polymer melt processing in different extrusion processes. A clear distinction is made between polymer melt fracture and polymer melt flow instability phenomena and only the first subject is examined in detail.

Chapters 5 and 6 present a comprehensive review of the literature on polymer melt fracture. In Chapter 5, we focus exclusively on the experimental findings and the techniques used to observe and measure polymer melt fracture. Chapter 6 looks into the influence of polymer architecture and polymer processing conditions on the onset and types of polymer melt fracture. This leads to Chapter 7, where we present the current understanding of polymer melt fracture and venture into a polymer melt fracture hypothesis. To capture that insight for practical use, it is important to develop models (mathematical equations). Models allow for simulations that may indicate potential solutions to manage polymer melt fracture. As an introduction, Chapter 8 is devoted to aspects of nonlinear constitutive equations and some microscopic theory. The advanced rheology presentation is required to understand some reported approaches and to develop a macroscopic model that may quantitatively describe polymer melt fracture phenomena. In Chapter 9, we explain the capabilities and limitations of such an approach. In Chapter 10, we consider it important to present an overview of pragmatic tools and approaches that have been used to prevent the appearance of polymer melt fracture.

Throughout the 10 chapters, only one class of viscoelastic fluids is considered, namely, thermoplastic polymer melts. Dilute or concentrated polymer solutions containing flexible or stiff polymers, liquid crystalline materials, or colloids are subjects that have not been dealt with in this book except through some relevant references.

The preface would not be complete without acknowledging the many people who supported (morally as well as actively) the completion of this book. First of all, we would like to acknowledge the patience of our families and friends. Throughout the many weekends, late evenings, and vacation days they always were very patient and lenient to the absent-minded writers. In addition, we would like to acknowledge many colleagues and friends for their unwavering support. As a first, thanks to Bob Sammler for reviewing and correcting the manuscript in detail. We appreciated very much his suggestions for textual improvements as well as his continued encouragement during difficult times. A particular word of appreciation is reserved for Dennis Liebman and Steve Martin. Their relentless support helped in putting the work together and meant a lot in the final analysis. The people who actually inspired RJK to embark on this endeavor were those participating in a daylong session discussing the issues of polymer melt fracture at the Dow Sarnia R&D labs in Canada. Among them Susan Hagan and her "sister" Deb Walker, and Tony Samurkas asked questions that were most difficult to answer, indicating that it was time to sort out what was really understood in this field of science and technology and, more importantly, how that understanding could be applied.

There have been many more people that contributed to this book as colleagues at The Dow Chemical Company or as university research partners. Their contributions came in many ways: in small ways through discussions, by providing test results, or sometimes by just being helpful, which allowed us to complete this book. As past and present Dow colleagues we would like to thank Marc Mangnus, Lizzy Vinjé, Larry Kale, Brian Dickie, Freddy van Damme, Monika Plass, Roger Michielsen, Bob Vastenhout, Flip Bosscher, Hans de Jonge, Sjoerd de Vries, Giel de Pooter, David Porter, Gunther Muggli, Paul Fisch, Gerard van de Langkruis, Cristina Serrat, Mehmet Demirors, Mark Vreys, Marc Dees, Bernadette Schelstraete, Marjan Sturm, Huguette Baete, Marlies Totté, Stéphane Costeux, Jérome Claracq, Jef van Dun, Joe Dooley, Kun Hyun, Joe Bicerano, Sandra Hofmann, Sarada Namhata, Sarah Patterson, Clive Bosnyak, Chris Christenson, Johan Thoen, Pat Andreozzi, Randy Collard, Hank Kohlbrand, Willem de Groot, Dana Gier, Thomas Allgeuer, Mark Murphy, and Nicolien Groosman.

Over the last 25 years, several research projects on the topic required us to collaborate with creative people, pragmatic industrialists, professors, and PhD and master students. Many of their names can be found in the referenced literature in this book. Thanks to Tony Daponte, Fons van de Ven, Stef van Eindhoven, Annemarie Aerts, Marcel Grob, Han Slot, Jean-François Agassant, Bruno Vergnes, Cécile Venet,

Virginie Durand, Jean-Pierre Villemaire, Jean-Michel Piau, Nadia El-Kissi, John Dealy, Savvas Hatzikiriakos, Stuart Kurtz, Malcolm Mackley, Meera Ranganathan, Karen Lee, Tom McLeish, Martin Laun, Manfred Wagner, Helmut Münstedt, Hans Hürlimann, Hans-Christian Öttinger, Joachim Meissner, Johan Dubbeldam, Roman Stepanyan, Michael Tchesnokov, Gérard Marin, Frédéric Leonardi, Ahmed Allal, Alexandrine Lavernhe, Armin Merten, Martin Jakob, Laurent Robert, Christelle Combeaud, Rudy Vallette, Yves Demay, Dawn Arda, Kalman Migler, Georgios Georgiou, Lourdes de Vargas, José Perez-Gonzalez, Oleg Kulikov, Suneel Kunamaneni, Steffen Berger, and Tobias Königer. For those collaborators who do not find their names, we apologize as they may have escaped our memories over the years. However, their contributions are equally valued, for they too have been fascinated by the subject of polymer melt fracture. We would like to thank them for their contributions.

Two of us, RJK and CFJdD, would like to express our gratitude to The Dow Chemical Company for providing us the opportunity to work on various occasions and in different capacities on the fascinating subject of polymer melt fracture, and for allowing us to use some pictures as well as supporting us through the entire publishing process.

Authors

Rudy Koopmans received his PhD in physical and macromolecular chemistry from the University of Antwerp, Antwerp, Belgium. He is a fellow in the Basic Plastics R&D organization of The Dow Chemical Company located in Horgen, Switzerland. Since he joined Dow in 1983, he has held various R&D positions in Europe and the United States. His main R&D focus is on materials development, polymer processing, and developing innovative technology solutions to market needs and identified market trends. In addition, he holds a visiting professorship at Leeds University (Leeds, United Kingdom) in the Department of Chemical Engineering. He has published more than 50 peer-reviewed papers in international journals and books, and is a holder of multiple patents.

Jaap den Doelder received his MSc in applied physics and applied mathematics at Eindhoven University of Technology, Eindhoven, the Netherlands. He received his PhD in applied mathematics at the same university in 1999 on the topic of polymer melt fracture. In the same year, he joined The Dow Chemical Company in Terneuzen, the Netherlands. He has since worked on a variety of topics related to materials science and modeling of polymers, connecting application requirements to molecular design. He is currently a research scientist in Dow's polyethylene business.

Jaap Molenaar studied mathematics and theoretical physics at Leiden University, Leiden, the Netherlands, and wrote a PhD thesis on the field of solid state physics. For more than a decade, he was involved in mathematics consulting. He received the Neways Award for his work on academic knowledge transfer to industry. Jaap specializes in the modeling of dynamical systems with a focus on differential equations and has published several books on these topics. His research focuses on fluid mechanics, in particular polymer melt flow. Recently, he has taken keen interest in systems biology. He is a full professor in applied mathematics and the head of the Department for Mathematical and Statistical Methods for the Life Sciences of Wageningen University and Research Centre, Wageningen, the Netherlands.

1 Polymer Melt Fracture Pictures

Polymer melt fracture is a visual experience on extrudate distortions. A book on polymer melt fracture is not complete without illustrating some of the many possible shapes and forms of distorted extrudates directly observable during polymer processing. Polymer melt extrusion of single extrudate strands or fibers, tubular blown films, thin cast films or thicker sheets, profiles, parisons, pipes, or tubes are prone to show visible defects under particular processing conditions. The defects are any deviation from a smooth, glossy, regular extrudate. In mold-filling extrusion operations, the defects may only become observable on the final solid part. For instance, injection-molded parts may show defects that reflect a polymer melt extrudate filling a mold inhomogeneously. Other forms of polymer processing such as calendering, compression, and rotational molding may induce, respectively, film, sheet, and part defects. Similarly, secondary processing techniques like thermoforming and lamination may give rise to part defects often resembling and associated with polymer melt fracture. The origin of these latter types of part defects, however, is often a confounded consequence of physical and mechanical processes. For some, polymer melt fracture could be one of the contributing factors.

What polymer melt fracture is precisely, how it comes about, and how it may be avoided is explored and explained in Chapters 5 to 10. But, before getting into the science and technology, let us have a look at intentionally created polymer melt fracture. The illustrations below show an amazing variety of distorted polymer melt extrudates. Even so, it is not intended to be exhaustive but merely illustrative of how polymer melts can "self-organize" to form new topologies under the influence of work in an "open system."

First, a series of optical microscopy pictures illustrates the various forms of surface and volume distortions in order of increasing severity. A picture of the extrudate distortion is shown together with its schematic representation as used also in Chapter 6 (Courtesy of C. Venet). The picture series is composed out of extrudates from different types of commercially available polyethylene (PE) grades. PE grades are selected as they manifest the widest range of different extrudate distortions. Typically, one polymer grade does not show all possible extrudate distortions. The illustrations have varying magnifications indicated for each picture. The extrudates are obtained with a constant-rate capillary Rhéoplast® rheometer (Courbon SA, St Etienne, France).

Second, scanning electron microscopy (SEM) pictures illustrate extrudates, magnified 40 times, for a selection of commercially available PE, polypropylene (PP), polystyrene (PS), and polycarbonate (PC) grades. For each grade, the extrudates are obtained at several apparent wall shear rates $\dot{\gamma}_a$ by using a constant-rate Göttfert 2001® capillary rheometer (Göttfert Werkstoff-Prüfmaschine GmbH, Buchen, Germany). The selected piston speed is kept approximately similar for the various tests. For all tests, the same capillary is used having a die aspect ratio defined as length L over diameter, that is, two times the capillary radius R—($L/2R = 20/1$ mm, 180° entry angle) except for one PE where $L/2R = 10/1$ mm in view of its significantly different rheological behavior. In all experiments, the polymer melt temperature is 180°C, except for PC (280°C). In some pictures, small dust particles (not related to polymer melt fracture) are attached to the surface of the extrudates as a result of the SEM sample pretreatment. The procedure includes mounting the solid extrudates on an aluminum sample holder, drying at room temperature, and sputtering it with gold to make the surface conductive. The SEM instrument is a Philips SEM505 operating at 30 kV. The picture legends provide further detail on the capillary experiment. The flow direction is from right to left or from top to bottom.

1.1 OPTICAL MICROSCOPY

Various forms of polymer melt fracture are illustrated in Figures 1.1 to 1.15. At increasing shear rates smooth extrudate surfaces become gradually more distorted. At even higher shear rates the extrudate shows a combination of surface distortions and smooth (Figure 1.9). Eventually exotic extrudate shapes are produced of which only a few are shown (Figures 1.10 to 1.15).

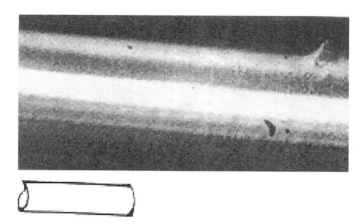

FIGURE 1.1 Transparent extrudate with smooth surface. Metallocene-catalyzed very-low-density ethylene-octene copolymer ($\rho = 870$ kg/m³; MFR (190°C/2.16 kg) = 1.0 dg/min)—magnification ×120; $L/2R = 16/1.39$ mm; entry angle 180°; 190°C; $\dot{\gamma}_a = 7$ s⁻¹. (From Venet, C., Propriétés d'écoulement et défauts de surface de résins polyéthylènes, PhD thesis, Ecole des Mines de Paris (CEMEF), Sophia Antipolis, France, 1996.)

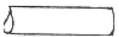

FIGURE 1.2 Transparent extrudate with "orange peel" surface. Ziegler–Natta-catalyzed linear low-density ethylene-octene copolymer ($\rho = 920\,kg/m^3$; MFR (190°C/2.16 kg) = 1.0 dg/min)—magnification ×120; $L/2R = 0$ (~very small)/1.39 mm; entry angle 180°; 190°C; $\dot{\gamma}_a = 99\,s^{-1}$. (From Venet, C., Propriétés d'écoulement et défauts de surface de résins polyéthylènes, PhD thesis, Ecole des Mines de Paris (CEMEF), Sophia Antipolis, France, 1996.)

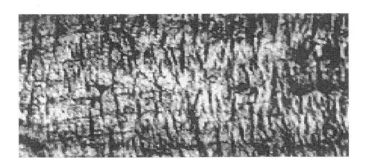

FIGURE 1.3 Matte extrudate with high-frequency low-amplitude "wavelets" surface. Ziegler–Natta-catalyzed linear low-density ethylene-octene copolymer ($\rho = 920\,kg/m^3$; MFR (190°C/2.16 kg) = 1.0 dg/min)—magnification ×120; $L/2R = 0$ (~very small)/1.39 mm; entry angle 180°; 150°C; $\dot{\gamma}_a = 244\,s^{-1}$. (From Venet, C., Propriétés d'écoulement et défauts de surface de résins polyéthylènes, PhD thesis, Ecole des Mines de Paris (CEMEF), Sophia Antipolis, France, 1996.)

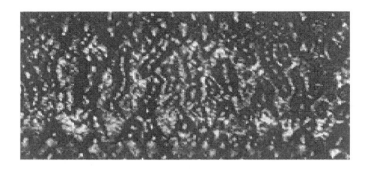

FIGURE 1.4 Matte extrudate with "pitted, cracked" surface. Metallocene-catalyzed high-density ethylene polymer containing long chain branching ($\rho = 957\,kg/m^3$; MFR $(190°C/2.16\,kg) = 1.0\,dg/min$)—magnification ×120; $L/2R = 0$ (~very small)/1.39 mm; entry angle 180°; 150°C; $\dot\gamma_a = 57\,s^{-1}$. (From Venet, C., Propriétés d'écoulement et défauts de surface de résins polyéthylènes, PhD thesis, Ecole des Mines de Paris (CEMEF), Sophia Antipolis, France, 1996.)

FIGURE 1.5 Transparent extrudate with "pitted, sharkskin-feel" surface. Metallocene-catalyzed very-low-density ethylene-octene copolymer ($\rho = 870\,kg/m^3$; MFR $(190°C/2.16\,kg) = 1.0\,dg/min$)—magnification ×120; $L/2R = 16/1.39\,mm$; entry angle 180°; 150°C; $\dot\gamma_a = 14\,s^{-1}$. (From Venet, C., Propriétés d'écoulement et défauts de surface de résins polyéthylènes, PhD thesis, Ecole des Mines de Paris (CEMEF), Sophia Antipolis, France, 1996.)

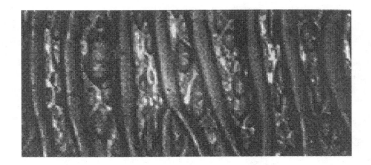

FIGURE 1.6 Translucent extrudate with "high-frequency low-amplitude ridged" surface. Metallocene-catalyzed very-low-density ethylene-octene copolymer ($\rho = 870\,\mathrm{kg/m^3}$; MFR ($190°\mathrm{C}/2.16\,\mathrm{kg}$) = $1.0\,\mathrm{dg/min}$)—magnification ×120; $L/2R = 0$ (~very small)/$1.39\,\mathrm{mm}$; entry angle 180°; 150°C; $\dot{\gamma}_a = 43\,\mathrm{s^{-1}}$. (From Venet, C., Propriétés d'écoulement et défauts de surface de résins polyéthylènes, PhD thesis, Ecole des Mines de Paris (CEMEF), Sophia Antipolis, France, 1996.)

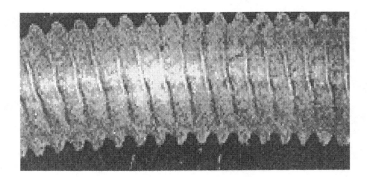

FIGURE 1.7 Translucent extrudate with "screw-treaded" surface. Metallocene-catalyzed very-low-density ethylene-octene copolymer ($\rho = 870\,\mathrm{kg/m^3}$; MFR ($190°\mathrm{C}/2.16\,\mathrm{kg}$) = $1.0\,\mathrm{dg/min}$)—magnification ×11; $L/2R = 16/1.39\,\mathrm{mm}$; entry angle 180°; 150°C; $\dot{\gamma}_a = 76\,\mathrm{s^{-1}}$. (From Venet, C., Propriétés d'écoulement et défauts de surface de résins polyéthylènes, PhD thesis, Ecole des Mines de Paris (CEMEF), Sophia Antipolis, France, 1996.)

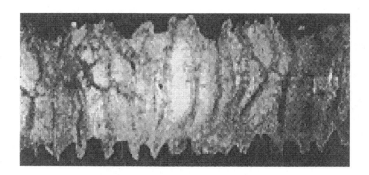

FIGURE 1.8 Translucent extrudate with "distorted, irregular screw-treaded" surface. Metallocene-catalyzed very-low-density ethylene-octene copolymer ($\rho = 870\,\text{kg/m}^3$; MFR (190°C/2.16 kg) = 1.0 dg/min)—magnification ×11; $L/2R = 16/1.39$ mm; entry angle 180°; 150°C; $\dot{\gamma}_a = 183\,\text{s}^{-1}$. (From Venet, C., Propriétés d'écoulement et défauts de surface de résins polyéthylènes, PhD thesis, Ecole des Mines de Paris (CEMEF), Sophia Antipolis, France, 1996.)

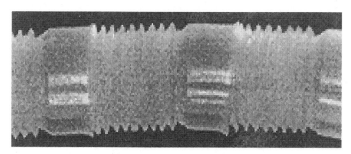

FIGURE 1.9 Transition zone "spurt" extrudate. Ziegler–Natta-catalyzed linear low-density ethylene-octene copolymer ($\rho = 920\,\text{kg/m}^3$; MFR (190°C/2.16 kg) = 1.0 dg/min)—magnification ×11; $L/2R = 16/1.39$ mm; entry angle 180°; 190°C; $\dot{\gamma}_a = 763\,\text{s}^{-1}$. (From Venet, C., Propriétés d'écoulement et défauts de surface de résins polyéthylènes, PhD thesis, Ecole des Mines de Paris (CEMEF), Sophia Antipolis, France, 1996.)

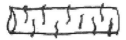

FIGURE 1.10 Surface distorted extrudate without volume distortions beyond the transition zone (see Chapter 5). Ziegler–Natta-catalyzed linear low-density ethylene-octene copolymer ($\rho = 920\,\text{kg/m}^3$; MFR (190°C/2.16 kg) = 1.0 dg/min)—magnification ×11; $L/2R = 16/1.39$ mm; entry angle 180°; 150°C; $\dot{\gamma}_a = 1370\,\text{s}^{-1}$. (From Venet, C., Propriétés d'écoulement et défauts de surface de résins polyéthylènes, PhD thesis, Ecole des Mines de Paris (CEMEF), Sophia Antipolis, France, 1996.)

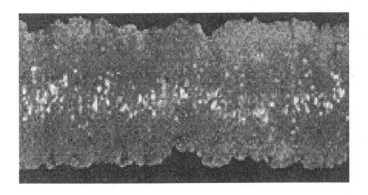

FIGURE 1.11 Surface- and volume-distorted extrudate. Ziegler–Natta-catalyzed linear low-density ethylene-octene copolymer ($\rho = 920\,\text{kg/m}^3$; MFR (190°C/2.16 kg) = 1.0 dg/min)—magnification ×11; $L/2R = 16/1.39$ mm; entry angle 180°; 190°C; $\dot{\gamma}_a = 3200\,\text{s}^{-1}$. (From Venet, C., Propriétés d'écoulement et défauts de surface de résins polyéthylènes, PhD thesis, Ecole des Mines de Paris (CEMEF), Sophia Antipolis, France, 1996.)

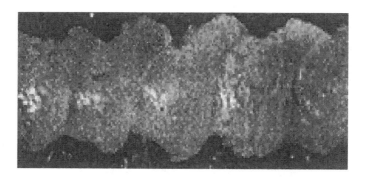

FIGURE 1.12 Helical volume-distorted extrudate. Metallocene-catalyzed very-low-density ethylene-octene copolymer ($\rho = 870\,\mathrm{kg/m^3}$; MFR ($190°C/2.16\,\mathrm{kg}$) = $1.0\,\mathrm{dg/min}$)—magnification ×11; $L/2R$ = $16/1.39\,\mathrm{mm}$; entry angle 180°; 150°C; $\dot{\gamma}_a = 320\,\mathrm{s^{-1}}$. (From Venet, C., Propriétés d'écoulement et défauts de surface de résins polyéthylènes, PhD thesis, Ecole des Mines de Paris (CEMEF), Sophia Antipolis, France, 1996.)

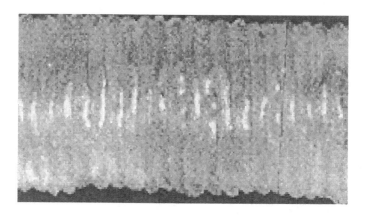

FIGURE 1.13 "Sausage-like" volume-distorted extrudate. Metallocene-catalyzed very-low-density ethylene-octene copolymer ($\rho = 870\,\mathrm{kg/m^3}$; MFR ($190°C/2.16\,\mathrm{kg}$) = $1.0\,\mathrm{dg/min}$)—magnification ×11; $L/2R$ = 0 (~very small)/$1.39\,\mathrm{mm}$; entry angle 180°; 150°C; $\dot{\gamma}_a = 763\,\mathrm{s^{-1}}$. (From Venet, C., Propriétés d'écoulement et défauts de surface de résins polyéthylènes, PhD thesis, Ecole des Mines de Paris (CEMEF), Sophia Antipolis, France, 1996.)

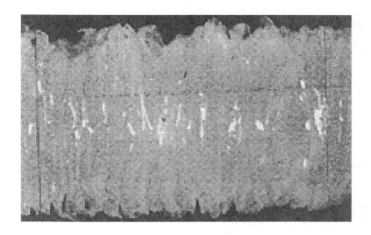

FIGURE 1.14 Surface- and volume-distorted extrudate. Ziegler–Natta-catalyzed linear low-density ethylene-octene copolymer ($\rho = 920$ kg/m³; MFR (190°C/2.16 kg) = 1.0 dg/min)— magnification ×11; $L/2R = 16/1.39$ mm; entry angle 180°; 190°C; $\dot{\gamma}_a = 4270$ s⁻¹. (From Venet, C., Propriétés d'écoulement et défauts de surface de résins polyéthylènes, PhD thesis, Ecole des Mines de Paris (CEMEF), Sophia Antipolis, France, 1996.)

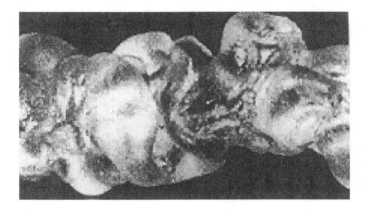

FIGURE 1.15 Irregular, "chaotic" volume-distorted extrudate. Metallocene-catalyzed very-low-density ethylene-octene copolymer ($\rho = 870$ kg/m³; MFR (190°C) = 1.0 dg/min)— magnification ×11; $L/2R = 0$ (~very small)/1.39 mm; entry angle 180°; 150°C; $\dot{\gamma}_a = 5720$ s⁻¹. (From Venet, C., Propriétés d'écoulement et défauts de surface de résins polyéthylènes, PhD thesis, Ecole des Mines de Paris (CEMEF), Sophia Antipolis, France, 1996.)

1.2   SCANNING ELECTRON MICROSCOPY

Scanning Electron Microscopy of distorted extrudates as generated using a capillary rheometer for different commercially available plastics reveal an amazing variety of shapes. The subsequent Figures 1.16 to 1.29 show polymer melt fracture at increasing shear rates going from (a) to (f).

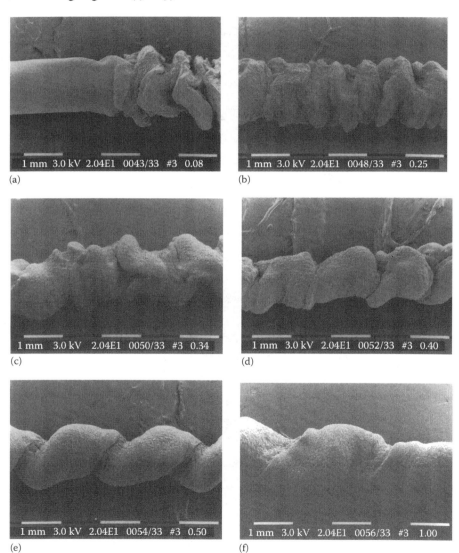

FIGURE 1.16   SEM of the capillary extrudates of a Ziegler–Natta-catalyzed linear high-density PE ($\rho = 965\,kg/m^3$; MFR (190°C/2.16 kg) = 0.2 dg/min)—magnification ×40; $L/2R = 20/1$ mm; entry angle 180°; 180°C; $\dot{\gamma}_a = 144$ (a), 450 (b), 612 (c), 720 (d), 900 (e), and 1800 s^{-1} (f). (Courtesy of M. Mangnus from The Dow Chemical Company.)

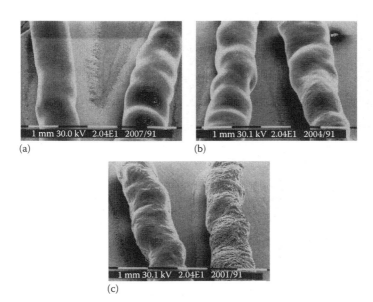

(a)   (b)

(c)

FIGURE 1.17   SEM of the capillary extrudates of a low-density PE ($\rho = 920\,kg/m^3$; MFR (190°C/2.16 kg) = 1.0 dg/min)—magnification ×40; $L/2R = 20/1$ mm; entry angle 180°; 180°C; $\dot{\gamma}_a = 144, 450, 620, 900, 1800$, and 4500 s^{-1}. (Courtesy of M. Mangnus from The Dow Chemical Company.)

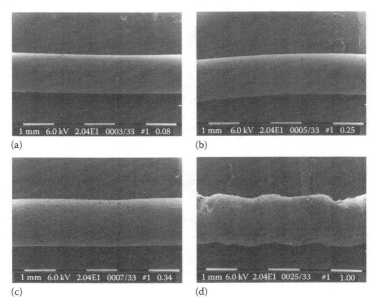

(a)   (b)

(c)   (d)

FIGURE 1.18   SEM of the capillary extrudates of a Ziegler–Natta-catalyzed linear low-density ethylene-octene copolymer ($\rho = 920\,kg/m^3$; MFR (190°C/2.16 kg) = 1.0 dg/min)—magnification ×40; $L/2R = 20/1$ mm; entry angle 180°; 180°C; $\dot{\gamma}_a = 144$ (a), 450 (b), 612 (c), and 1800 s^{-1} (d). (Courtesy of M. Mangnus from The Dow Chemical Company.)

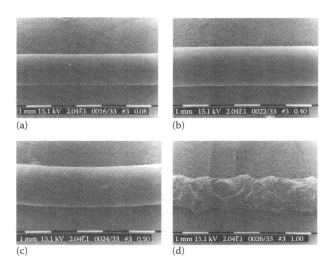

FIGURE 1.19 SEM of the capillary extrudates of a Ziegler–Natta-catalyzed linear low-density ethylene-octene copolymer ($\rho = 903\,\text{kg/m}^3$; MFR ($190°C/2.16\,\text{kg}$) = $1.0\,\text{dg/min}$)—magnification ×40; $L/2R = 20/1$ mm; entry angle 180°; 180°C; $\dot{\gamma}_a = 144$ (a), 720 (b), 900 (c), and $1800\,\text{s}^{-1}$ (d). (Courtesy of M. Mangnus from The Dow Chemical Company.)

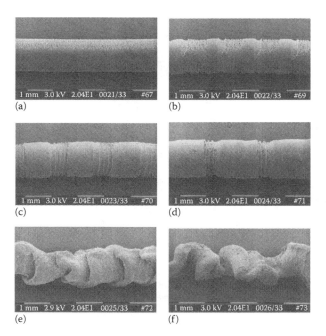

FIGURE 1.20 SEM of the capillary extrudates of a Ziegler–Natta-catalyzed linear low-density ethylene-octene copolymer ($\rho = 923\,\text{kg/m}^3$; MFR ($190°C/2.16\,\text{kg}$) = $0.7\,\text{dg/min}$)—magnification ×40; $L/2R = 10/1$ mm; entry angle 180°; 180°C; $\dot{\gamma}_a = 270$ (a), 450 (b), 612 (c), 720 (d), 2700 (e), and $4500\,\text{s}^{-1}$ (f). (Courtesy of M. Mangnus from The Dow Chemical Company.)

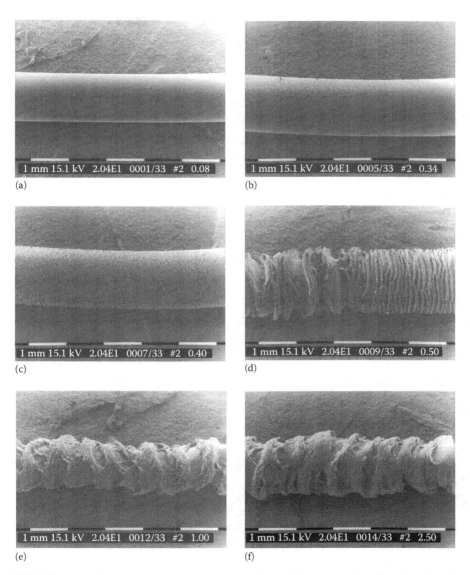

(a)

(b)

(c)

(d)

(e)

(f)

FIGURE 1.21   SEM of the capillary extrudates of a metallocene-catalyzed low-density ethylene-octene copolymer ($\rho = 905\,kg/m^3$; MFR $(190°C/2.16\,kg) = 1.0\,dg/min$)—magnification
×40; $L/2R = 20/1$ mm; entry angle 180°; 180°C; $\dot{\gamma}_a = 144$ (a), 612 (b), 720 (c), 900 (d), 1800 (e),
and $4500\,s^{-1}$ (f). (Courtesy of M. Mangnus from The Dow Chemical Company.)

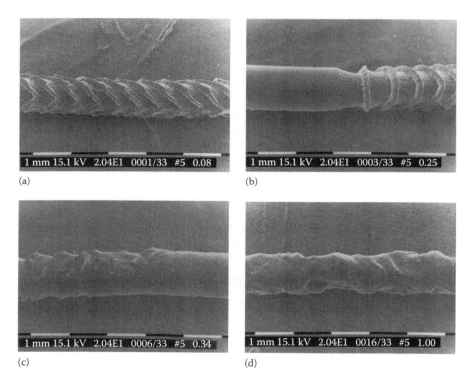

(a)

(b)

(c)

(d)

FIGURE 1.22 SEM of the capillary extrudates of a metallocene-catalyzed low-density ethylene-butene copolymer ($\rho = 905\,\text{kg/m}^3$; MFR ($190°\text{C}/2.16\,\text{kg}$) = 1.0 dg/min)—magnification ×40; $L/2R = 20/1$ mm; entry angle 180°; 180°C; $\dot{\gamma}_a = 144$ (a), 450 (b), 612 (c), and 1800 s^{-1} (d). (Courtesy of M. Mangnus from The Dow Chemical Company.)

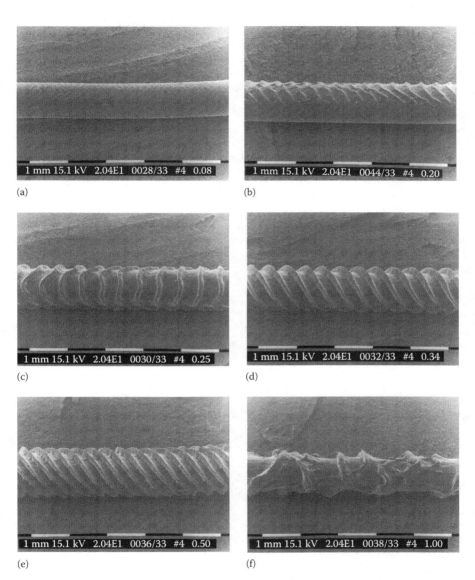

FIGURE 1.23 SEM of the capillary extrudates of a metallocene-catalyzed low-density ethylene-butene polymer ($\rho = 885\,kg/m^3$; MFR ($190°C/2.16\,kg) = 2.0\,dg/min$)—magnification ×40; $L/2R = 20/1$ mm; entry angle 180°; 180°C; $\dot{\gamma}_a = 144$ (a), 360 (b), 450 (c), 612 (d), 900 (e), and 1800 s^{-1} (f). (Courtesy of M. Mangnus from The Dow Chemical Company.)

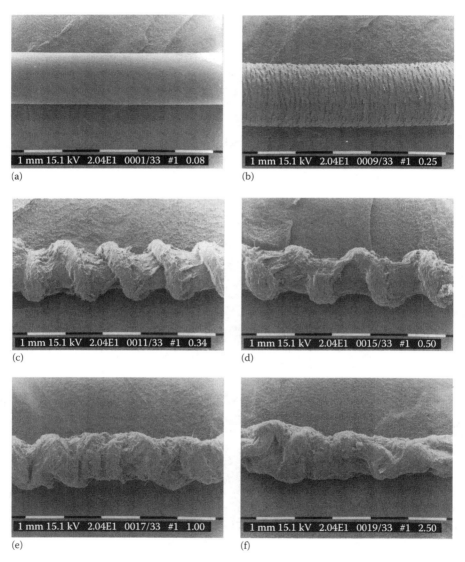

FIGURE 1.24 SEM of the capillary extrudates of a metallocene-catalyzed very-low-density ethylene-octene polymer ($\rho = 870\,kg/m^3$; MFR ($190°C/2.16\,kg$) $= 1.0\,dg/min$)—magnification ×40; $L/2R = 20/1$ mm; entry angle 180°; 180°C; $\dot{\gamma}_a = 144$ (a), 450 (b), 612 (c), 900 (d), 1800 (e), and 4500 s^{-1} (f). (Courtesy of M. Mangnus from The Dow Chemical Company.)

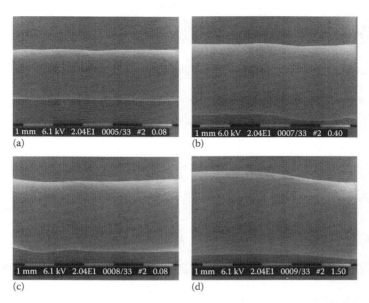

FIGURE 1.25 SEM of the capillary extrudates of an isotactic PP homopolymer ($\rho = 905$ kg/m^3; MFR (230°C/2.16 kg) = 3.0 dg/min)—magnification ×40; $L/2R = 20/1$ mm; entry angle 180°; 180°C; $\dot{\gamma}_a = 144$ (a), 720 (b), 1440 (c), and 2700 s^{-1} (d). (Courtesy of M. Mangnus from The Dow Chemical Company.)

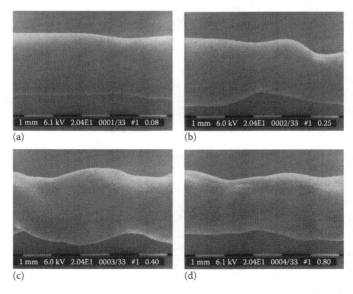

FIGURE 1.26 SEM of the capillary extrudates of an electron-beam irradiated isotactic PP homopolymer ($\rho = 905$ kg/m^3; MFR (230°C/2.16 kg) = 3.0 dg/min)—magnification ×40; $L/2R = 20/1$ mm; entry angle 180°; 180°C; $\dot{\gamma}_a = 144$ (a), 450 (b), 720 (c), and 1440 s^{-1} (d). (Courtesy of M. Mangnus from The Dow Chemical Company.)

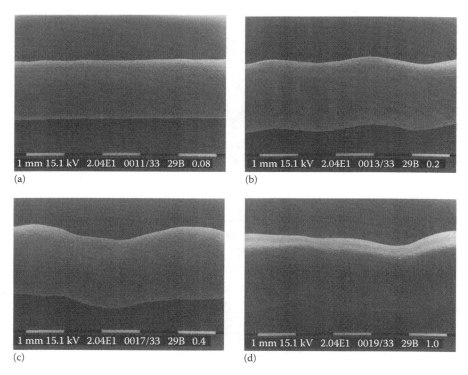

FIGURE 1.27   SEM of the capillary extrudates of a random ethylene-propylene copolymer ($\rho = 905\,kg/m^3$; MFR $(230°C/2.16\,kg) = 2.0\,dg/min$)—magnification ×40; $L/2R = 20/1\,mm$; entry angle 180°; 180°C; $\dot\gamma_a = 144$ (a), 360 (b), 720 (c), and $1800\,s^{-1}$ (d). (Courtesy of M. Mangnus from The Dow Chemical Company.)

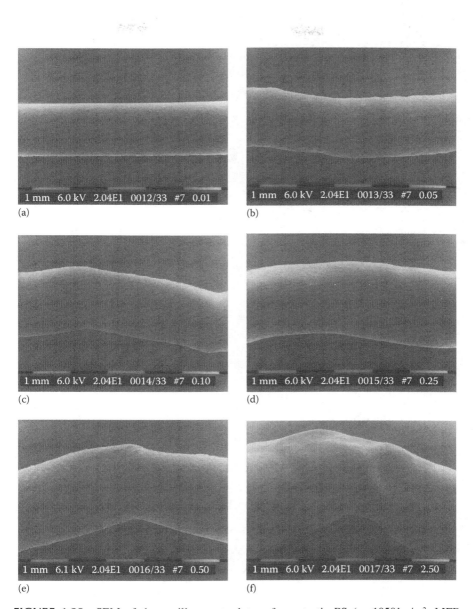

FIGURE 1.28 SEM of the capillary extrudates of an atactic PS ($\rho = 1050 \, \text{kg/m}^3$; MFR ($200°C/5 \, \text{kg}) = 8.0 \, \text{dg/min}$)—magnification ×40; $L/2R = 20/1 \, \text{mm}$; entry angle 180°; 180°C; $\dot{\gamma}_a = 18$ (a), 90 (b), 180 (c), 450 (d), 900 (e), and 4500 s^{-1} (f). (Courtesy of S. Namhata from The Dow Chemical Company.)

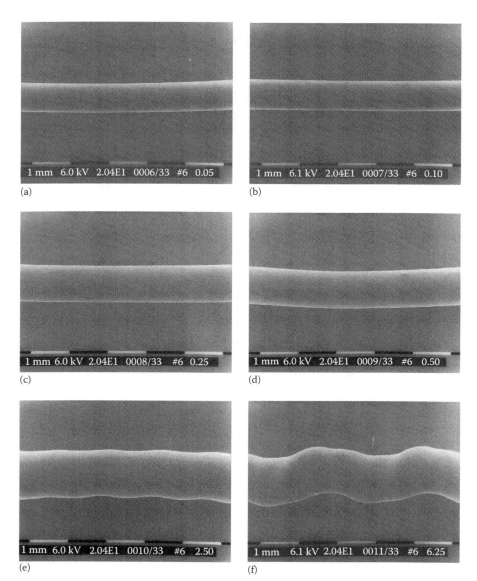

FIGURE 1.29 SEM of the capillary extrudates of a PC ($\rho = 1200\,kg/m^3$; MFR ($300°C/1.2\,kg) = 10\,dg/min$)—magnification ×40; $L/2R = 20/1\,mm$; entry angle 180°; 180°C; $\dot{\gamma}_a = 90$ (a), 180 (b), 450 (c), 900 (d), 4500 (e), and $10530\,s^{-1}$ (f). (Courtesy of S. Namhata from The Dow Chemical Company.)

REFERENCE

1. Venet, C., Propriétés d'écoulement et défauts de surface de résins polyéthylènes. PhD thesis, Ecole des Mines de Paris (CEMEF), Sophia Antipolis, France, 1996.

2 Polymer Characteristics

In 1920, Herman Staudinger [1], a German chemist residing as a professor of organic chemistry at the Federal Institute of Technology (ETH) in Zürich, Switzerland, formulated a theory about the chemical nature of a group of materials of synthetic and natural origin. He hypothesized that these materials consisted of *hochmolekulare Verbindungen*, that is, macromolecules. Only many years later, the concept was sufficiently validated, accepted, and finally rewarded with a Nobel Prize in 1953. Today, the science of macromolecules is a part of everyday life.

Macromolecules are very large molecules. One class of macromolecules is commonly known as *polymers*. The term "poly-mer" is derived from the Greek language, meaning many-parts (i.e., many-"mer" units). The mer-unit is the smallest possible combination of atoms forming a reactive chemical entity, that is, a molecule. Many such identical molecules have sufficient chemical reactivity to covalently connect with each other (typically >1000) to form a macromolecule. Molecules with only a few mer-units are called *oligomers*. In contrast to polymers, the physical properties of an oligomer change with the addition or removal of one or more mer-units from the molecule. The mer-unit is also called the *repeat unit* (Figure 2.1).

Many biological materials such as proteins, carbohydrates, deoxyribonucleic acid (DNA), and ribonucleic acid (RNA) are macromolecules, often referred to as biopolymers. Their chemical configuration typically consists, however, of many different mer-units and is vastly more complex than man-made synthetic polymers. The latter are commonly referred to as plastics or more to the point, as expressed in the Germanic languages, *Kunststoffen* (meaning artificial or synthetic materials). From a scientific perspective, there is a subtle but important difference between polymers and plastics, as will be pointed out below.

Synthetic polymers are frequently pictured as flexible strands or strings. The picture is a gross oversimplification of the nature of polymers. However, it suggests two key molecular features, namely, "very long" and "flexible." Typically, two types of synthetic polymers are defined: thermoplastics and thermosets (Figure 2.2).

The distinguishing feature between these two types of synthetic polymers is based on a physical property. Upon heating, thermoplastics preserve the identity of the macromolecule, and the solid–liquid transition is thermo-reversible. The primary covalent bonds remain stable, while a great many intermolecular secondary bonds are broken and reorganized. Common thermoplastics are polyethylene (PE), polypropylene (PP), polyvinylchloride (PVC), polystyrene (PS), polyethyleneterephthalate (PET), polycarbonate (PC), and polyamide (PA, better know as nylon) (Table 2.1). In thermosets, the giant network of covalently interconnected molecules decomposes upon heating. The primary bonds rupture and the polymer degrades. A well-known thermoset is Bakelite®, a phenol-formaldehyde synthetic polymer. The more commonly available thermosets are polyurethane (PUR) and epoxy resins. Thermosets

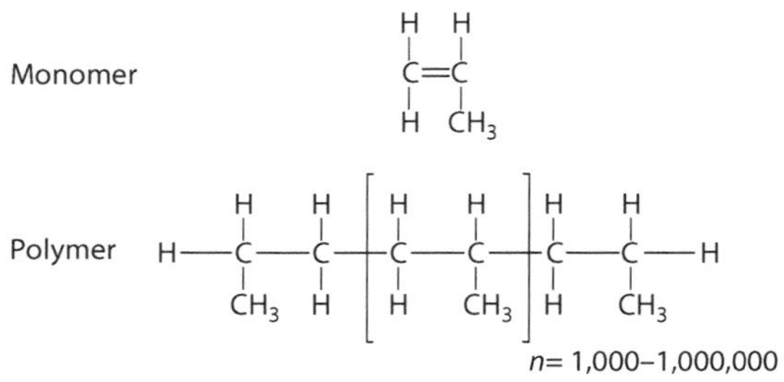

Monomer

$$\begin{array}{cc} H & H \\ | & | \\ C = C \\ | & | \\ H & CH_3 \end{array}$$

Polymer

$$H - \begin{array}{cc} H & H \\ | & | \\ C - C \\ | & | \\ CH_3 & H \end{array} \left[\begin{array}{cc} H & H \\ | & | \\ C - C \\ | & | \\ H & CH_3 \end{array} \right] \begin{array}{cc} H & H \\ | & | \\ C - C \\ | & | \\ H & CH_3 \end{array} - H$$

$n= 1,000–1,000,000$

FIGURE 2.1 The chemical structure representation of a polymer (PP) with the repeat unit shown in brackets. The chemical structure representation of the monomer (here propylene) is closely reflected in that of the repeat unit. For PP, the mer-unit can be added to a polymerizing macromolecules in so-called head–tail or head–head position, as illustrated by the position of the methyl group in the left and right side of the PP formula.

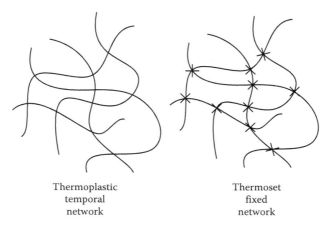

Thermoplastic
temporal
network

Thermoset
fixed
network

FIGURE 2.2 Synthetic polymers are often represented as flexible strands. Two types of polymers are defined: thermoplastics (e.g., PE) and thermosets (e.g., phenol-formaldehyde). The first is characterized by a temporal network of physically interacting macromolecules, the latter by a chemically fixed network, that is, one giant macromolecule.

are typically produced by mixing very low viscosity liquids. The components react in situ to form a giant molecule of (mostly) covalently interconnected molecules. Although defects in or distortions of the final part may appear in thermoset products, polymer melt fracture is typically not an issue. Accordingly, this book focuses on thermoplastics.

2.1 POLYMERS

Polymer melt fracture is a very critical issue when processing thermoplastics. This peculiar behavior of the polymer in the fluid state during flow is related to the polymers being "very long" and "flexible." Any chemical or physical change that affects these two features will impact polymer melt fracture. Therefore, some basic aspects

TABLE 2.1
Chemical Structure Representation of the Repeat Unit for Some Common Synthetic Polymers

Polymer Name, Abbreviation	Repeat Unit	Polymer Name, Abbreviation	Repeat Unit
Polyethylene, PE	$-CH_2-$	Polystyrene, PS	$-CH_2-CH(C_6H_5)-$
Polypropylene, PP	$-CH_2-CH(CH_3)-$	Poly(1,4-butadiene)	$-CH_2-CH=CH-CH_2-$
Poly(1-butene), PB	$-CH_2-CH(CH_2CH_3)-$	Polyisoprene, PI	$-CH_2-CH=C(CH_3)-CH_2-$
Poly(4-methyl-1-pentene)	$-CH_2-CH(CH_2-CH(CH_3)-CH_3)-$	Polytetrafluoro-ethylene, PTFE	$-CF_2-CF_2-$
Poly(1-octene)	$-CH_2-CH(CH_2CH_2CH_2CH_2CH_2CH_3)-$	Polyhexafluoro-propylene, PHFP	$-CF_2-CF(CF_3)-$
Polyvinylchloride, PVC	$-CH_2-CHCl-$	Poly(vinylidene-difluoride), PVDF	$-CH_2-CF_2-$
Poly(vinylidene dichloride), PVDC	$-CH_2-CCl_2-$	Polyacrylate	$-CH_2-CH(C(=O)-O-CH_3)-$

(continued)

TABLE 2.1 (continued)

Chemical Structure Representation of the Repeat Unit for Some Common Synthetic Polymers

Polymer Name, Abbreviation	Repeat Unit	Polymer Name, Abbreviation	Repeat Unit
Polyacrylonitrile, PAN	$-\text{C}-\text{C}-$ with H, H above and H, CN below	Polymethylmethacrylate, PMMA	$-\text{C}-\text{C}-$ with H, CH_3 above; H, $\text{C}-\text{O}-CH_3$ (with =O) below
Poly(vinyl alcohol), PVAL	$-\text{C}-\text{C}-$ with H, H above and H, OH below	Polyamide n, PA n	$-\overset{\text{O}}{\text{C}}-(CH_2)_{n-1}-\overset{\text{H}}{\text{N}}-$
Poly(vinyl acetate), PVAC	$-\text{C}-\text{C}-$ with H, H above and H, $\text{O}-\text{C}-CH_3$ (with =O) below	Poly(ethylene terephthalate), PET	$-\text{O}-\text{C}-\text{C}-\text{O}-\overset{\text{O}}{\text{C}}-\bigcirc-\overset{\text{O}}{\text{C}}-$ with H,H above and below each C
Polydimethylsiloxane, PDMS	$-\overset{CH_3}{\underset{CH_3}{\text{Si}}}-\text{O}-$	Poly(bisphenol-A-carbonate), PC	$-\text{O}-\overset{\text{O}}{\text{C}}-\text{O}-\bigcirc-\overset{CH_3}{\underset{CH_3}{\text{C}}}-\bigcirc-$

of the physicochemical polymer nature, that is, polymer architecture, need introduction. The focus of attention is carbon-based synthetic polymers. In particular, primary covalently induced polymer architectures are considered. For the topic at hand, secondary or higher order configurational and intermolecular interactions are of less importance and only briefly mentioned where relevant.

Polymers are macromolecules characterized by their chemical makeup. In contrast to small molecules, they are not only defined by their molar mass (MM) but also by their molar mass distribution (MMD). It implies that polymers, in their most simple form, are in fact a mix of chemically identical macromolecules but with varying size as defined by MM or the number of repeat units. These fundamental characteristics determine the intrinsic physicochemical polymer properties and material behavior.

2.1.1 POLYMER ARCHITECTURE

The molecular composition of the repeat unit defines the basic polymer architecture. In the case of PE, the ethylene monomer $\{CH_2=CH_2\}$ is polymerized to give a macromolecule with $\{-CH_2-\}$ as repeat unit (and not $\{-CH_2-CH_2-\}$!). For most other polymers, the repeat unit is a close copy of the monomer (Table 2.1). The chemical composition of the polymer is further defined by the degree of polymerization (DP) or the number of monomers in the macromolecule. The DP is a direct measure for the length of the macromolecule and its MM. The MM of the macromolecule equals the MM of the repeat unit times DP. DP and MM are used interchangeably, provided, the macromolecule is linear and has only one type of repeat unit. For linear PE and PP, such an assumption is an acceptable estimate. However, in the case of a PA, with repeat unit $\{-NH-(CH_2)_n-CO-\}$, water (H_2O) is abstracted from the monomer $\{NH_2-(CH_2)_n-COOH\}$ during the step-growth polymerization reaction. The difference between MM and DP cannot be ignored anymore.

The monomers associated with the repeat units represented in Table 2.1 are called *bifunctional*, indicating the number of covalent bonds by which a monomer is incorporated into a polymer. Polymers from bifunctional monomers form, strictly speaking, linear macromolecules.

Polymers synthesized from only one type of monomer A are called *homopolymers*. When two chemically different monomers A and B are used, a copolymer is formed. Similarly, when three chemically different monomers A, B, and C are used, a terpolymer is formed. Each additional monomer creates new possibilities to modify the polymer architecture. The relative concentration and distribution of the comonomer may change. It gives rise to several classes of copolymers (Figure 2.3): random copolymers (the monomers' order follows no regular sequence), alternating copolymers (the monomers have an alternating sequence distribution), and block copolymers (the monomers are grouped together in "blocks"). In block copolymers, the number of monomers in a block may change from two to as much as comprising

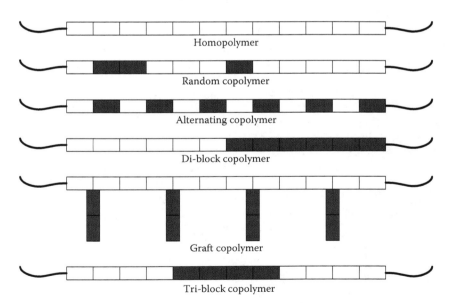

FIGURE 2.3 Schematic macromolecule representation using a flexible strand for homopolymers, random-, alternating-, and di- and tri-block copolymers. Each "rectangle" represents either the main monomer A (white) or the second (co-) monomer B (black). In polymers, a distribution of such macromolecules of similar chemical mer-units makeup, but different in MM should be imagined.

50 mol% of the total polymer. The number of monomers per block is frequently referred to as dyads (two), triads (three), tetrads (four), and pentads (five) [2] to define the distribution sequences. The combinations of different block lengths are possible in the same macromolecule. Several block copolymers can be envisioned (Figure 2.3).

It is possible to link a block of B monomers to a much longer macromolecule $\{-A_n-\}$ anywhere in between the end groups. In this case, the block of B monomers is said to be grafted to the main macromolecule, and one speaks of a *graft* copolymer. The main macromolecule is identified as the longest possible sequence of identical repeat units. The grafting introduces a branch in the polymer. Typically, branches are mer-units attached to the main macromolecule. Branches can be composed of the same monomer as the main macromolecule or can have a different chemical composition. Chemistry provides different pathways to introduce branches into a polymer. One approach is grafting, another is the use of higher functional monomers during polymerization. However, tri- and tetra-functional monomers tend to form interlinked macromolecules (thermosets) instead of highly branched ones when the chemistry is not controlled properly. A special type of polymer with a well-controlled branch structure, and polymerized from higher functional monomers is called a *dendrimer*. Typically, a trifunctional monomer connects to three trifunctional monomers to yield six new reactive attachment sites or branch points. The subsequent reactions of the six trifunctional monomers increases the number of branch points by a factor of two. Each increase or doubling of branch points is called a *generation*, starting with generation zero, the monomer. If the polymerization process is steered away from the functionality doubling, the branching becomes irregular and so-called *hyperbranched polymers* are produced.

Bifunctional monomers can also give rise to branched polymers. Side reactions during chain-growth or step-growth polymerization may yield one or more reactive centers, leading to one or more branches on the main macromolecule [3]. Typically, a distinction is made between short chain branches (SCB) that have an oligomeric nature, and long chain branches (LCB) that have a much longer length. How long is long remains a point of discussion and, in the literature, no absolute criterion is defined as a measure. For example, in PE, SCBs are associated with the pendant groups (methyl, ethyl, butyl, hexyl, etc.) introduced either via side reactions or via the addition of co- or termonomers (e.g., propene-1, butene-1, 4-methyl-pentene-1, pentene-1, hexene-1, and octene-1). For most homopolymers, the single pendant atom or group of atoms, such as methyl in PP, chlorine in PVC, and benzyl in PS, is not considered an SCB (although they are in the strictest sense). The maximum length of the SCB could be defined as anything shorter than LCB. For polymer melts, the length of a long branch is typically defined as the length of a macromolecular segment with a MM, M_e, being equal to the average MM between entanglement coupling points [4]. The MM, M_e should be considered as a measure of the spacing between topological constraints (or the entanglement coupling points) (Figure 2.4). Although this definition may have no clear physical meaning, it represents a number that can be calculated from the rheological properties of the polymers using the "rubber elasticity" theory. For PE, M_e is reported as 1250 g/mol at 190°C [4]. With a mer-unit MM of 14 g/mol, this would define the length of an LCB to be about 90 {–CH$_2$–} units long. By definition, the maximum length of a single LCB cannot be longer than half the length of the main macromolecule.

Next to the length of branches, it is possible to distinguish between at least four different types of branched polymers (Figure 2.5).

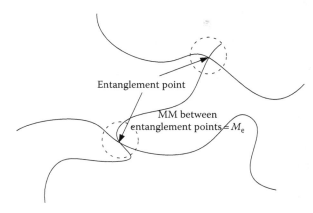

Entanglement point

MM between entanglement points $= M_e$

FIGURE 2.4 Many macromolecules together will intertwine forming loops and points of connection between different entities. These are typically called entanglement points or, in terms of molecular dynamics, topological constraints, as such regions restrict the free motion of a macromolecule. The schematic representation of two such entanglement points suggests a distance characterized by the MM M_e of that part of the macromolecule that is constrained.

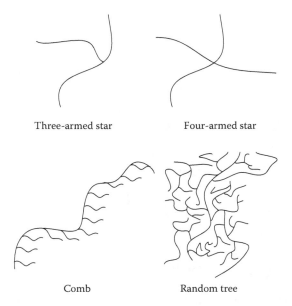

Three-armed star   Four-armed star

Comb   Random tree

FIGURE 2.5 Schematic representation of different branch types in polymers. A comb, a three-armed and a four-armed star, as well as a random tree type of branching are shown.

A regular comb refers to many branches of equal length attached to branch points equally spaced along the macromolecule. In principle, PP or any other linear homopolymer except for PE could be considered to have a comb-type branch structure. The sequence order of comb-type branches can also be irregular, and the length of the branches may vary (also in one macromolecule). Polymers in which the branches are of equal length and attached to a single branch point are called regular three- or four-armed stars. The branch points are tri- or tetra-functional. The arm length can be irregular. When the branches vary in length and are randomly organized on the main macromolecule as well as on the branches, one speaks of a random tree resembling the branching pattern of a tree.

2.1.2 MOLAR MASS DISTRIBUTION

Up till now, the terms "macromolecule" and "polymer" were used interchangeably. However, there is more to a polymer than just being a very long, flexible (macro) molecule. Chemically, small molecules, for example, water and benzene, are characterized by a single MM (respectively, about 18 and 78 g/mol), which implies uniform molecules. The synthesis of polymers always yields a collection of macromolecules of different MM, that is, nonuniform macromolecules. This results for all polymers in an MMD characterized by a mean and a spread of MM. Therefore, the term polymer needs to be understood as being a collection of nonuniform macromolecules. The typical MMD of polymers is determined by the polymerization boundary conditions (pressure, temperature) and the probability function that governs the polymerization mechanism (Figure 2.6).

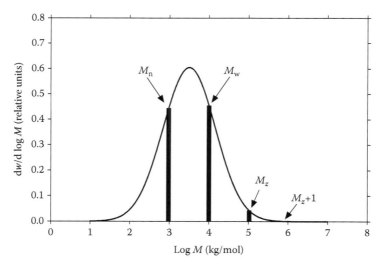

FIGURE 2.6 A typical representation of a polymer's MMD. The normalized weight w (indicating a relative amount) is presented versus the logarithm of the MM. The MM averages (M_n, M_w, M_z, M_{z+1}) or moments of this simulated distribution function are 1,000, 10,000, 100,000, and 800,000 kg/mol, respectively.

TABLE 2.2

Definitions of Molar Mass Moments of Polymers Can Be Presented in Terms of the Number of Macromolecules n_i of Molar Mass M_i or the Weight w_i of Macromolecules of Molar Mass M_i

Number of Macromolecules n_i of Molar Mass M_i	Weight of Macromolecules w_i of Molar Mass M_i
$$M_n = \frac{\sum_i n_i M_i}{\sum_i n_i}$$	$$M_n = \frac{\sum_i w_i}{\sum_i w_i / M_i}$$
$$M_w = \frac{\sum_i n_i M_i^2}{\sum_i n_i M_i}$$	$$M_w = \frac{\sum_i w_i M_i}{\sum_i w_i}$$
$$M_z = \frac{\sum_i n_i M_i^3}{\sum_i n_i M_i^2}$$	$$M_z = \frac{\sum_i w_i M_i^2}{\sum_i w_i M_i}$$
$$M_{z+1} = \frac{\sum_i n_i M_i^4}{\sum_i n_i M_i^3}$$	$$M_{z+1} = \frac{\sum_i w_i M_i^3}{\sum_i w_i M_i^2}$$

Note: The units of the MM is g/mol or Dalton (Da).

As a consequence, the MM of a polymer can only be expressed in terms of average numbers or so-called MM moments (Table 2.2). The most commonly used moments are the number-average MM, M_n, and the weight-average MM, M_w. The ratio M_w/M_n is used as a measure defining the width of the distribution and is called the MM dispersity [5]. A ratio M_w/M_n equal or nearly equal to one, refers to a uniform or nearly uniform MMD. An important special case, the "symmetric monomodal" MMD, for example, a normal or lognormal distribution, is fully defined by only two numbers that can be associated with M_n and M_w, or M_w and M_w/M_n. For most polymers, the MMD is not symmetric indicating that an average MM and an MM dispersity cannot define the entire MMD in sufficient detail (Table 2.2). Each MM moment provides its own fractional information about the composition of the MMD. Such limitations severely restrict correlating detailed polymer architecture to material behavior, in particular rheological features.

It is possible to express the MM moments in terms of the number of macromolecules n_i of mass M_i, but it is often more convenient to work with the weight of the macromolecule. The average that accounts for the fact that large molecules hold more of the total MM M_w is often more representative than the mere average of the number of MMs, M_n. The weight w_i of the macromolecules of mass M_i is related to the number of macromolecules n_i:

$$n_i = \frac{N_A w_i}{M_i} \qquad (2.1)$$

where N_A is the number of Avogadro (602×10^{21} mol^{-1}).

Example 2.1:   Moments of a Molar Mass Distribution

The information contained in the MM moments of an MMD can be understood by blending 90 kg of polymer A having MM 200 kg/mol and 10 kg of polymer B of MM 20 kg/mol, and 90 kg of polymer A with 10 kg of polymer C of MM 2000 kg/mol.

By using the equations in Table 2.2, the MM moments of the blend can be calculated. The implicit assumption is that both polymers are uniform. The addition of polymers with a low or a high MM in equal quantities has a different influence on the calculated MM averages (Table 2.3). Adding a small amount of low MM B (20 kg/mol) to a polymer with $M_n = M_w = M_z = M_{z+1} = 200$ kg/mol has a small influence on M_w (a reduction from 200 to 182 kg/mol) and practically none on M_z and M_{z+1}. In contrast, M_n is reduced to half (from 200 to 105 kg/mol)! Adding a small amount of high MM C (2000 kg/mol) has a strong influence on M_w (an increase from 200 to 380 kg/mol) and an even stronger influence on M_z and M_{z+1}. M_n shows only a slight increase (from 200 to 220 kg/mol). The exercise shows that M_n is very sensitive to low MM changes while M_w and the higher molecular moments are very sensitive to high MM changes.

TABLE 2.3
Molar Mass Moments of Two Polymer Blends Obtained by Combining (1) 90 kg of Polymer A of Molar Mass 200 kg/mol and 10 kg of Polymer B of Molar Mass 20 kg/mol, and (2) 90 kg of Polymer A with 10 kg of Polymer C of Molar Mass 2000 kg/mol

	90 kg A + 10 kg B	90 kg A + 10 kg C
M_n	105	220
M_w	182	380
M_z	198	1147
M_{z+1}	200	1851
M_w/M_n	1.09	3.02

Example 2.2: Molar Mass Moments of a Blend

Calculating the MM moments for blends of two very narrowly distributed, nearly uniform MMDs (M_w = 200 (A) and 20 (B) kg/mol each with M_w/M_n = 1.1) shows how they are affected when the relative blend composition varies (Figure 2.7). Polymer A has M_n = 182 kg/mol, M_z = 222 kg/mol, and M_{z+1} = 242 kg/mol. The addition of the low MM component practically does not influence the M_z and M_{z+1} MM moment.

The combination of MMD variation with the many polymer architecture options offers many possibilities to design different polymers and leads to the definition of polymer subclasses.

First, it is possible to distinguish between homogeneous and heterogeneous polymers. Homogeneous polymers have the same macromolecular architecture independent of the MM. For example, a homogeneous ethylene–propylene copolymer will have an identical amount of propylene comonomer on a mol% basis in the low and high MM macromolecules. Mol% is defined as the number of monomers in a macromolecule composed of 100 monomer molecules. Heterogeneous polymers have a varying mol% comonomer depending on the macromolecules' MM. In fact, heterogeneous polymers are a blend of two or more different polymer architectures, each of which may have their own MMD.

Second, each industrial producer of polymers that has a proprietary polymerization process technology and operational discipline will make, in fact, a "different" polymer.

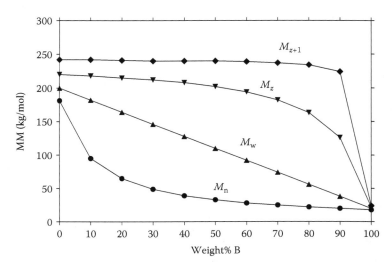

FIGURE 2.7   The influence on the MM moments as functions of the changing relative amounts when mixing two very narrowly distributed MMD of M_w/M_n = 1.1, and having a M_w of (A) 200 and (B) 20 kg/mol, respectively.

2.1.3 POLYMERIZATION PROCESSES

The general chemistry routes for polymer synthesis are well documented [6]. Polymers are synthesized by either step-growth (polycondensation) or chain-growth (polyaddition) polymerization.

In step-growth polymerization, two different monomers are combined while small molecules are released (e.g., H_2O or CO_2). In the case of PA, nylon 6,6, for example, an organic acid (adipic acid) is combined with an amine (hexamethyldiamine) to form a polymer and water. The kinetics of polymerization is not affected by the size of the reacting species. The degree of polymerization ($DP = M/M_0$, with M_0, the monomer MM) is defined by the reaction time and the reaction kinetics of the monomers (assuming that the monomers are not depleted in this process). The MM of the macromolecules is fairly uniform and the MMD relatively narrow. The MM of step-growth polymers is often much lower than that of chain-growth polymers, that is, an indicative M_w between 10^4 and 10^5 versus 10^5 and 10^6 for the latter.

A chain-growth polymerization involves three consecutive stages: initiation, propagation, and termination. Each macromolecule is individually initiated and grows (propagates) very rapidly to a high MM until its growth is terminated. Depending on the type of initiation, it is possible to distinguish between free radical, cationic, anionic, and coordination complex polymerization (mostly an anionic surface-active system). In the respective cases, the reacting carbon atom either contains a free radical, is positively charged, negatively charged, or forms a coordination complex with a transition metal (e.g., titanium (Ti) or chromium (Cr)). In principle, relatively narrow MMDs ($M_w/M_n = 1.1$–2) can be obtained. However, all kinds of side reactions can happen at each stage, depending on the reaction mechanism, reaction temperature, pressure and time, catalyst type, cocatalyst, and the purity of the monomers. Therefore, chain-growth polymerization often yields polymers with broad MMDs ($M_w/M_n > 2$).

The differentiation and generation of subclasses for polymers with the same repeat unit is also a consequence of the strive for what is economically desirable and technically feasible. On a commercial level, a process technology choice will be a commitment for many years in view of the very large amount of capital required. Large-volume, low-cost processes are driven by maximizing the output for the lowest possible investment while still producing an acceptable product. Low-volume, high-cost processes are typically aiming for more technically demanding polymer applications. Irrespective of size and efficiency, operational cost minimization continuously drives the search for improved engineering solutions. This has led to the development of several process methodologies for polymerization. The most widely practiced in terms of polymer volume are bulk, solution, gasphase, suspension, and emulsion polymerization. In bulk polymerization, a liquid monomer is initiated to react and produce the desired polymer. In the case of an exothermic free radical polymerization, the heat buildup becomes a problem, and the removal of all monomers in the final product is a challenge. Solution and gasphase technology are ways to circumvent heat transfer and residual monomer issues. Suspension polymerization is in fact a "microscale" bulk polymerization. Small droplets of monomer in an immiscible liquid react to form powders of >100 μm. In emulsion polymerization, water is the heat transfer medium that contains a water-soluble initiator and reactive immiscible monomer droplets stabilized with surfactants.

Most technologies are proprietary and protected in patents. The key differentiation often lies more with the applied catalyst than the process engineering. Strictly, catalysts accelerate the reaction rate but do not affect the chemical reaction. Industrial catalysts, however, are sophisticated systems that do affect the nature of the macromolecule being grown. Particularly, coordination complex catalysis changes the boundary conditions for monomers to be initiated and for "micro"-molecules to grow into macromolecules. Heterogeneous coordination complex catalysis is mainly controlled by surface phenomena. The catalyst is a solid, or attached to a solid, suspended in the reaction medium. Homogeneous catalysis uses typically similar coordination complexes, but as organometallic compounds soluble in the reaction medium. The industrially synthesized polymers originating from different producers and/or processes are understandably different. The polymer architectural and MMD differences are reflected in the physicochemical and processing properties producers offer into the market. The variety of polymer subclasses that can be produced may be illustrated when considering the polymerization of ethylene monomer into PE.

Ethylene gas is produced by heating a petroleum derivative (liquefied petroleum gas [LPG] or naphtha) in a furnace (a steam-cracker) at a high temperature (~800°C) for a very short time (<s). Together with ethylene, other useful polymerizable monomers are produced, propene-1, butene-1, and benzene. The ethylene gas is commercially converted into PE using either low or high pressure polymerization process technology (Table 2.4) [7–9].

The high-pressure, bulk process yields low-density PE (LDPE) made in either an autoclave or a tubular reactor. At very high temperatures (~200°C–300°C) and pressures (~1000–3000 bar (×10^5 Pa)), the ethylene monomer is polymerized according to a free-radical process initiated by air, oxygen, or peroxide catalysts. During a very short residence time, ranging from 20 to 50 s, about 15%–30% of ethylene is converted to polymer. The difference between an autoclave (high-pressure vessel) and a tubular reactor (very long high-pressure tube) has mainly a historical reason as it reflects the evolution in process technology. However, autoclave reactors tend to have lower ethylene conversion and yield more LCB containing random tree–type polymers as compared to tubular LDPE. In either case, the polymers contain a high number of SCB (ethyl and butyl groups) as a consequence of side reactions and the use of polymerization-controlling agents also known as *chain transfer agents*. The SCB restricts the density between 915 and 930 kg/m³. The MMD of LDPE is usually very broad.

The low-pressure processes can be subdivided into solution, gas phase, and slurry technology. The polymerization conditions of pressure and temperature are bounded by the solubility of the polymer (solution process) or its melting point (gas phase and slurry process). Low-pressure processes have a much higher catalyst efficiency and ethylene conversion (up to 95% or higher). They offer a higher degree of flexibility to synthesize ethylene at lower pressures and temperatures and tend to be more economic to operate than the high-pressure processes. Comonomers are extensively used in low-pressure technology to control the solid-state density of polymers. The PE density is indicative for the comonomer concentration of low-pressure processes.

TABLE 2.4

Overview of the Main Ethylene Polymerization Technologies Leading to a Subclassification of PE according to Polymer Density

	High-Pressure		Low-Pressure		
Process	Autoclave	Tubular	Solution	Gas Phase	Slurry
P (bar) ($\times 10^5$ Pa)	1000–3000	1000–3000	20–100	20–100	1–20
T (°C)	>200	>200	100–200	50–100	50–100
Catalyst	Air peroxides	Air peroxides	Ziegler–Natta metallocene	Ziegler–Natta metallocene	Ziegler–Natta metallocene
Reaction	Radical	Radical	Coordination	Coordination	Coordination
Polymer	Random tree SCB+LCB	Random tree SCB+LCB	Linear, comb Homopolymer Copolymers SCB+LCB	Linear, comb Homopolymer Copolymers SCB+LCB	Linear, comb Homopolymer Copolymers SCB+LCB
M_w/M_n	10–30	10–30	2–10	2–35	5–30
Density (kg/m³)	915–935	915–935	865–965	890–965	935–965
Name	LDPE	LDPE	ULDPE VLDPE LLDPE MDPE HDPE	VLDPE LLDPE MDPE HDPE	MDPE HDPE UHMPE

It has led to a loosely employed subclassification of PE in terms of high-density—(HDPE) (950–965 kg/m³), medium-density—(MDPE) (935–950 kg/m³), linear-low-density—(LLDPE) (915–935 kg/m³), very-low-density—(VLDPE) (890–915 kg/m³), and ultra-low-density PE (ULDPE) (860–890 kg/m³). Depending on the PE producer, different comonomers (butene-1, pentene-1, 4-methyl-pentene-1, hexene-1, and octene-1) are preferred. The low pressure processes use heterogeneous Ziegler–Natta or homogeneous metallocene catalysts.

A Ziegler–Natta catalyst consists of organo-transition metal (mainly Ti and Cr) molecules, typically fixed (supported) onto an inorganic particle (e.g., $MgCl_2$ or SiO_4). Each catalyst-containing particle has many active catalyst molecules (sites) that each polymerizes a different macromolecule. Such multisite catalysts yield heterogeneous, nonuniform macromolecules. Each site has a different polymerization activity related to sterical hindrance effects, monomer diffusion limitations, and comonomer reactivity differences. The MMD for Ziegler–Natta-catalyzed polymers can vary from an M_w/M_n of 3 to as much as 30.

In contrast, metallocene catalysts are so-called *single-site catalysts*. They can also be linked to a supporting inert medium. The unsupported version is soluble in the reaction medium and allows for a homogeneous catalysis. The macromolecules synthesized with unsupported metallocene catalysts yield fairly homogeneous, nearly uniform (M_w/M_n of about 2) macromolecules with a nearly random distribution of

SCB (in case of copolymers). Supported metallocene products can be heterogeneous but still have a very narrow MMD.

Irrespective of the catalyst and comonomer, additional branching can be formed in low-pressure processes either because of side reactions during polymerization or thermo-mechanical treatment during the finishing steps in the process (e.g., solvent removal or densification of powder into granules with extrusion). Side reactions generally induce unsaturated carbon–carbon bonds or unreacted free radicals that either react forming further branches during the polymerization or during subsequent thermo-mechanical treatments. To prevent the latter reactions, additional chemicals (additives) are mixed into nearly all commercially available PE (and other types of polymers).

Besides the companies' economic benefits of product differentiation and sustainable competitiveness, the multitude of polymerization technologies places a major challenge for scientists for the accurate characterization of the polymer architecture and MMD. Moreover, "polymers" become "plastics" when commercially traded. For various reasons, polymers are loaded with all kinds of additives either as a consequence of the polymerization technology, for example, catalyst or solvent residues, or as defined by the application performance needs, for example, antioxidants, processing aids, or fillers. This procedure applies to all polymers with few exceptions, that is, those used for medical or pharmaceutical applications.

2.2 POLYMER CHARACTERIZATION

The multitude of polymer classes and the compositional variety within one polymer class challenge the characterization resources for defining exactly what the polymerization process has actually made. A key issue for understanding and preventing the occurrence of polymer melt fracture is defining which polymer characteristic induces the phenomenon and which one prevents it. A single, absolute polymer characterization tool that allows a complete and unambiguous definition of the polymer architecture and MMD does not exist. Even if available, the challenge of finding a quantitative relation for the onset of polymer melt fracture would remain.

Polymer characterization is made possible through a wide variety of experimental techniques. Most require highly specialized analysts mainly for interpreting the experimental results in relation to the validity of the hypothesis and the supporting theory, the absolute or relative character of the findings, and the precision and accuracy of the method applied. Commonly used techniques can be roughly divided into those probing the average chemical composition of an ensemble of macromolecules and those that measure average bulk/material properties.

2.2.1 POLYMER ARCHITECTURE

A detailed reconstruction of the polymer architecture and MMD from a liquid, paste, powder, or granule focuses primarily on identifying the polymer repeat unit. It entails direct experimental observations and interpretations using established theoretical understanding. It is a deductive process combining findings at various levels of time and length scale—from the molecular to the macroscopic. Advances in analytical technology continue to bring more detailed and accurate characterization [10–13]. Here, only a few of the many possible analytical techniques are mentioned in view of their relevance to the subject matter of this book.

2.2.1.1 Spectrometry: Ultraviolet, Visible, Infrared, and Raman

Spectrometric techniques measure the intensity and wavelength (or frequency) of the absorption and emission of electromagnetic radiation. Molecules change their energy state when interacting with electromagnetic radiation ("light") either by increasing (absorption) or lowering (emission) their energy state.

When absorption happens at wavelengths between 100 and 1000 nm, ultraviolet-visible light-sensitive molecules are present. The UV-VIS technique is typically used for characterizing aromatic molecule containing polymers, for example, PC, PS [14].

All atoms in molecules vibrate at temperatures above absolute zero (0 K). The vibration frequencies are characteristic and unique for different groups of atoms referred to as *functional groups*. These frequencies can be detected using infrared (IR) light that falls in the wavelength band between 2.5 and 30 μm or the wave number (the number of waves per centimeter) ranging from 4000 to 330 cm^{-1} (Figure 2.8). Near-IR (2.5–0.8 μm) and far-IR (1000–30 μm) may provide additional information.

Traditionally, IR instruments were dispersive and double-beam using a prism or grating to expose the sample to isolated frequencies. By scanning the consecutive frequencies, an IR-absorption spectrum was obtained. Today, Fourier transform infrared (FTIR) spectrometers equipped with a Michelson interferometer [15] measure an interferogram using a laser beam transmitting through the sample. A Fourier transformation provides the IR spectrum.

Raman spectrometry complements FTIR but in fact is a scattering technique. The sample is illuminated with a very-high-intensity laser beam with frequency in the visible, near-IR or near-ultraviolet range. The part of the scattered laser light that has undergone a frequency change is detected and is representative to the characteristic vibration frequency of the functional group.

These spectrometric techniques provide characteristic absorption bands corresponding to a well-tabulated functional group vibration mode. However, a reference or calibration curve is required to obtain quantitative results. The preferred mode of operation for the quick identification of a polymer is "fingerprinting." Each IR spectrum is characteristic of the polymer architecture. Comparing experimental findings with IR-spectrum libraries containing thousands of reference spectra allows for a first-pass identification polymer characterization [16].

2.2.1.2 Nuclear Magnetic Resonance Spectrometry

Nuclear magnetic resonance (NMR) spectrometry, also known as magnetic resonance imaging (MRI) [17–20], is a key analytical tool to quantitatively elucidate the detailed molecular composition of a polymer. Without getting into the detailed physics of the technique, NMR exploits differences in the environment surrounding each atom in the macromolecule. To that purpose, strong magnetic fields are used in combination with a radio frequency tuning operation. In this fashion, the concentration of, for example, carbon (C) atoms surrounded by only two or more carbon atoms is identified as different. However, this is only possible for isotopes with non-equal

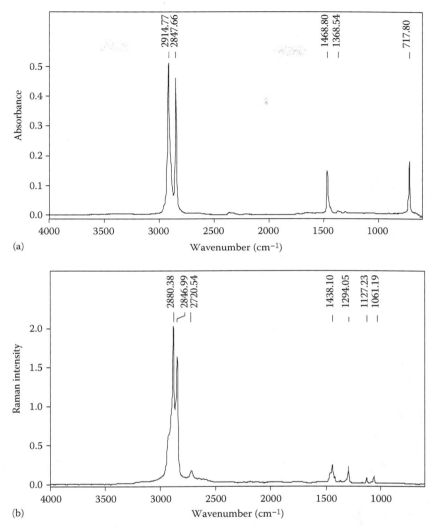

FIGURE 2.8 A typical FTIR-spectrum (a) and Raman spectrum of a 1 mm thick poly(ethylene-*co*-1-octene) LLDPE disk presented as absorbance respectively Raman intensity versus wave number (b). Raman spectrometry provides more detailed structural information in the 1600–600 cm⁻¹ range. (Courtesy of M. Plass from The Dow Chemical Company.)

numbers of protons and neutrons. Hydrogen (^{1}H) and carbon (^{13}C) are commonly used for synthetic polymers. A typical ^{13}C NMR spectrum of a poly(ethylene-*co*-1-octene) (LLDPE) shows peaks associated with ^{13}C carbons, each experiencing a different environment (Figure 2.9).

The position of each peak, expressed as the chemical shift parameter (δ) (in unit parts per million (ppm)), is identified relative to a reference signal of a well-known standard material, often tetramethylsilane (TMS). Recording the ^{13}C NMR spectra

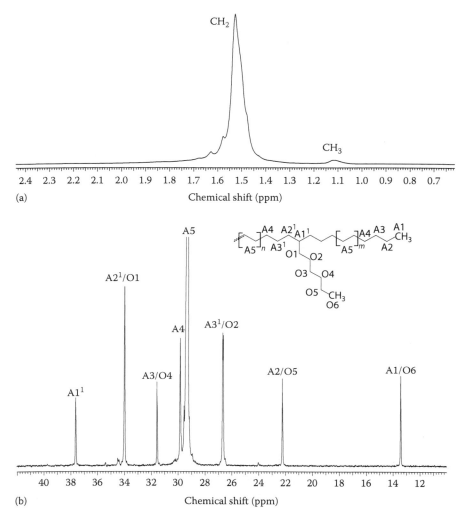

FIGURE 2.9 ^{1}H (a) and ^{13}C NMR (b) spectra of a poly(ethylene-*co*-1-octene) LLDPE dissolved in tetrachlorobenzene. The peak sizes are the quantitative measures of concentration, and the peak position or chemical shift is defined by the surrounding atoms. (Courtesy of N. Ni Bhriain from The Dow Chemical Company.)

of polymers can require very long sampling times (from 48 up to 72 h) because of the low abundance of sensitive isotopes in nature. NMR can be applied to polymer solutions and polymer solids at ambient and higher temperatures. This technique is quantitative and can be applied to determine the mean concentrations of SCB and LCB, as well as branching distribution and other polymer architectural details [19]. For PE, NMR is capable of quantifying the type, the length, and the distribution of branches induced by comonomers or side reactions [20]. Distinct peaks for branch point carbons that contain fewer than about six carbons can be detected. The quantitative detection of macromolecular end groups allows for obtaining a number-average MM.

2.2.1.3 Density

Although density [19] is a material "bulk" property, it is extensively used as a characterization tool—although an indirect one—of the polymer architecture. For all polymers, density (ρ) is a standard measure. The property indicates the ratio of the mass (M) and volume (V) and has the units of kg/m^3. The reciprocal density is called the specific volume ($v = \rho^{-1}$). Mass is a fundamental material property and can be determined unambiguously. Volume is more difficult to define as it depends on the packing quality of polymers. Packing is a complex property influenced by the electronic structure of atoms, by the type of bonding forces, and by structural and spatial variations in the macromolecules [21]. Density is usually reported at room temperature and measured according to standard test methods such as ASTM D 792 (Water displacement method) [22], ASTM D 1505 (density gradient column method) [23], and ASTM D 4883 (ultrasound technique) [24]. It is a very sensitive measure to differentiate between polymers (e.g., PE and PVC) but also to identify different PE subclasses. Density is a property that changes with pressure and temperature.

2.2.1.4 Thermal Analysis

Beyond molecular spectrometry and chemical techniques, the elucidation of the polymer architecture is mainly an effort of exploiting physical principles at the macroscopic scale. Density is one example implicitly providing information on how the polymer architecture affects the packing of the macromolecules. Similarly, the analysis of the thermal characteristics of polymers can provide such information. There are multiple approaches but differential scanning calorimetry (DSC), crystallization fractionation, and thermal decomposition are the most representative.

Macromolecules will, depending on the thermodynamic constraints, organize themselves and stack up to form "crystals" or remain unorganized and "amorphous." Polymers containing crystals will have a "melting point" while amorphous polymers only have a glass transition temperature T_g—a temperature below which polymers have glass-like properties, that is, hard and brittle. DSC [12,13,19,25] is an analytical technique that measures the heat (enthalpy) exchanges of the sample with its environment and referenced versus a material of known heat capacity. A temperature scan (typically −30°C to 250°C) at an arbitrarily selected rate (e.g., 10°C/min) is performed of the sample (Figure 2.10). A deviation from the reference indicates a thermal process in the sample of either using heat or generating heat. The melting of the "crystals" uses an amount of heat that is quantitatively proportional to the amount of "crystals," that is, the amount of macromolecules capable of organizing themselves. For PE, this has a direct bearing on the nature and amount of SCB. In block copolymers, it may be indicative of the block size and concentration. Similarly, DSC can detect the T_g. In combination with the theoretical understanding of tabulated T_g and chemical structure parameters [21], polymer architecture can be deduced.

Crystallization fractionation (CrysTAF) [26] is a technique to fractionate macromolecules based on their ability to organize and form "crystals." The less SCB a macromolecule contains, the better its ability to form crystals. This technique is

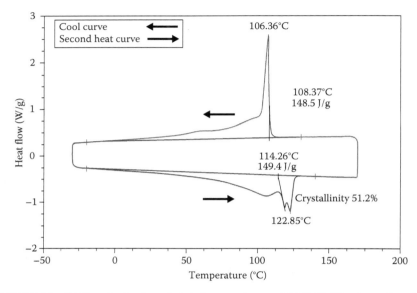

FIGURE 2.10 DSC analysis of a poly(ethylene-co-1-octene) LLDPE. The thermal analysis is typically done at a rate of 10°C/min and encompasses a first heating (not shown) above the melting point of the semicrystalline polymer followed by a cooling (top) and a second heating (bottom). The latter thermogram indicates the heterogeneous semicrystalline nature of LLDPE polymers as reflected in the multiple melting peaks. (Courtesy of C. de Zwart from The Dow Chemical Company.)

particularly useful for distinguishing PE subclasses. A DSC apparatus is used, but now the polymer sample is immersed into a solvent. A temperature scan is started and the dissolution heat is detected. The assumption is that macromolecules with the least crystal forming ability, that is, those having the most SCB, will dissolve first (Figure 2.11).

A more invasive approach to polymer architecture determination is to decompose it. Heating the polymer above its decomposition temperature and detecting the degradation products with FTIR, chromatographic techniques, or mass spectrometry may help defining the chemistry of the macromolecules. This can be done under an inert or oxidative atmosphere to provide additional details of the chemical configuration. The approach can be very instructive for step-growth thermoplastics containing various functional groups. Pyrolysis and gas analysis [27], or thermogravimetry-combined gas analysis [28] qualify as analytical techniques.

2.2.2 MOLAR MASS, MOLAR MASS DISTRIBUTION

Common MM and MMD determination techniques make use of dilute polymer solutions. Typically, and for most polymers, they are very reliable and accurate if the polymer can be easily dissolved—mostly synthetic amorphous polymers (e.g., PS, PC, and PMMA). Synthetic, high-MM, semicrystalline polymers are far more challenging to dissolve (e.g., PE and PP). The experimental MM moments or MMD

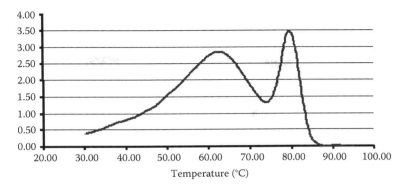

Temperature (°C)

FIGURE 2.11 A crystallization fractionation graph of a poly(ethylene-co-1-octene) LLDPE as obtained using DSC. A relative response indicative of the heat of dissolution is presented as the function of temperature. The information obtained is comparable to TREF. The two peaks, respectively, at low and high temperature represent high comonomer-containing macromolecules and very low to no comonomer-containing macromolecules (cf. Figure 2.15). (Courtesy of C. de Zwart from The Dow Chemical Company.)

are more difficult to reproduce. MM and MMD results should be considered with care and expert knowledge on how the polymer was synthesized [29,30]. Of particular concern is the representativeness of the very small sample size—a few (or less) micrograms of polymer in solution—for the larger mass of the polymer. Additionally, the use of different and multiple detectors generates different MM moments. The accuracy of the findings needs reporting in combination with the technique used for the proper reference and use of the data. Alternatively, melt flow characterization techniques are used to measure properties from which, indirectly, MM and MMD information can be extracted.

2.2.2.1 Dilute Solution Viscosity

A polymer dissolved in a solvent increases the viscosity of the solution. Dilute solution viscosity is a measure of the average size of the longest contiguous macromolecule in solution. It is implied with dilute solutions that the macromolecules are completely surrounded by solvent molecules and that there are no intermacromolecular interactions. The macromolecules take the shape of a random coil (Figure 2.12). The volume occupied by the macromolecule in the solvent is referred to as the *hydrodynamic volume*, and is characterized by a radius of gyration. The latter is defined as the *root-mean-square average* of the distances between molecular segments from the centre of gravity of the coil and represented as ($\langle S^2 \rangle^{1/2}$). The coil size of a linear PE with MM of 70,000 g/mol is about 20 nm in solution but may vary depending on the solvent [31].

FIGURE 2.12 A schematic representation of a macromolecule in a solution that forms a random coil of dimensions defined by the radius of gyration $\langle S^2 \rangle^{1/2}$. The hydrodynamic volume is suggested to occupy a sphere (dotted line).

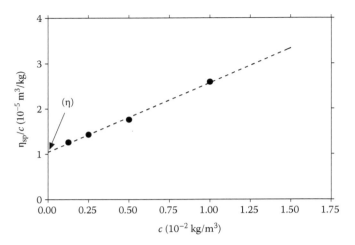

FIGURE 2.13   The specific viscosity of dilute polymer solutions of various polymer concentrations allows determining the limiting viscosity number [η] via an extrapolation to zero-concentration.

If the concentration of the polymer in solution is c, yielding a viscosity η, and the viscosity of the solvent is η_s, it is possible to define a quantity called specific viscosity η_{sp} as

$$\eta_{sp} = \frac{\eta - \eta_s}{\eta_s} \tag{2.2}$$

The measurement of the specific viscosity as the function of concentration yields a linear relationship that allows for an extrapolation to "zero-concentration" (Figure 2.13). The limiting viscosity number (in units of dL/g) defined in Equation 2.3 is also known as intrinsic viscosity [η]. The specific viscosity is typically measured by detecting the flow rate of a specified volume of solvent and polymer solution at a well-defined temperature using standardized glass capillaries (ASTM D445) [32]. Alternatively, a differential viscometer can be used to measure the specific viscosity directly. This is accomplished with a balanced network of four capillaries arranged in a manner analogous to a "Wheatstone bridge." A differential pressure proportional to the specific viscosity is measured across the "bridge" when a solution is injected into one of the capillaries, while solvent flows continuously over the other three capillaries [33,34]:

$$[\eta] = \lim_{c \to 0} \eta_{sp} = \lim_{c \to 0} \left(\frac{\eta - \eta_s}{\eta_s} \right) \tag{2.3}$$

Mark, Houwink, and Sakurada [35,36] established an empirical relationship between the limiting viscosity number and the MM for a set of (nearly) uniform polymers:

$$[\eta] = KM^a \tag{2.4}$$

For (nonuniform) polymers, the MM in Equation 2.4 is replaced by an average number: the viscosity average MM M_v. The values of K and a for various polymers can be found in *Polymer Handbook* [35,36]. The exponent a is commonly between 0.5 and 0.8 for ideal solvents.

The hydrodynamic volume of a macromolecule in solution depends on repeat unit type, comonomer level (in copolymers), branching level, and polymer/solvent interaction. Branched polymers have a smaller hydrodynamic volume than linear ones of the same MM. Therefore, the limiting viscosity number, which is smaller for branched polymers, can be used as a measure for branching. If it holds that branching only affects the radius of gyration, a branching factor, g, can be defined as the ratio of the root-mean-square radius of gyration for branched ($\langle S^2 \rangle_b^{1/2}$) and linear ($\langle S^2 \rangle_l^{1/2}$) polymers of the same MM:

$$g = \frac{\langle S^2 \rangle_b^{1/2}}{\langle S^2 \rangle_l^{1/2}}\Bigg|_M \tag{2.5}$$

The branching factor is a dimensionless quantity between 0 and 1, often related to a similar ratio g' based on the limiting viscosity number:

$$g' = \frac{[\eta_b]}{[\eta_l]}\Bigg|_M \approx \left(\frac{\langle S^2 \rangle_b^{1/2}}{\langle S^2 \rangle_l^{1/2}}\right)^{3/2} = g^{3/2} \tag{2.6}$$

Zimm and Stockmayer derived relationships between the branching factor and the number and type of branch points [37].

2.2.2.2 Light Scattering

The theory of light scattering photometry dates back to 1946 when Debye first used this technique for determining absolute MMs [38]. When a light beam of a certain wavelength illuminates a dilute polymer solution, light scatters in all directions. The intensity of the scattered light depends on the polymer concentration (c) and the scattering angle (Θ). The main parameter in the light-scattering theory is the Rayleigh ratio (R_Θ), which is defined as the ratio of the intensity (I_s) of the scattered light at an angle Θ and the intensity (I) of the incident beam times the scattering volume (V):

$$R_\Theta = \frac{I_s}{IV} \tag{2.7}$$

The scattering volume is essentially the volume of light measured in a conical space between the sample and the detector. The Rayleigh ratio is related to the absolute MM M, the polymer concentration c, and a constant K that contains all the instrumental and physical constants of the system:

$$R_\Theta = KcM \tag{2.8}$$

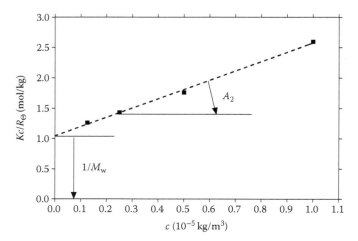

FIGURE 2.14 A typical LALLS plot to derive the absolute M_w and the second virial coefficient A_2.

For nonuniform polymers, Equation 2.8 is integrated over all MMs and the corresponding concentrations and leads to the determination of M_w. For most practical dilute solutions, Equation 2.8 must be adapted to include additional correction terms such as the second virial coefficient A_2. (Equation 2.8 is rewritten as a power series in terms of the variable c. The coefficients of the higher order terms of the series expansion are called virial coefficients.) After including the M_w and rearranging the factors, Equation 2.8 becomes

$$\frac{Kc}{R_\Theta} = \frac{1}{M_w} + 2A_2c + \cdots \tag{2.9}$$

The second virial coefficient can be obtained by plotting Kc/R_Θ versus concentration for different scattering angles and concentrations. An extrapolation to "zero-scattering angle" and "zero-concentration" defines M_w and A_2 (Figure 2.14).

The use of laser sources for light scattering and advanced detection systems have turned this extremely cumbersome analytical technique into a routine methodology for absolute MM determination. The high-intensity laser light source allows to measure at very low angles ($\Theta \sim 7°$) and very low concentration, and still obtains a reliable scattering signal. This technique is referred to as low angle laser light scattering (LALLS) [39–41]. Other light-scattering technologies measure the scattered light under different angles simultaneously ($\Theta \sim 15°–160°$), and are referred to as multiple angle laser light scattering (MALLS) [42].

2.2.2.3 Colligative Property Measurement Techniques

Vapor pressure lowering, boiling point elevation, freezing point depression, and osmotic pressure are the colligative properties of solutions. These properties depend solely on the number of particles in the solutions and not on their identity (e.g., size and mass). Analytical techniques measuring these properties of dilute polymer solutions obtain the number average MM of polymers, M_n. For typical high MM polymers, the changes to be detected are very small. It puts a high demand on the quality of the detection system and limits the use of these approaches for many polymers.

2.2.2.4 Gel Permeation Chromatography

Gel permeation chromatography (GPC), also known as size exclusion chromatography (SEC), is the most widely used analytical method to determine MMD and MM averages [43]. By injecting a dilute polymer solution in a separation column packed with a porous cross-linked gel, macromolecules are separated according to their hydrodynamic volume. Large macromolecules have little access to the pores of the gel and they are eluted after only a short period of time. Smaller macromolecules can permeate the porous gel better, follow a more tortuous path, and stay longer in the column. The accuracy and the degree of separating the macromolecules depend on the size distribution of the pores. The concentration passing through the column is recorded continuously as a function of time. The resulting GPC chromatogram represents the MMD in coordinates of a retention time or the equivalent, elution volume (mL/s; 10^{-6} m³/s) as function of a relative detector response (Figure 2.15).

GPC is a relative methodology, and a calibration curve is required to correlate the retention time (elution volume) to the MM of the different macromolecules. The nearly uniform PS of different MM is typically used as calibration standard to provide a highly accurate calibration curve. Beyond PS, that is, for other polymers, a calibration curve can only be derived by assuming that the size of the macromolecule is the only factor controlling GPC separation. Theoretical analysis shows that for nearly uniform polymers, the product of the MM (M) and the limiting viscosity number [η] is proportional to the hydrodynamic volume of the macromolecule in solution [44–46]. A graph of [η]M versus the elution volume provides a universal

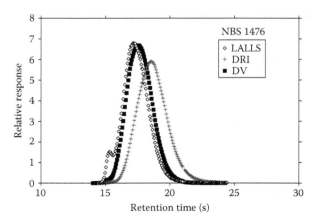

FIGURE 2.15 GPC chromatograms for a dilute National Bureau of Standards (NBS) 1476 PE solution using different polymer concentration detectors: a differential refractive index (DRI) device, differential viscosimeter (DV), and low angle laser light scattering (LALLS) system. (Courtesy of F. Bosscher from The Dow Chemical Company.)

calibration curve that is valid for all polymers [45,47,48]. Accordingly, for any elution volume, a correlation with the PS standards can be established, as shown in Equation 2.10:

$$[\eta]M = [\eta]_{ps} M_{ps} \qquad (2.10)$$

By combining Equation 2.10 and the Mark–Houwink–Sakurada equation (2.4), the MM of an unknown polymer using the PS calibration curve can be calculated:

$$[\eta]M = KM^{a+1} \qquad (2.11)$$

$$KM^{a+1} = K_{ps}M^{a_{ps}+1} \qquad (2.12)$$

The parameters K and a must be known for the unknown polymer and for PS (K_{ps}, a_{ps}) in the solvent used, and under the same experimental conditions.

The GPC device is often used in combination with more than one concentration detector. Most common is a DRI detector that determines the solvent-to-polymer solution ratio as the function of the polymer concentration eluting from the columns. A DRI detector is often used [49–51] in combination with a LALLS or MALLS detector and a DV (Figure 2.16). The DRI, LALLS, and DV signals may be different because of differences in high MM polymers detection sensitivity (Figure 2.15).

The combination of detectors allows determining the branching content, more specifically, the LCB in PE and other branched polymers. For heterogeneous polymers, the hydrodynamic volume of a low MM linear and a higher MM branched macromolecule can be the same. The absolute character of the LALLS/MALLS detector allows separating the macromolecule of identical hydrodynamic volume according to true MM. In combination with the DRI signal, a measure for LCB can be found [52,53].

2.2.2.5 Fractionation

GPC fractionates homogeneous polymers according to hydrodynamic volume. Heterogeneous polymers often require, in addition, other fractionation techniques that rely on differences in solubility, crystallization, and diffusion between macromolecules to determine the precise polymer architecture. Partial or successive precipitation fractionation, successive solution fractionation (SSF), and temperature rise elution fractionation (TREF) are a few techniques often used for this purpose [19].

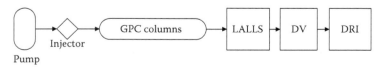

FIGURE 2.16 Schematic setup of a GPC device connected to multiple online detectors: DV, (multiple) LALLS, and DRI.

Most fractionation work is very time consuming and very labor intensive. A faster method for the fractionation of LDPE was developed by Desreux and Spiegel [54] based on a column elution technique. The principle of the technique is to precipitate polymer on a support packed in a column. By eluting the column with a solvent at gradually higher temperatures, individual fractions can be collected and analyzed. This (analytical) temperature rise elution fractionation ((A)TREF) technique is very helpful to define the heterogeneous comonomer distribution in LLDPE [55,56]. TREF, in combination with an IR detector, defines the number of SCB via the detection of the methyl end groups of the comonomer (Figure 2.17). TREF can also be combined with LALLS/MALLS detectors to obtain an indication of the MM of the branched macromolecules. The technique is still very time consuming, in particular, the column preparation. Crystallization fractionation using a DSC provides a less detailed but faster alternative.

2.2.2.6 Rheology

Rheology is the scientific discipline that studies the deformation of fluids. The fluid behavior during deformation is in part a reflection of the polymer architecture and MMD. Rheological material functions can be correlated to polymer parameters [21,57]. The challenge is the precision and accuracy of the experimental data to be correlated. Alternatively, theoretical considerations on polymer architecture have led to considerable insight into the rheological behavior of polymers. It has opened a pathway for a more accurate and practical application of rheology as a tool to direct polymer architecture design [31,58].

2.2.2.6.1 Melt Flow Rate

Probably the most widely used and simplest rheological test method is the melt flow rate (MFR). It is a measure for the ease of flow of polymer melts. A high MFR indicates easy flow or low viscosity fluid, while a low MFR indicates a slow flow or high viscosity fluid. Each polymer has its own specified condition for measuring MFR. It is specified in standard test methods (ASTM D1238) [59]. MFR is defined as the amount of polymer that emerges out of a capillary die of specified dimension in 10 min under the influence of a standard weight at a given temperature. For most commercial polymers, an empirical correlation with the M_n [60] and M_w has been reported, or can be derived easily [61,62]. By measuring MFR under different conditions, the ratio of the findings is often associated to the MM dispersity of the MMD with higher ratios reflecting a broader MMD. It is implied, however, that the correlation is only valid within a polymer class or subclass. It means that, for example, HDPE, LDPE, PP, and PS polymers can have the same MFR but their M_w and MMD will be very different. Furthermore, the MM moments of the calibration standards have to be known with sufficient accuracy to arrive at a meaningful quantitative correlation between MFR and MM. In all other cases, at best, indicative and empirical information is obtained. Extrapolations beyond the datasets are speculative.

2.2.2.6.2 Zero-Shear Viscosity

For dilute polymer solutions, the empirical Mark–Houwink–Sakurada relationship (Equation 2.4) provides a direct correlation between the limiting viscosity number

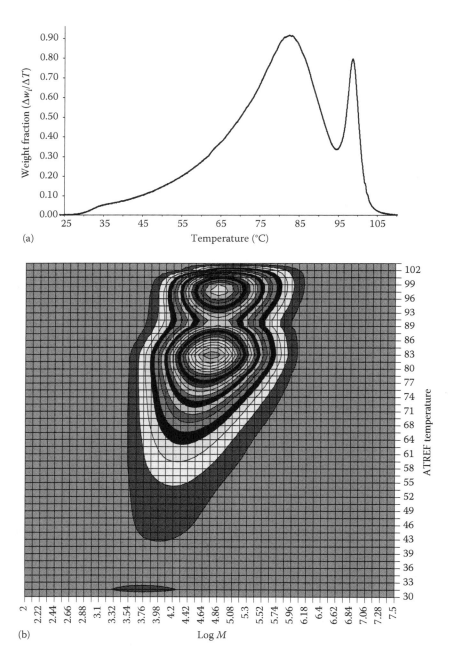

FIGURE 2.17 A TREF curve for an LLDPE (a) [56]. The lower temperature peak indicates the presence of macromolecules containing plenty of comonomers, while the higher temperature peak refers to the linear macromolecules. TREF, combined with LALLS, provides a response surface representing an MMD for each fraction (b). (Courtesy of F. Bosscher and L. Hazlitt from The Dow Chemical Company.)

and the MM of a uniform polymer—but only when the implicit assumptions are valid (cf. Section 2.2.2.1).

For polymers in an amorphous fluid state, that is, in the absence of any degree of order induced through crystallization or any other form of organization of the macromolecules, the empirical Mark–Houwink–Sakurada relationship is applied [63]. However, the limiting viscosity number is replaced by zero-shear viscosity η_0 and the MM by M_w (Equation 2.13):

$$\eta_0 = KM_w^{3.4} \tag{2.13}$$

Berry and Fox [64] demonstrated the validity of the relationship for many polymer melts. For all linear polymer melts, the coefficient a of Equation 2.4 becomes 3.4, provided the MM is above a critical value M_c and below which a becomes equal to 1. Any deviation from that number is attributed to either the presence of branching containing macromolecules, microstructure formation (e.g., through aggregation), or nonuniform polymers with a very broad, often skewed, MMD [65].

Over the years, several empirical correlations have been established to include these deviations by adapting the Mark–Houwink–Sakurada/Berry–Fox relation [3,65]. The applicability of such equations is hampered by the accuracy of the MM moments and the difficulty of experimentally defining a zero-shear viscosity for most commercially available polymers.

2.2.2.6.3 Rheology and Polymer Characteristics

Rheological experiments have the appeal of being fast, sensitive to polymer characteristics, and highly reproducible with a minimum sample size. Measurement of material functions (e.g., viscosity, modulus) in various modes of deformation—shear, extension, and compression—holds the promise of extracting polymer architecture and MMD information. Unfortunately, this is an ill-posed mathematical inversion problem. Many polymer design solutions are possible that fit a given material function. Without additional information to guide the inversion algorithm, it is not possible to find a unique solution. Many approaches have been reported typically for linear polymers only and with mixed success. The limited applicability of the approach is often related to the incompleteness of the material function or the absence of experimental data over a wide-enough timescale window. Even in view of the apparent success of some methods for a number of well-defined polymers, the complexity of the polymer characteristics defies the best of inversion techniques as yet [66–71]. Chapters 3 and 8 will address further the challenges of rheology and polymer characteristics.

2.3 GENERAL OBSERVATION

Polymers are versatile materials. Their macromolecular nature is both challenging to comprehend and fascinating to study. To begin understanding polymer melt fracture, it is essential to understand the basic elements of polymer architecture. Of particular importance is knowing how to elucidate polymer characteristics with what tools and within which constraints. This chapter can only be a (very) brief summary of the many scientific insights and theories that have been published over the years. It aims, however, to point out the need to comprehend the governing hypothesis and critically assess experimental data and the conclusions drawn from them.

REFERENCES

1. Staudinger, H., *From Organic Chemistry to Macromolecules: A Scientific Autobiography Based on My Original Papers*. Wiley-Interscience, New York, 1970.
2. Harwood, H. J., *Characterization of Materials and Research: Ceramics and Polymers*. Syracuse University Press, Syracuse, NY, 1975.
3. Vasile, C. and R. B. Seymour, *Handbook of Polyolefins: Synthesis and Properties*. Marcel Dekker, New York, 1993.
4. Ferry, J. D., *Viscoelastic Properties of Polymers*. John Wiley & Sons Inc., New York, 1980.
5. IUPAC recommendations, Dispersity in polymer science. *Pure Appl. Chem.*, **81(2)**:351–353 (2009).
6. Walton, D. J. and J. P. Lorimer, *Polymers*. Oxford University Press, New York, 2000.
7. Krentsel, B. A., Y. V. Kissin, V. I. Kleiner, and L. L. Statskaya, *Polymers and Copolymers of Higher Alpha-Olefins*. Hanser Publishers, New York, 1997.
8. Didier, C., Le polyethylene haute densite. *Inform. Chem.*, **256**:193–203 (1984).
9. International Chem Systems, *LLDPE/LDPE*. Chem Systems Inc., Tarrytown, NY, 1997.
10. Hunt, B. J. and M. I. James, *Polymer Characterisation*. Chapman & Hall, Glasgow, U.K., 1993.
11. Campbell, D., R. A. Pethrick, and J. R. White, *Polymer Characterization*, 2nd edn. Stanley Thornes Publishing Ltd., Cheltenham, U.K., 2000.
12. Stuart, B. H., *Polymer Analysis*. John Wiley & Sons, Chichester, U.K., 2002.
13. Seidel A. (Ed.), *Characterization and Analysis of Polymers*. John Wiley & Sons, Hoboken, NJ, 2008.
14. Williams, D. H. and I. Fleming, *Spectroscopic Methods in Organic Chemistry*, 2nd edn. McGraw Hill, London, U.K., 1973.
15. Griffiths, P. R. and J. A. De Haseth, *Fourier Transform Infrared Spectroscopy*. Wiley Interscience, New York, 1986.
16. Hummel, D. O., *Infrared Spectra of Polymers in the Medium and Long Wavelength Regions*. Wiley Interscience, New York, 1966.
17. Ibbett, R. N., *NMR Spectroscopy of Polymers*. Chapman & Hall, Glasgow, U.K., 1993.
18. Kitayama, T. and K. Hatada, *NMR Spectroscopy of Polymers*. Springer Laboratory, Series XII. Springer Verlag, Munchen, Germany, 2004
19. Rabek, J. F., *Experimental Methods in Polymer Chemistry*. John Wiley & Sons, New York, 1980.
20. Randall, J. C., Polymers and irradiation treatment method. US. WO 84/01156, 1984.
21. Van Krevelen, D. W., *Properties of Polymers*, 2nd edn. Elsevier, Amsterdam, the Netherlands, 1990.
22. ASTM D 792, Standard test methods for density and specific gravity (relative density) of plastics by displacement, 2008.
23. ASTM D 1505, Standard test method for density of plastics by the density-gradient technique, 2003.
24. ASTM D 4883, Standard test method for density of polyethylene by the ultrasound technique, 2008.
25. Bershtein, V. A. and V. M. Egorov, *Differential Scanning Calorimetry of Polymers: Physics, Chemistry, Analysis, Technology*. Ellis Horwood Series in *Polymer Science and Technology*. Ellis Horwood Ltd., New York, 1994.
26. Monrabal, B., Crystallization analysis fractionation: A new technique for the analysis of branching distribution in polyolefins. *J. Appl. Polym. Sci.*, **52(4)**:491–499 (1994).

27. Koopmans, R. J., R. van der Linden, and E. F. Vansant, The characterisation of newly developed and promising hydrolyzed ethylene vinyl acetate copolymers. *J. Adhesion*, **11**:191–202 (1980).

28. Earnest, C. M., *Compositional Analysis of Thermogravimetry*. ASTM Special Technical Publication. ASTM, Philadelphia, PA, 1988.

29. Prochazka, O. and P. Kratochvil, An analysis of the accuracy of determining molar mass averages of polymers by GPC with an on line light scattering detector. *J. Appl. Polym. Sci.*, **34**:2325–2336, 1987.

30. Grinshpun V., K. F. O'Driscoll, and A. Rudin, On the accuracy of SEC analyses of molecular weight distributions of polyethylenes. *J. Appl. Polym. Sci.*, **29**:1071–1077 (1984).

31. Greassley, W. W., *Polymeric Liquids and Networks: Structure and Properties*. Garland Science, New York, 2004.

32. ASTM D445, Standard test method for kinematic viscosity of transparent and opaque liquids (and calculation of dynamic viscosity), 2006.

33. Haney, M. A., The differential viscosimeter. I. A new approach to the measurement of specific viscosities of polymer solutions. *J. Appl. Polym. Sci.*, **30**:3023–3036 (1985).

34. Haney, M. A., The differential viscometer. II. On-line viscosity detector for size exclusion chromatography. *J. Appl. Polym. Sci.*, **30**:3037–3049 (1985).

35. Brandrup, J. and E. H. Immergut, *Polymer Handbook*. Wiley Interscience, New York, 1989.

36. Wagner, H. J., The Mark–Houwink–Sakurada equation for the viscosity of linear polyethylene. *J. Phys. Chem. Ref. Data*, **14**:611–617 (1985).

37. Zimm, B. H. and W. H. Stockmayer, The dimensions of chain molecules containing branches and rings. *J. Chem. Phys.*, **17**:1301–1314 (1949).

38. Debye, P. P., A photoelectric instrument for light scattering measurements and a differential refractometer. *J. Appl. Phys.*, **17**:392–398 (1946).

39. Hjertberg, T., L.-I. Kulin, and E. Sorvik, Laser light scattering as GPC detector. *Polym. Test.*, **3**:267–289 (1983).

40. Pope, J. W. and B. Chu, A laser light scattering study on molecular weight distribution of linear polyethylene. *Macromolecules*, **17**:2633–2640 (1984).

41. Chu, B., M. Onclin, and J. R. Ford, Laser light scattering characterisation of polyethylene in 1,2,4-trichlorobenzene. *J. Phys. Chem.*, **88**:6566–6575 (1984).

42. Podzimek, S., The use of GPC coupled with a multiangle laser light scattering photometer for the characterization of polymers. On the determination of molecular weight, size, and branching. *J. Appl. Polym. Sci.*, **54**:91–103 (1994).

43. Moore, J. C., Gel permeation chromatography. I. A new method for molecular weight distribution of high polymers. *J. Polym. Sci.*, *A2*, 835–843 (1964).

44. Benoit, H., Z. Grubisic, P. Rempp, D. Decker, and J. G. Zilliox, Study by liquid phase chromatography of linear and branched polystyrene of known structure. *J. Chem. Phys.*, **63**:1507–1514 (1966).

45. Grubisic, Z., P. Rempp, and H. Benoit. A universal calibration for gel permeation chromatography. *Polym. Lett.*, **5**:753–759 (1967).

46. Weiss, A. R. and E. Cohn-Ginsberg, A note on the universal calibration curve for gel permeation chromatography. *Polym. Lett.*, **7**:379–381 (1969).

47. Coll, H. and D. K. Gilding, Universal calibration in GPC: A study of polystyrene, poly-alpha-methylstyrene, and polypropylene. *J. Polym. Sci. A2*, **8**:89–109 (1970).

48. Sanayei, R. A., S. Pang, and A. Rudin, A new approach to establishing universal calibration curves for size exclusion chromatography. *Polymer*, **34**:2320–2323 (1993).

49. Gedde, U. W., Molecular structure of crosslinked polyethylene as revealed by ^{13}C nuclear magnetic resonance and infrared spectroscopy and gel permeation chromatography. *Polymer*, **27**:269–274 (1986).

50. Pang, S. and A. Rudin, Characterization of polyolefins by size exclusion chromatography with low angle light scattering and continuous viscometer detectors. *Polymer*, **33**:1949–1952 (1992).

51. Goedhart, D. and A. Opschoor, Polymer characterization by coupling gel permeation chromatography and automated viscometry. *J. Appl. Polym. Sci.*, **8**:1227–1233 (1970).

52. Drott, E. E. and R. A. Mendelson, Determination of polymer branching with gel permeation chromatography. I. Theory. *J. Polym. Sci.*, *A2*, **8**:1361–1371 (1970).

53. Drott, E. E. and R. A. Mendelson, Determination of polymer branching with gel permeation chromatography. II. Experimental results for polyethylene. *J. Polym. Sci., A-2*, **8**:1373–1385 (1970).

54. Wild, L., Temperature Rising Elution Fractionation. In *Separation Techniques Thermodynamics Liquid Crystal Polymers* (L. Wild and G. Glockner, Eds.), Advances in Polymer Science Series, 98, 1991, pp. 1–47.

55. Wild, L., T. R. Ryle, D. C. Knobeloch, and I. R. Peat, Determination of branching distributions in polyethylene and ethylene copolymers. *J. Polym. Sci. Polym. Phys. Ed.*, **20**:441–455 (1982).

56. Hazlitt, L. G., Determination of short-chain branching distributions of ethylene copolymers by automated analytical temperature rising elution fractionation. *J. Polym. Sci. Appl., Polym. Symp.*, **45**:25–37 (1990).

57. Porter, D., *Group Interaction Modelling of Polymer Properties*. Marcel Dekker Inc., New York, 1995.

58. Rubenstein, M. and R. H. Colby, *Polymer Physics*. Oxford University Press, New York, 2003.

59. ASTM D1238, Standard test method for melt flow rates of thermoplastics by extrusion plastometer, 2004.

60. Sperati, C. A., W. A. Franta, and J. H. W. Starkweather, The molecular structure of polyethylene. V. The effect of chain branching and molecular weight on physical properties. *J. Am. Chem. Soc.*, **75**:6127–6133 (1953).

61. Shenoy, A. V., S. Chattopadhyay, and V. M. Nadkarni, From melt flow index to rheogram. *Rheol. Acta*, **22**:90–101 (1983).

62. Shenoy, A. V. and D. R. Saini, *Thermoplastics Melt Rheology and Processing*. Marcel Dekker, New York, 1996.

63. Bueche, F., Influence of rate of shear on the apparent viscosity of A—Dilute polymer solutions and B—Bulk polymers. *J. Chem. Phys.*, **22**:1570–1576 (1954).

64. Berry, G. C. and T. G. Fox, The viscosity of polymers and their concentrated solutions. *Adv. Polym. Sci.*, **5**:261–357 (1968).

65. Santamaria, A., Influence of long chain branching in melt rheology and processing of low density polyethylene. *Mater. Chem. Phys.*, **12**:1–28 (1985).

66. Carrot, C. and J. Guillet, From dynamic moduli to molecular weight distribution: A study of various polydisperse linear polymers. *J. Rheol.*, **41**:1203–1220 (1997).

67. Tuminello, W. H., Determination of molecular weight distribution (MWD) from melt rheology: A review. In *SPE ANTEC 87*, Los Angeles, CA, 1987, pp. 990–995.

68. Malkin, A. Y. and A. E. Teishev, Flow curve molecular weight distribution: Is the solution of the inverse problem possible? *Polym. Eng. Sci.*, **31**:1590–1596 (1991).

69. Wasserman, S. H. and W. W. Graessley, Effects of polydispersity on linear viscoelasticity in entangled polymer melts. *J. Rheol.*, **36**:543–572 (1992).

70. Tuminello, W. H. and N. Cudre-Mauroux, Determining molecular weight distributions from viscosity versus shear rate curves. *Polym. Eng. Sci.*, **31**:1496–1507 (1991).

71. Macskási, L., A critical analysis of the models connecting molecular mass distribution and shear viscosity functions. *eXPRESS Polym. Lett.*, **3(6)**:385–399 (2009).

3 Polymer Rheology

The major growth of the polymer industry in the second half of the twentieth century challenged scientists and engineers to develop a better understanding of how polymer melts flow [1]. Polymeric fluids exhibit a variety of flow features not seen in low-molar-mass fluids (e.g., extrudate swell, rod climbing, melt fracture). These observations are attributed to the viscoelastic nature of polymeric fluids in the sense that during flow, the stress experienced by the fluid depends upon the history of the deformation experienced [2]. The Navier–Stokes equations [3] that describe how the velocity, pressure, temperature, and density of a moving fluid are related are insufficient to quantify the viscoelastic behavior and, certainly, the polymer melt fracture. The conventional transition from laminar to turbulent fluid flow, as defined by Osborne Reynolds [4,5] in 1883, does not apply to polymer melts. Polymer melt fracture is a laminar flow phenomenon observed at very low Reynolds numbers. However, the macroscopic continuum mechanical modeling approach remains a valid avenue for developing insight into the viscoelastic flow behavior associated with polymer melt fracture. Alternatively, microscopic modeling, taking into account polymer conformation distributions and polymer–polymer or polymer–die interaction approaches, are developed to understand the macromolecular origin of polymer melt fracture. This chapter provides a brief introduction to the basic mathematics of continuum mechanics and some of the macroscopic concepts developed for quantifying polymer melt fracture. The coupling to the microscopic scale will be made in Chapter 8. Both chapters provide necessary background material for the mathematical modeling of polymer melt fracture, as presented in Chapter 9.

3.1 CONTINUUM MECHANICS

Polymers can be considered as one of the most intricate "many-particle" systems. Not only do these "particles" show an enormous diversity in architecture (see Chapter 2), but their interactions are also strongly coupled. Such complexity gives rise to a special flow behavior that is not yet fully understood. Finding interrelations between the microscopic details and macroscopic flow remains an active area of ongoing research [6–11].

The theory developed to describe solids and fluids on the macroscopic scale is continuum mechanics [12–15]. The properties are represented by functions depending on time t and position $\mathbf{x}$. Their interrelations are expressed in terms of algebraic and differential equations. For example, the function $\rho(\mathbf{x}, t)$ represents the density at position $\mathbf{x}$ and time t. From a microscopic point of view, it is not meaningful to speak about the mass at one particular point. However, in continuum mechanics this makes sense, thanks to a commonly accepted interpretation. The density in point $\mathbf{x}$ is read as the amount of mass m contained in a box around $\mathbf{x}$ divided by the volume of the box. This box should, on the one hand, contain so many polymer chains that the average density value does not depend on its detailed form, and on the other hand, be so small that it can be considered as a point from a macroscopic perspective. Since the functions representing system properties depend on the continuous position variable $\mathbf{x}$, they are usually called *fields* [15]. Other terms used in the literature are *density*, *distribution*, and *profile*.

3.2 SCALARS, VECTORS, AND TENSORS

In continuum mechanics, the independent variables are position $\mathbf{x}$, which has coordinates $\mathbf{x} = (x_1, x_2, x_3)$ with respect to a fixed Cartesian coordinate system (Figure 3.1), and time t. Dependent variables may be scalar fields, such as mass, density $\rho(\mathbf{x}, t)$, temperature $T(\mathbf{x}, t)$, and energy $U(\mathbf{x}, t)$, or vector fields with, as a typical example, velocity $\mathbf{v} = (v_1(\mathbf{x}, t), v_2(\mathbf{x}, t), v_3(\mathbf{x}, t))$, which have components that are themselves scalar fields. Scalar and vector fields are not appropriate to represent the stress in a fluid or solid, that is, the forces acting in such materials. For this purpose, the concept of a tensor field has been introduced. Although the concept of a tensor is, in general, related to its properties under coordinate transformations, in rheology, only a restricted use is made of this concept. Here, a tensor field $\mathbf{T}$ is identified with a 3×3 matrix of nine elements $T_{ij}(\mathbf{x}, t)$ that are themselves scalar fields. The indices i and j refer to the three coordinate axes. A tensor $\mathbf{T}$ has the matrix form

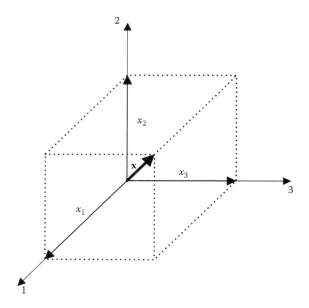

FIGURE 3.1 Projection of a position vector $\mathbf{x}$ in a fixed Cartesian coordinate system. The components of $\mathbf{x}$ are the scalars x_1, x_2, and x_3.

$$\mathbf{T} = \begin{bmatrix} T_{11} & T_{12} & T_{13} \\ T_{21} & T_{22} & T_{23} \\ T_{31} & T_{32} & T_{33} \end{bmatrix}. \tag{3.1}$$

In contrast to scalars, vectors, and tensors are denoted in bold. Scalars have no index, vectors have one index, and tensors two indices, running from 1 to 3. As is also usual in rheology, no distinction is made between column and row vectors.

For the multiplication of vectors and tensors, the usual operations apply. A well-known operation is the scalar (also called inner) product of two vectors $\mathbf{u} = (u_1, u_2, u_3)$ and $\mathbf{v} = (v_1, v_2, v_3)$, which is defined as

$$\mathbf{u} \cdot \mathbf{v} = u_1 v_1 + u_2 v_2 + u_3 v_3 = \sum_k u_k v_k. \tag{3.2}$$

The scalar product of two vectors yields a scalar. Note that the scalar product of a vector $\mathbf{v}$ with itself, $\mathbf{v} \cdot \mathbf{v}$, yields the length of $\mathbf{v}$ squared.

The scalar product of a tensor $\mathbf{T}$ and a vector $\mathbf{v}$, $\mathbf{T} \cdot \mathbf{v}$ is a vector with elements

$$(\mathbf{T} \cdot \mathbf{v})_i = \sum_k T_{ik} v_k. \tag{3.3}$$

The scalar product of two tensors $\mathbf{S}$ and $\mathbf{T}$, $\mathbf{S} \cdot \mathbf{T}$ yields again a tensor with elements given by

$$(\mathbf{S} \cdot \mathbf{T})_{ij} = \sum_k S_{ik} T_{kj}. \tag{3.4}$$

A different kind of product of two vectors is the dyadic product. This operation starts with two vectors $\mathbf{u}$ and $\mathbf{v}$ and yields a tensor. The elements of this tensor, denoted by $\mathbf{uv}$, are given by the simple rule

$$(\mathbf{uv})_{ij} = u_i v_j. \tag{3.5}$$

Note that the difference in notation between the scalar product and the dyadic product is only the dot, but the outcome of both operations is quite different! In continuum mechanics, frequent use is made of the nabla operator ∇

$$\nabla = \left(\frac{\partial}{\partial x_1}, \frac{\partial}{\partial x_2}, \frac{\partial}{\partial x_3} \right). \tag{3.6}$$

When this vector operator is applied to a scalar field, it yields the *gradient* of that field. The scalar product of $\mathbf{V}$ and a vector field $\mathbf{v} = (v_1(\mathbf{x}, t), v_2(\mathbf{x}, t), v_3(\mathbf{x}, t))$, is given by

$$\mathbf{V} \cdot \mathbf{v} = \frac{\partial v_1}{\partial x_1} + \frac{\partial v_2}{\partial x_2} + \frac{\partial v_3}{\partial x_3}, \tag{3.7}$$

and is called the *divergence* of $\mathbf{v}$. It measures whether the distance between neighboring particles decreases or increases when they move with the flow. If the divergence vanishes (becomes equal to zero), the distances apparently remain the same and the flow shows locally incompressible behavior. The dyadic product of $\mathbf{V}$ and the velocity $\mathbf{v}$ yields the tensor

$$\mathbf{V}\mathbf{v} = \begin{bmatrix} \dfrac{\partial v_1}{\partial x_1} & \dfrac{\partial v_2}{\partial x_1} & \dfrac{\partial v_3}{\partial x_1} \\ \dfrac{\partial v_1}{\partial x_2} & \dfrac{\partial v_2}{\partial x_2} & \dfrac{\partial v_3}{\partial x_2} \\ \dfrac{\partial v_1}{\partial x_3} & \dfrac{\partial v_2}{\partial x_3} & \dfrac{\partial v_3}{\partial x_3} \end{bmatrix}. \tag{3.8}$$

The relation between the scalar product of $\mathbf{V} \cdot \mathbf{v}$ and the dyadic product $\mathbf{V}\mathbf{v}$ is given by

$$\mathbf{V} \cdot \mathbf{v} = Tr(\mathbf{V}\mathbf{v}) \tag{3.9}$$

with *Tr* referring to *Trace*, that is, the sum of the diagonal elements.

It is important to remember that, contrary to scalar fields, the functional form of the components of vector and tensor fields depends on the coordinate system in use. For example, if one changes from a Cartesian to a cylindrical coordinate system, vectors and tensors have to be transformed too, according to specific transformation rules. However, each 3×3 tensor $\mathbf{T}$ has three scalar properties that are independent of the coordinate system. These scalar properties are referred to as the first, second, and third invariants:

$$I_T = Tr(\mathbf{T}) = T_{11} + T_{22} + T_{33}, \tag{3.10}$$

$$II_T = \frac{1}{2}(I_T^2 - Tr(\mathbf{T}^2)), \tag{3.11}$$

$$III_T = Det(\mathbf{T}) \tag{3.12}$$

where *Det* stands for *Determinant*. Any expression containing only these invariants will itself be invariant under coordinate transformations. The invariance of the third property is clear, since the determinant is, apart from the sign, given by the volume of the parallelepiped spanned by the column vectors of $\mathbf{T}$, which is coordinate independent. The invariance of all three quantities directly follows from the fact that they act as coefficients in the so-called *characteristic equation* of **T**. The roots of the characteristic equation

$$Det(\mathbf{T} - \lambda \mathbf{I}) = 0, \tag{3.13}$$

with **I** the 3×3 unit tensor, are the eigenvalues of **T**. After the evaluation of the determinant in Equation 3.13, a polynomial in λ of order 3 is obtained:

$$\lambda^3 - I_T \lambda^2 + II_T \lambda - III_T = 0. \tag{3.14}$$

An important property is that each tensor satisfies its own characteristic equation. This Cayley–Hamilton theorem implies in the present case that

$$\mathbf{T}^3 - I_T \mathbf{T}^2 + II_T \mathbf{T} - III_T \mathbf{I} = 0. \tag{3.15}$$

It follows that $\mathbf{T}^3$ can be expressed in terms of the lower powers of **T**. In turn, this implies that all $\mathbf{T}^n$ with $n > 2$ can be expressed in terms of $\mathbf{T}^2$, **T**, and **I**. An alternative formulation is obtained by multiplying Equation 3.15 by the inverse $\mathbf{T}^{-1}$ of **T**:

$$\mathbf{T}^2 = I_T \mathbf{T} - II_T \mathbf{I} + III_T \mathbf{T}^{-1}. \tag{3.16}$$

Alternatively, all $\mathbf{T}^n$ with $n \geq 2$ can be expressed in terms of **T**, $\mathbf{T}^{-1}$, and **I**.

Example 3.1: Invariants of a 3 × 3 Tensor

Consider the tensor that, with respect to some coordinate system, is represented by the matrix

$$\mathbf{T} = \begin{bmatrix} 1 & 1 & 0 \\ 0 & 1 & 0 \\ 0 & 0 & 1 \end{bmatrix}. \tag{3.17}$$

Elementary matrix manipulations yield

$$\mathbf{T}^2 = \begin{bmatrix} 1 & 2 & 0 \\ 0 & 1 & 0 \\ 0 & 0 & 1 \end{bmatrix}, \quad \mathbf{T}^3 = \begin{bmatrix} 1 & 3 & 0 \\ 0 & 1 & 0 \\ 0 & 0 & 1 \end{bmatrix}, \quad \mathbf{T}^{-1} = \begin{bmatrix} 1 & -1 & 0 \\ 0 & 1 & 0 \\ 0 & 0 & 1 \end{bmatrix}. \tag{3.18}$$

The resulting invariants have the values

$$I_T = 3, \quad II_T = 3, \quad III_T = 1. \tag{3.19}$$

The characteristic equation reads as

$$\lambda^3 - 3\lambda^2 + 3\lambda - 1 = (\lambda - 1)^3 = 0, \qquad (3.20)$$

giving three identical roots

$$\lambda_1 = \lambda_2 = \lambda_3 = 1. \qquad (3.21)$$

Equations 3.15 and 3.16 can easily be checked for this case.

3.3   STRESS TENSOR

The tensor concept is required when describing the effect of local forces acting in a material. This can be understood as follows. At position $\mathbf{x}$ and time t, an imaginary surface element of area A is placed in the fluid or solid. The shape of this surface element is not relevant, that is, it can be a circle or a square. Area A should be small compared to the macroscopic dimensions of the medium, but large compared to the molecular dimensions. The position in space of the surface element is determined by its normal $\mathbf{n}$, which is assumed to have unit length. The material on the side of the surface element opposed to $\mathbf{n}$ exerts forces on the material at the same side as $\mathbf{n}$. Compared with the macroscopic size of the flow as a whole, the microscopic forces have short ranges. Thus, only forces between molecules very close to the surface are relevant. The sum of all these forces through the surface element is denoted as $\mathbf{F}_n$. If A is small enough, $\mathbf{F}_n$ will be proportional to area A. In the limit of smaller and smaller A, the ratio

$$\sigma_n = \frac{\mathbf{F}_n}{A} \qquad (3.22)$$

will become independent of A. This normalized force can be decomposed into a component in the plane of the surface element, orthogonal to $\mathbf{n}$, and a component parallel to $\mathbf{n}$. The first component is called the *shear stress* and the second one the *normal stress*. If the surface element would really exist, the shear stress tends to drag it parallel to itself, whereas the normal stress tends to move it in the direction of $\mathbf{n}$. By definition, the tensor that relates $\mathbf{n}$ and σ_n is called the *stress tensor*. In formula:

$$\sigma_n = \sigma \cdot \mathbf{n} \qquad (3.23)$$

If the direction of the normal $\mathbf{n}$ varies, σ_n would also vary, but the stress tensor itself is independent of $\mathbf{n}$. If the stress tensor is known, the forces in the material can be calculated for any direction of $\mathbf{n}$.

In the case of fluids composed of small molecules, normal stresses appear to be independent of direction and shear stresses vanish. Such fluids are called *isotropic* fluids. In such fluids, anisotropy would be immediately cancelled out by a fast rearrangement of the molecules. In contrast, for fluids consisting of very large molecules such as polymer melts, anisotropy is found and both shear and normal stresses play an important role. The macromolecules strongly constrain each other's movement during flow inducing different conformational arrangements depending on the flow direction.

In an isotropic medium at rest, the stress tensor can be expressed as

$$\sigma = -p\mathbf{I}, \tag{3.24}$$

where
  $\mathbf{I}$ is the unit tensor
  p the (hydrostatic) pressure

It is common to decompose the stress in an arbitrary medium in an isotropic and an anisotropic part, and to write

$$\sigma = -p\mathbf{I} + \tau. \tag{3.25}$$

The tensor τ is called the *extra* or *viscous stress tensor.*

3.4   STRAIN TENSORS

In solid-state materials (metals, ceramics), each particle remains in the vicinity of its rest position. At positive temperature, the particles vibrate around these positions. Forces cause the material to deform locally, but the particles move away from their rest positions only over relative small distances. In a first-order approach, it is assumed that the deviations from the rest positions are linearly proportional to the restoring forces. Such behavior is called *elastic* or *Hookean*, after the English scientist Robert Hooke, who introduced this linear concept in 1687 for springs.

In fluids like polymer melts, the notion of rest position is not useful. The particles move along with the flow over arbitrary distances. Accordingly, only the relative particle positions are important. The deformation of the material (the change in distances between particles) is called *strain*. Strain variations are intimately coupled to stress variations. The relation between strain and stress is described by constitutive equations. In polymer melts, these relations are nonlinear and extremely complicated. In order to formulate constitutive equations, it is necessary to have appropriate tensor descriptions not only of stress, but also of strain. Two complementary approaches are discussed, since the constitutive equations in the literature make use of both. Either one focuses on the change in particle position or on the change in particle velocity.

3.4.1   Finger Tensor

The path or trajectory in space $\mathbf{x}(\mathbf{x'}, s; t)$ of a particle at time t will depend on its initial position $\mathbf{x'}$ at time s. In this notation, the dependence of the path on the initial time and position is made explicit. By definition $\mathbf{x} = \mathbf{x'}$ if $t = s$. In principle, the path $\mathbf{x}$ can be calculated from the initial particle configuration at time s. For that purpose, a set of differential equations (the equations of motions) needs to be solved. This can be a huge task if the constitutive equations that couple these differential equations are complicated. If $\mathbf{x}(\mathbf{x}', s; t)$ is known for all $\mathbf{x}'$, these paths contain all information about the deformation of the material. It is useful to introduce the tensor

$$\mathbf{A}(s,t) = \nabla'\mathbf{x}(\mathbf{x}', s; t), \tag{3.26}$$

where the ∇' operator applies to the argument $\mathbf{x}'$ in $\mathbf{x}(\mathbf{x}', s; t)$. The tensor $\mathbf{A}$ measures the change in $\mathbf{x}$ relative to the change in initial position $\mathbf{x}'$. $\mathbf{A}$ depends on both the initial and the final particle configurations as expressed through the arguments (s, t).

In Example 3.2, it is shown that $\mathbf{A} = \mathbf{I}$, the unit tensor, if the flow velocity is the same everywhere so that no deformation takes place. At first sight, the tensor $\mathbf{A}$ seems an appropriate measure for the rate of deformation. Its deviation from $\mathbf{I}$ could be used for this purpose. However, in Example 3.3, it is shown that this measure contains too much information. In a rotating flow, where the relative distances of the particles remain the same and thus no deformation occurs, $\mathbf{A}$ still deviates from $\mathbf{I}$. For that reason, the Finger tensor $\mathbf{B}$ is introduced, which is defined as

$$\mathbf{B}(s,t) = \mathbf{A}^{\mathrm{T}}(s,t) \cdot \mathbf{A}(s,t). \tag{3.27}$$

The Finger tensor $\mathbf{B}$ (and its inverse $\mathbf{B}^{-1}$, the Cauchy strain tensor) is symmetric as directly follows from its construction.

Example 3.2:   Finger Tensor in Steady and Nonsteady Laminar Flow

In the laminar flow (also called simple-shear flow), the material planes slide over each other. In Cartesian coordinates, the sliding planes are considered parallel to the (x_1, x_3) plane and moving in the x_1 direction (Figure 3.2). In steady laminar flow, the velocity v_1 in the x_1 direction only depends on x_2, so $v_1 = v_1(x_2)$. A material point at position (x_1', x_2', x_3') at time s moves to position (x_1, x_2, x_3) at time t according to

$$x_1 = x_1' + (t - s)v_1(x_2'),$$

$$x_2 = x_2', \tag{3.28}$$

$$x_3 = x_3'.$$

Consequently, the strain tensor $\mathbf{A}$ is given by

$$\mathbf{A}(s,t) = \begin{bmatrix} 1 & 0 & 0 \\ (t-s)\dot{\gamma} & 1 & 0 \\ 0 & 0 & 1 \end{bmatrix}, \tag{3.29}$$

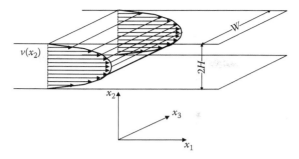

FIGURE 3.2 Velocity profile in a laminar simple-shear flow between parallel plates separated by a distance $2H$.

with the shear rate defined as

$$\dot{\gamma}(x_2) = \frac{dv_1}{dx_2}. \tag{3.30}$$

The shear rate and **A** depend on position via x_2. For a uniform velocity, one has **A = I**, since v_1 is independent of x_2. The shear rate notation $\dot{\gamma}$ is very common in the literature. The dot refers to a time derivative. This can be understood when replacing v_1 by dx_1/dt in Equation 3.30, and interchanging the derivatives with respect to t and x_2:

$$\dot{\gamma} = \frac{d}{dt}\left(\frac{dx_1}{dx_2}\right). \tag{3.31}$$

The shear rate has the dimension of reciprocal time. In steady flow, the quantity

$$\gamma(s, t, x_2) = (t - s)\dot{\gamma}(x_2) \tag{3.32}$$

is called the shear.

In nonsteady flow, $v_1 = v_1(x_2, t)$, and the shear rate definition requires the use of partial derivatives:

$$\dot{\gamma} = \frac{\partial v_1}{\partial x_2}. \tag{3.33}$$

The corresponding shear is given by

$$\gamma(s, t) = \int_s^t \dot{\gamma}(t') \, dt'. \tag{3.34}$$

In both steady and nonsteady laminar flow, the Finger tensor is given by

$$\mathbf{B}(s,t) = \begin{bmatrix} 1+\gamma^2(s,t) & \gamma(s,t) & 0 \\ \gamma(s,t) & 1 & 0 \\ 0 & 0 & 1 \end{bmatrix}. \tag{3.35}$$

The inverse of the Finger tensor, the Cauchy (strain) tensor, is

$$\mathbf{B}^{-1}(s,t) = \begin{bmatrix} 1 & -\gamma(s,t) & 0 \\ -\gamma(s,t) & 1+\gamma^2(s,t) & 0 \\ 0 & 0 & 1 \end{bmatrix}. \tag{3.36}$$

In rotational flow, the Finger tensor equals the unit tensor. This is shown in Example 3.3. The equality to the unit tensor also follows from the property that every 3×3 matrix represents a linear transformation of space and can be written as the product of a pure rotation matrix $\mathbf{A_r}$, and a stretching matrix $\mathbf{A_s}$: $\mathbf{A} = \mathbf{A_r} \cdot \mathbf{A_s}$. The rotation matrix $\mathbf{A_r}$ is orthogonal, that is, $(\mathbf{A_r})^{-1} = (\mathbf{A_r})^T$, which means that the inverse of the rotation is given by its transposed matrix. Then, it follows that

$$\mathbf{B} = (\mathbf{A_r} \cdot \mathbf{A_s})^T \cdot \mathbf{A_r} \cdot \mathbf{A_s} = \mathbf{A_s^T} \cdot \mathbf{A_s}. \tag{3.37}$$

The Finger tensor thus measures the strain due to local stretching, but does not account for rotations.

Example 3.3: Finger Tensor in Purely Rotational Flow

Consider a flow that rotates with angular velocity ω. For convenience, the x_3-axis is chosen as the rotation axis. Particles then move around in planes parallel to the (x_1, x_2) plane (Figure 3.3). A particle at position (x_1', x_2', x_3') at time s moves to (x_1, x_2, x_3) at time t.

The position transformation is given by

$$x_1 = x_1' \cos \omega(t-s) + x_2' \sin \omega(t-s),$$

$$x_2 = -x_1' \sin \omega(t-s) + x_2' \cos \omega(t-s), \tag{3.38}$$

$$x_3 = x_3'.$$

This leads to the strain tensor

$$\mathbf{A}(s,t) = \begin{bmatrix} \cos \omega(t-s) & -\sin \omega(t-s) & 0 \\ \sin \omega(t-s) & \cos \omega(t-s) & 0 \\ 0 & 0 & 1 \end{bmatrix}. \tag{3.39}$$

It is easily verified from substitution into Equation 3.27 that the corresponding Finger tensor $\mathbf{B}(s, t)$ equals the unit tensor $\mathbf{I}$.

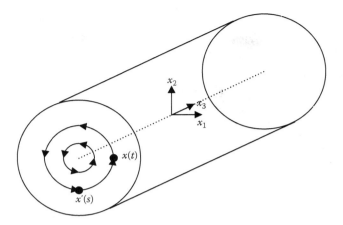

FIGURE 3.3   Movement of a particle in a flow rotating around the x_3 axis.

3.4.2 RATE OF DEFORMATION AND VORTICITY TENSOR

An alternative for the Finger tensor is obtained when considering the flow velocities. If the velocity $\mathbf{v}(\mathbf{x}, t)$ is spatially uniform, no deformation takes place. This suggests taking the tensor

$$\mathbf{C}(\mathbf{x},t) = \nabla\mathbf{v}(\mathbf{x},t), \tag{3.40}$$

introduced in Equation 3.8, as a measure of the deformation. The tensor $\mathbf{C}$ vanishes if there is no deformation. Its deviation from zero can be used as an indication of the rate of deformation. To separate between stretch and rotation, the tensor $\mathbf{C}$ is split up into its symmetric and antisymmetric parts:

$$\mathbf{C} = \mathbf{D} + \mathbf{V}. \tag{3.41}$$

The rate of deformation tensor $\mathbf{D}$ is defined as

$$\mathbf{D} = \tfrac{1}{2}(\mathbf{C} + \mathbf{C}^{\mathrm{T}}). \tag{3.42}$$

The vorticity tensor $\mathbf{V}$ is defined as

$$\mathbf{V} = \tfrac{1}{2}(\mathbf{C} - \mathbf{C}^{\mathrm{T}}). \tag{3.43}$$

For a pure rotation, $\mathbf{D} = 0$, and thus $\mathbf{C} = \mathbf{V}$. If $\mathbf{V} = 0$ and $\mathbf{C} = \mathbf{D}$, the flow is said to be irrotational.

Example 3.4: Rate of Deformation Tensor in Laminar Flow

As seen in Example 3.2 and Figure 3.2, the velocity in laminar flow has the form

$$\mathbf{v} = (v_1(x_2, t), 0, 0). \tag{3.44}$$

Accordingly the tensor $\mathbf{C}$ has the form

$$\mathbf{C}(t) = \begin{bmatrix} 0 & 0 & 0 \\ \dot{\gamma} & 0 & 0 \\ 0 & 0 & 0 \end{bmatrix}. \tag{3.45}$$

The corresponding $\mathbf{D}$ and $\mathbf{V}$ tensors are given by

$$\mathbf{D}(t) = \frac{1}{2}\begin{bmatrix} 0 & \dot{\gamma} & 0 \\ \dot{\gamma} & 0 & 0 \\ 0 & 0 & 0 \end{bmatrix}, \quad \mathbf{V}(t) = \frac{1}{2}\begin{bmatrix} 0 & -\dot{\gamma} & 0 \\ \dot{\gamma} & 0 & 0 \\ 0 & 0 & 0 \end{bmatrix}. \tag{3.46}$$

Since neither $\mathbf{D}$ nor $\mathbf{V}$ is vanishing, laminar flow is neither purely rotational nor irrotational. The flow is irrotational only if v_1 is not dependent on x_2. The invariants of $\mathbf{D}$ are given by

$$I_D = III_D = 0 \tag{3.47}$$

and

$$II_D = -\frac{1}{4}\dot{\gamma}^2 \tag{3.48}$$

Example 3.5: Rate of Deformation Tensor in Purely Rotational Flow

The velocities in purely rotational flow are given by

$$v_1 = \omega x_2$$

$$v_2 = -\omega x_1 \tag{3.49}$$

$$v_3 = 0$$

This can be obtained from differentiating Equation 3.38 with respect to t. The corresponding tensor $\mathbf{C}$ reads as

$$\mathbf{C} = \begin{bmatrix} 0 & -\omega & 0 \\ \omega & 0 & 0 \\ 0 & 0 & 0 \end{bmatrix}. \tag{3.50}$$

From Equations 3.42 and 3.43, it follows that $\mathbf{D} = 0$ and $\mathbf{V} = \mathbf{C}$.

3.4.3 RELATION BETWEEN FINGER TENSOR B AND RATE OF DEFORMATION TENSOR D

As both **B** and **D** can be used as deformation measures, it makes sense to understand the relationship between the two. In the definition of **B**, the position vector is used, whereas in the definition of **D**, the velocity vector is used. Since the velocity is the derivative of position with respect to time, it can be expected that also **B** and **D** are related through a time derivative, as shown by

$$\frac{\partial}{\partial t}\mathbf{B} = \frac{\partial}{\partial t}(\mathbf{A}^{\mathrm{T}} \cdot \mathbf{A}) = \left[\frac{\partial}{\partial t}\mathbf{A}^{\mathrm{T}}\right] \cdot \mathbf{A} + \mathbf{A}^{\mathrm{T}} \cdot \left[\frac{\partial}{\partial t}\mathbf{A}\right] \tag{3.51}$$

Next, an expression is needed for the derivative of tensor $\mathbf{A}(s, t)$ with respect to t. From the definition in Equation 3.26 and the interchange of derivatives with respect to time and position, one finds

$$\frac{\partial}{\partial t}\mathbf{A}(s,t) = \frac{\partial}{\partial t}(\nabla'\mathbf{x}(\mathbf{x}',s;t)) = \nabla'\mathbf{v}(\mathbf{x}',s;t). \tag{3.52}$$

with **v** the velocity of a particle at time t that took position **x**′ at time s. The ∇' operator operates on the position in the past. Using the chain rule, the right-hand side of Equation 3.52 can be expressed in terms of **A** and **C**:

$$\nabla'\mathbf{v}(t,s,\mathbf{x}') = \nabla'\mathbf{x} \cdot \nabla\mathbf{v}(t,\mathbf{x}) = \mathbf{A}(s,t) \cdot \mathbf{C}(t). \tag{3.53}$$

Substituting this result into Equations 3.51 and 3.52 gives

$$\frac{\partial}{\partial t}\mathbf{B}(s,t) = \mathbf{C}^{\mathrm{T}}(t) \cdot \mathbf{B}(s,t) + \mathbf{B}(s,t) \cdot \mathbf{C}(t). \tag{3.54}$$

Since $\mathbf{B}(t, t) = \mathbf{I}$, it follows that

$$\lim_{s \to t} \frac{\partial}{\partial t}\mathbf{B}(s,t) = \mathbf{C}^{\mathrm{T}}(t) + \mathbf{C}(t) = 2\mathbf{D}(t). \tag{3.55}$$

The tensor $\mathbf{B}(s, t)$ contains the integrated effect of all deformations in the time interval $[s, t]$. The tensor $\mathbf{D}(t)$ measures the change in strain at time t. Both tensors effectively contain the same information and they are used in a specific situation depending on convenience. However, from Equation 3.54, it is clear that **D** cannot simply be interpreted as the derivative of **B** with respect to time.

3.5 EQUATIONS OF MOTION

In fluid mechanics, the Navier–Stokes equations govern the dynamics. These equations express the conservation of mass, momentum, and energy. Since energy conservation is usually not considered in the discussions on polymer melt fracture, only the conservation of mass and momentum will be described.

3.5.1 TRANSPORT THEOREM

All conservation laws in flowing media can be derived from one general unifying principle, the transport theorem. For the general derivation of the transport theorem, consider a property $f(\mathbf{x}, t)$, with f being a scalar field, for example, mass, heat, and energy, or a vector field, for example, velocity. The nature of f does not need specification, although in this context, it will be identified with mass and momentum. During flow, the property f is transported through the medium. The corresponding flux field is denoted by $\mathbf{Q}(\mathbf{x}, t)$. Property f can be produced and/or annihilated. The rate at which this happens is given by the source field $S(\mathbf{x}, t)$ (Figure 3.4).

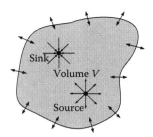

FIGURE 3.4 Fluxes across the boundary of a fixed volume V, including sinks and sources within V.

Considering an arbitrary volume V in the medium and applying the Eulerian approach, the volume will have a fixed position with respect to an external coordinate frame. The amount of f in V, $F(t)$, is given by the three-dimensional (3D) integral over volume V:

$$F(t) = \iiint_V f(\mathbf{x}, t)\, dV. \tag{3.56}$$

Since the volume is fixed, the change in $F(t)$ is given by

$$\frac{dF}{dt} = \iiint_V \frac{\partial f}{\partial t}\, dV. \tag{3.57}$$

The conservation of f implies that, per unit of time, this change in F equals the flux through the surface A of V plus the production of f in V. This change is given by a two-dimensional (2D) integral over the surface A and a 3D integral over the volume:

$$\frac{dF}{dt} = -\iint_A \mathbf{Q} \cdot \mathbf{n}\, dA + \iiint_V S\, dV. \tag{3.58}$$

The integral over the surface A contains an inner product of the flux $\mathbf{Q}$ with the outward normal $\mathbf{n}$ to A. This inner product yields the component of $\mathbf{Q}$ orthogonal to A. By integrating this component over the surface A, the total flux through A is obtained. The minus sign represents that an outward flux of f through A leads to the loss of f inside V. The integral over the volume V represents the total production/annihilation of f in V. The theorem of Gauss states for a vector density such as $\mathbf{Q}$ that

$$\iint_A \mathbf{Q} \cdot \mathbf{n}\, dA = \iiint_V \nabla \cdot \mathbf{Q}\, dV, \tag{3.59}$$

The theorem allows rewriting the surface integral into a volume integral. Bringing all terms together, it can be concluded that f is conserved only if at any time and for any volume V it holds that

$$\iiint_V \left\{ \frac{\partial f}{\partial t} + \nabla \cdot \mathbf{Q} - S \right\} dV = 0. \tag{3.60}$$

This is the global or integral form of the transport theorem. Since the volume V is chosen arbitrarily and the integrand is assumed to be continuous in space, it can be concluded that the integrand must vanish. This yields the local or differential form of the transport theorem:

$$\frac{\partial f}{\partial t} + \nabla \cdot \mathbf{Q} = S. \tag{3.61}$$

3.5.2 MASS BALANCE

To obtain the mass balance, property f is identified with mass having mass density $\rho(\mathbf{x}, t)$. The corresponding mass flux is obtained by multiplying the mass density with the velocity field: $\mathbf{Q} = \rho \mathbf{v}$. Substitution in Equation 3.61 leads to the continuity equation

$$\frac{\partial \rho}{\partial t} + \nabla \cdot (\rho \mathbf{v}) = S. \tag{3.62}$$

Using the identity

$$\nabla \cdot (\rho \mathbf{v}) = \rho \nabla \cdot \mathbf{v} + \mathbf{v} \cdot \nabla \rho, \tag{3.63}$$

which directly follows from the chain rule of differentiation, Equation 3.63 can be rewritten as

$$\frac{\partial \rho}{\partial t} + \mathbf{v} \cdot \nabla \rho + \rho \nabla \cdot \mathbf{v} = S. \tag{3.64}$$

The first two terms can be taken together by introducing the notion of material time derivative, also referred to as total time derivative:

$$\frac{D}{Dt} := \frac{\partial}{\partial t} + \mathbf{v} \cdot \nabla. \tag{3.65}$$

This time derivative measures the change in a property at the position of an observer who moves with the flow and thus follows a streamline, that is, the path of a material point in time. This leads to an alternative form of the continuity equation

$$\frac{D\rho}{Dt} + \rho\nabla \cdot \mathbf{v} = S. \tag{3.66}$$

If no material is created or annihilated, and $S=0$, the continuity equation reads as

$$\frac{D\rho}{Dt} = -\rho\nabla \cdot \mathbf{v}. \tag{3.67}$$

A special case is that of incompressible flow. Then, the amount of mass in an arbitrary volume is conserved when the volume is transported with the flow, although the volume itself may change its position and shape in time. In an incompressible flow

$$\frac{D\rho}{Dt} = 0. \tag{3.68}$$

Combining Equations 3.67 and 3.68 incompressibility implies that the divergence of the velocity field vanishes everywhere, i.e.,

$$\nabla \cdot \mathbf{v} = 0. \tag{3.69}$$

Example 3.6:   Pipeline Transport

Consider a pipe through which immiscible fluids, for example, oil and water, are pumped sequentially. Since fluids are considered incompressible, they all travel at the same speed. Let the fluids have densities $\rho_1, \rho_2, \dots$. The mass density $\rho(\mathbf{x}, t)$ along the pipe is constant in each fluid and changes across fluid interfaces. For a fixed position x along the pipe, the derivative of ρ with respect to t vanishes as long as a fluid passes x, but sharply peaks when an interface passes x. For an observer running along the pipe at the same speed as the fluids, ρ does not change with time. This is expressed by $D\rho/Dt=0$. Note that for this pipeline transport the condition $\nabla \cdot \mathbf{v} = 0$ indeed holds (Figure 3.5).

3.5.3   MOMENTUM BALANCE

The property f in Equation 3.61 is now identified with momentum, which is defined as the product of mass density and velocity: $\mathbf{f} = \rho\mathbf{v}$. The flux of this momentum has

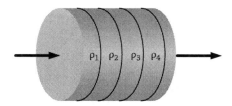

FIGURE 3.5   Pipeline transport of immiscible fluid batches.

two directions: momentum itself has a direction, and the direction in which it is transported. The resulting flux $\mathbf{Q}$ is a tensor with two indices, given by the dyadic product of the momentum with the velocity $\mathbf{v}$:

$$\mathbf{Q} = \rho \mathbf{v}\mathbf{v}. \tag{3.70}$$

When applying the transport theorem in Equation 3.61, the role of the sink/source term $\mathbf{S}$ needs consideration. Newton's second law states that the derivative of momentum with respect to time equals the sum of the exerted forces on the material within V. In this case, two kinds of forces may act on the material within the volume V. First, the contact force $\mathbf{F}_c$, due to the interaction with the surroundings. The contact force acts on the surface A of V and its total contribution is given by an integral over A. According to Equation 3.23, the contact force in point $\mathbf{x}$ on A is obtained by the matrix-vector product of the stress tensor σ and the outward normal $\mathbf{n}$ in $\mathbf{x}$ to the surface. The total contact force is then given by the surface integral

$$\mathbf{F}_c(t) = \iint_A \sigma(\mathbf{x},t) \cdot \mathbf{n}(\mathbf{x},t) \, dA. \tag{3.71}$$

It is convenient to transform this surface integral into a volume integral by way of the Gauss divergence theorem:

$$\mathbf{F}_c(t) = \iiint_V \nabla \cdot \sigma(\mathbf{x},t) \, dV. \tag{3.72}$$

Second, body forces may apply. They represent the effect of external force fields on the material within V. The body forces act on each point in V. Examples are gravity and electromagnetic forces. In the rheological practice, electromagnetic forces are seldom relevant. Gravity is always present, although it may often be neglected, since the dynamics of flows is usually dominated by the internal interactions. Denoting all external fields together as the force field $\mathbf{f}_b(\mathbf{x}, t)$, the total body force $\mathbf{F}_b$ is given by the integral

$$\mathbf{F}_b(t) = \iiint_V \mathbf{f}_b(\mathbf{x},t) \, dV. \tag{3.73}$$

In case of gravity, $\mathbf{f}_b = \rho \mathbf{g}$, with $\mathbf{g}$ the (constant) gravitational acceleration direction. Newton's second law states that

$$\frac{d\rho_v}{dt} = \mathbf{F}_c + \mathbf{F}_b. \tag{3.74}$$

Applying the transport theorem with $\mathbf{S}$ given by the sum of Equations 3.72 and 3.73 leads to

$$\iint_V \left\{ \frac{\partial(\rho \mathbf{v})}{\partial t} + \nabla \cdot (\rho \mathbf{v v}) \right\} dV = \iint_V \{ \nabla \cdot \boldsymbol{\sigma} + \mathbf{f}_b \} dV. \tag{3.75}$$

Since V is arbitrarily chosen, the integrands must be identical so that the integral form is equivalent to the differential form

$$\frac{\partial(\rho \mathbf{v})}{\partial t} + \nabla \cdot (\rho \mathbf{v v}) = \nabla \cdot \boldsymbol{\sigma} + \mathbf{f}_b. \tag{3.76}$$

Equation 3.75 can be rewritten as

$$\frac{\partial(\rho \mathbf{v})}{\partial t} + \rho \mathbf{v} (\nabla \cdot \mathbf{v}) + \mathbf{v} \cdot \nabla(\rho \mathbf{v}) = \nabla \cdot \boldsymbol{\sigma} + \mathbf{f}_b. \tag{3.77}$$

In terms of the material time derivative (Equation 3.65)

$$\frac{D(\rho \mathbf{v})}{Dt} + \rho \mathbf{v} \nabla \cdot \mathbf{v} = \nabla \cdot \boldsymbol{\sigma} + \mathbf{f}_b. \tag{3.78}$$

Using the continuity equation, Equation 3.67 can be rewritten in the form

$$\rho \frac{D\mathbf{v}}{Dt} = \nabla \cdot \boldsymbol{\sigma} + \mathbf{f}_b. \tag{3.79}$$

Substituting the decomposition equation of the stress tensor (Equation 3.25) an alternative expression is obtained:

$$\rho \frac{D\mathbf{v}}{Dt} = -(\nabla p)\mathbf{I} + \nabla \cdot \boldsymbol{\tau} + \mathbf{f}_b. \tag{3.80}$$

If the system is in at rest, i.e., $\mathbf{v} = \mathbf{0}$ the momentum equation (Equation 3.78) reduces to

$$\nabla \cdot \boldsymbol{\sigma} = -\mathbf{f}_b. \tag{3.81}$$

In solids, the body forces are usually negligible. The geometry of a system in static equilibrium and without the presence of body forces is thus governed by

$$\nabla \cdot \boldsymbol{\sigma} = 0, \tag{3.82}$$

an equation that should be solved in combination with the applied boundary conditions.

Example 3.7: Momentum Conservation in Laminar Flow

Since in laminar flow, such as in Figure 3.2, the velocity has only a nonvanishing component v_1 in the x_1-direction, the momentum equation (Equation 3.80) reduces to a scalar equation for this component

$$\rho\left(\frac{\partial v_1}{\partial t} + v_1 \frac{\partial v_1}{\partial x_1}\right) = -(\nabla p)_1 + (\nabla \cdot \boldsymbol{\tau})_1 + (\mathbf{f}_b)_1. \tag{3.83}$$

On the right-hand side of Equation 3.83, the subscripts denote that the first components are considered. Since v_1 does not depend on x_1, the second term on the left-hand side vanishes. So,

$$\rho \frac{\partial v_1}{\partial t} = -\frac{\partial p}{\partial x_1} + (\nabla \cdot \boldsymbol{\tau})_1 + (\mathbf{f}_b)_1. \tag{3.84}$$

In narrow slit dies and capillaries, the pressure p depends linearly on x_1 only. If in addition the body forces are ignored, the momentum equation becomes

$$\rho \frac{\partial v_1}{\partial t} = \frac{P}{L} + (\nabla \cdot \boldsymbol{\tau})_1, \tag{3.85}$$

where
  P is the pressure difference over the die
  L is the die length

This equation will be worked out further in Example 3.9.

3.6 CONSTITUTIVE EQUATIONS

Essential in modeling the rheological behavior of polymers is the specification of the constitutive relation, that is, an explicit or implicit expression for the relation between stress and strain. A few, relatively simple models that are frequently used are introduced. These models are very useful for understanding the concepts of viscoelastic polymer flows in general, but cannot describe the full complexity of polymer flow. In Chapter 8, more sophisticated models will be considered including those derived from microscopic polymer dynamics considerations.

3.6.1 ELASTIC BEHAVIOR

An elastic model comprises the concept introduced by Robert Hooke for a spring. Hooke observed that for the small deformations of a spring, the deformation and the force are linearly related. Analogously, in an elastic constitutive model, the local stress is assumed to be a linear function of the local deformation. This is expressed by a linear relation between the extra stress tensor $\boldsymbol{\tau}$ and the Finger tensor $\mathbf{B}$:

$$\boldsymbol{\tau} = G\mathbf{B}. \tag{3.86}$$

The coefficient G is the elastic modulus. This modulus depends on the polymer architecture and has the same function as the spring constant for a linear spring. If G is constant (typically denoted as G_0), the model is called *Hookean*.

Example 3.8:   Laminar Flow of a Hookean Fluid

The form of the Finger tensor **B** for laminar flow, given in Equation 3.35, shows that the diagonal components of the stress tensor are

$$\tau_{11}(s,t) = G(1 + \gamma^2(s,t)),$$

$$\tau_{22} = \tau_{33} = G.$$

(3.87)

The other nonvanishing components are

$$\tau_{12}(s,t) = \tau_{21}(s,t) = G\gamma(s,t). \tag{3.88}$$

In the extensions of this basic model, G is made dependent on **B**. In an elastic medium, a local deformation gives rise to a local stress that tends to restore the deformation. The local stress continues to do that as long as the deformation is present.

3.6.2   VISCOUS BEHAVIOR

The standard mechanical analogy of viscous behavior is the dashpot. In such a system, the force necessary to displace the piston depends on the resulting velocity of the piston and is independent of its position. For a fast movement of the piston, a large force is needed. For viscous flows, any state of deformation can be reached, provided that the changes take place slow enough. Following a suggestion of Newton, viscous behavior is modeled by relating the local stress linearly to the rate of the deformation tensor **D**:

$$\tau = 2\eta\mathbf{D} \tag{3.89}$$

The coefficient η is called the viscosity. If η is a constant (typically denoted as η_0), the model is referred to as *Newtonian*. The relation was initially proposed for laminar flow. The tensor **D** in Equation 3.46, for laminar flow reduces to

$$\tau_{12} = \tau_{21} = \eta\dot{\gamma}. \tag{3.90}$$

It is the original definition of viscosity, being the ratio of shear stress and shear rate in laminar flow.

Example 3.9: Laminar Flow of a Newtonian Fluid

The form of the rate of the deformation tensor $\mathbf{D}$ for general laminar flow is given in Equation 3.46. Combination with Equations 3.85 and 3.89 gives

$$\rho \frac{\partial v_1}{\partial t} = \frac{P}{L} + 2\eta (\nabla \cdot \mathbf{D})_1 = \frac{P}{L} + \eta \frac{\partial \dot{\gamma}}{\partial x_2} = \frac{P}{L} + \eta \frac{\partial^2 v_1}{\partial x_2^2}. \tag{3.91}$$

In this last step, definition (3.33) is used. A special case of this expression arises when the time derivative on the left-hand side vanishes. The solution of the resulting second-order ordinary differential equation yields the stationary velocity profile. As indicated in Figure 3.2, the height of the slit die is equal to $2H$, thus $-H < x_2 < H$. The well-known parabolic Poiseuille profile is obtained by applying the no-slip-boundary condition $v_1(-H) = v_1(H) = 0$:

$$v_1(x_2) = \frac{P}{2\eta L}(H^2 - x_2^2). \tag{3.92}$$

For the extensions of the basic model, the so-called *generalized Newtonian model*, η is made dependent on $\mathbf{D}$. A scalar function like η is independent of coordinate transformations and can therefore only depend on the invariants of the tensor $\mathbf{D}$ as defined in Equations 3.10 through 3.12. For laminar flow, measuring η is a relatively easy experiment. In such flows, the first and third invariants vanish. From Equation 3.48, it can be concluded that

$$\eta = \eta(\dot{\gamma}). \tag{3.93}$$

For most polymers, it is observed that η decreases as shear rate increases. This phenomenon is called *shear thinning*. Its counterpart is shear thickening. A popular way to model shear thinning and thickening is the power law

$$\tau_{12} = k\dot{\gamma}^n. \tag{3.94}$$

This two-parameter model contains a pre-factor k and a dimensionless exponent n. The exponent is used to discriminate between shear thinning ($n < 1$) and shear thickening ($n > 1$). For $n = 1$, Equation 3.94 reduces to Equation 3.90 with k being the shear viscosity. The value of n is easily estimated from the slope of the τ_{12} dependence of $\dot{\gamma}$ plotted on double logarithmic axes. An alternative form of Equation 3.94 is

$$\eta = k\dot{\gamma}^{n-1} \tag{3.95}$$

Example 3.10:   Laminar Flow of a Power Law Fluid

In the case of a generalized Newtonian flow, Equation 3.84 reads

$$\rho \frac{\partial v_1}{\partial t} = \frac{P}{L} + \frac{\partial}{\partial x_2}(\tau_{12}). \tag{3.96}$$

Substitution of Equation 3.94 yields

$$\rho \frac{\partial v_1}{\partial t} = \frac{P}{L} + k \frac{\partial \dot{\gamma}^n}{\partial x_2} \tag{3.97}$$

The corresponding stationary shear rate profile is given by

$$\dot{\gamma} = \frac{dv_1}{dx_2} = \left(\frac{P|x_2|}{kL} \right)^{1/n}. \tag{3.98}$$

Note that $\dot{\gamma}(0)=0$, due to symmetry. Integrating this equation and using the no-slip-boundary condition $v_1(H)=0$, the equation one obtains for the velocity profile is

$$v_1(x_2) = \frac{n}{n+1} \left(\frac{P}{kL} \right)^{1/n} \left(H^{(n+1)/n} - |x_2|^{(n+1)/n} \right). \tag{3.99}$$

For Newtonian flow, where $n=1$, Equation 3.99 reduces to the Poiseuille flow. The total mass Q through a slit die is obtained by integrating the velocity profile over the die cross section. For a slit die of length L, height $2H$, and width W, this yields

$$Q = \int_0^W dx_3 \int_{-H}^H dx_2 v_1(x_2) = \frac{2n}{2n+1} WH^{(2n+1)/n} \left(\frac{P}{kL} \right)^{1/n}. \tag{3.100}$$

For Newtonian flow, where $n=1$, this reduces to

$$Q = \frac{2WH^3 P}{3\eta_0 L}. \tag{3.101}$$

In this case, the mass flux Q and the pressure difference P over the die are linearly related. The shear rate at the wall of a power law fluid can be expressed in terms of P and in terms of Q. The combination of Equations 3.98 and 3.100 yields

$$\dot{\gamma}_{wall} = -\left(\frac{PH}{kL} \right)^n = -\frac{2n+1}{2n} \frac{Q}{WH^2}. \tag{3.102}$$

Thus the shear rate at the wall of a power law fluid is linearly related to Q, and that is the reason why these quantities are often interchanged in the literature.

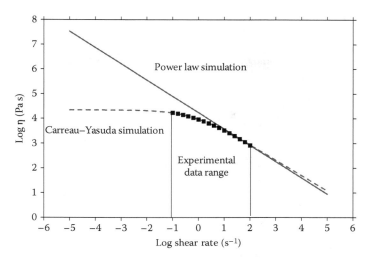

FIGURE 3.6 Typical representation of polymer melt viscosity versus shear rate data in the case of a shear-thinning fluid. The Carreau–Yasuda and power law equations may deviate from the true viscosity function when extrapolated beyond the experimental dataset.

A clear shortcoming of the power-law assumption is that η is not finite at low shear rates if $n \neq 1$. For shear thinning fluids, alternative models have been proposed to fit data better. An example is the five-parameter Carreau–Yasuda model [12] given by

$$\frac{\eta - \eta_\infty}{\eta_0 - \eta_\infty} = (1 + \lambda^a (\dot{\gamma})^2)^{(n-1)/a}, \tag{3.103}$$

where
η_0 is the zero-shear-rate viscosity
η_∞ the infinite-shear-rate viscosity
λ a time constant
n the power-law exponent
a a dimensionless parameter that describes the transition region between the zero-shear-rate region and the power-law region (Figure 3.6)

3.6.3 Viscoelastic Behavior

Elastic and viscous behaviors are often combined in polymer flow. For example, if a polymer rod is suddenly extended in a step strain, the stress shows a time-dependent behavior. At first, the stress increases to a maximum and after some time the stress decreases to an asymptotic value. This phenomenon is called *relaxation*. For most polymers, the asymptotic stress value will vanish. In other viscoelastic materials this asymptotic value may remain different from zero. A model for viscoelastic behavior can be obtained by combining Equations 3.86 and 3.89. In the following approach, the viscosity has the constant value η_0 and the elastic modulus has the constant value G_0.

The so-called *Maxwell model* is obtained by first differentiating both sides of Equation 3.86. From Equation 3.55, an alternative expression for elastic behavior can be found:

$$\dot{\tau} = 2G_0\mathbf{D}, \tag{3.104}$$

where $\dot{\tau}$ denotes partial differentiation with respect to time. Next, Equations 3.89 and 3.104 are combined in the following way:

$$\frac{1}{\eta_0}\tau + \frac{1}{G_0}\dot{\tau} = 2\mathbf{D}. \tag{3.105}$$

The usual form of this two-parameter Maxwell model is

$$\tau + \lambda_0\dot{\tau} = 2\eta_0\mathbf{D}, \tag{3.106}$$

where the time constant λ_0 has been defined as η_0/G_0. The Maxwell model is most easily interpreted in a one-dimensional (1D) system (e.g., a slender rod of material). In one dimension, the Maxwell model describes the behavior of the system as a linear spring and a linear dashpot in series (Figure 3.7).

If $\lambda_0 \gg 1$, the viscous part (dashpot) dominates. For $\lambda_0 \ll 1$, the elastic part (spring) is most important. Equation 3.106 can be integrated if multiplied by an integrating factor $\exp(\lambda_0 t)$ and the result

$$\tau(t) = \frac{2\eta_0}{\lambda_0} \int_{-\infty}^{t} e^{-(t-t')/\lambda_0}\mathbf{D}(t')\ dt' \tag{3.107}$$

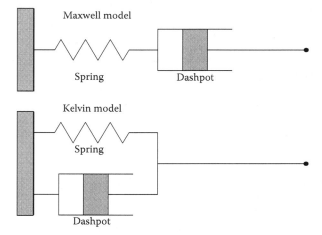

FIGURE 3.7 Linear spring and linear dashpot. If combined in series, the Maxwell model is represented, and if in parallel, the Kelvin model is obtained.

can be easily verified by substituting into Equation 3.106. Here, it is assumed that at $t = -\infty$, the material was strain-less. The integral shows that in a viscoelastic flow, the stress at time t is coupled to the strains at all earlier times: the system has a memory. The parameter λ_0 determines how fast the memory effects fade away and λ_0 is called the *relaxation time*.

Example 3.11: Maxwell Model for Laminar Flow

In laminar flow, the only nonvanishing components of the rate of deformation tensor **D** are $D_{12} = D_{21}$. The resulting equation for τ_{12} reads as

$$\tau_{12} + \lambda_0 \, \dot{\tau}_{12} = \eta_0 \dot{\gamma} \tag{3.108}$$

and similar for τ_{21}. The solution is

$$\tau_{12}(t) = \frac{\eta_0}{\lambda_0} \int_{-\infty}^{t} e^{-(t-t')/\lambda_0} \dot{\gamma}(t') \, dt'. \tag{3.109}$$

In steady laminar flow, $\dot{\gamma}$ is constant and Equation 3.109 reduces to

$$\tau_{12} = \eta_0 \dot{\gamma}. \tag{3.110}$$

The Maxwell model does not show shear thinning. The other stress components die away. For example, τ_{11} satisfies the equation

$$\tau_{11} + \lambda_0 \dot{\tau}_{11} = 0 \tag{3.111}$$

with solution

$$\tau_{11}(t) = \tau_{11}(0)e^{-t/\lambda_0}. \tag{3.112}$$

Equations 3.86 and 3.89 can also be combined differently. Lord Kelvin proposed the combination

$$\tau = G_0 \mathbf{B} + \eta_0 \dot{\mathbf{B}} = G_0 (\mathbf{B} + \lambda_0 \dot{\mathbf{B}}). \tag{3.113}$$

In one dimension, the Kelvin model describes the behavior of a spring and a dashpot in parallel, as depicted in Figure 3.7. In the Maxwell model, the strain can become infinitely large whereas in the Kelvin model, the repelling spring force will prohibit this. Equation 3.113 can be integrated in exactly the same manner as Equation 3.106. The result is that the strain **B** at time t is given by an integral over the stress at all times earlier than t. Also, here is the rate of memory determined by the parameter λ_0:

$$\mathbf{B}(-\infty, t) = \frac{1}{G_0 \lambda_0} \int_{-\infty}^{t} e^{-(t-t')/\lambda_0} \, \boldsymbol{\tau}(t') \, dt'. \qquad (3.114)$$

3.6.4 LINEAR VISCOELASTICITY

Equation 3.107 for the Maxwell model suggests for viscoelastic flow to write the stress as an integral over the deformation history:

$$\boldsymbol{\tau}(t) = 2 \int_{-\infty}^{t} G(t - t') \mathbf{D}(t') \, dt' \qquad (3.115)$$

This stress model is referred to as being linear, since the strains at all times in the past linearly contribute to the integral. This is the superposition principle proposed by Ludwig Boltzmann in 1876. The function $G(t)$ is called the relaxation function. It is to be expected that $G(t)$ is a monotonously decreasing function, since the memory of the system is fading.

Equation 3.115 has only limited applicability. Most polymer-based fluids respond nonlinearly, especially if the molecular interactions are complicated; as is the case for entangled macromolecules. However, Equation 3.115 remains an important tool, since all fluids tend to behave linearly as long as the deformations are small and slow. For this purpose, significant research is done in determining $G(t)$.

Example 3.12:   Relaxation Function in Laminar Flow

In Section 3.4.3, it is shown that the strain tensors **B** and **D** of laminar flow are not simply related by time differentiation. However, in laminar flow, such a relation still holds for the '12' and '21' components of these tensors (Examples 3.2 and 3.4):

$$B_{12}(s,t) = \gamma(s,t),$$
$$D_{12}(t) = \frac{1}{2}\dot{\gamma}(t) \qquad (3.116)$$

and similar for B_{21} and D_{21}. For these components Equation 3.115 can be rewritten by integrating by parts

$$\tau_{12}(t) = \int_{-\infty}^{t} G(t-t')\dot{\gamma}(t') \, dt' = \int_{-\infty}^{t} M(t-t')\gamma(t,t') \, dt' \qquad (3.117)$$

where use is made of $\gamma(t, t) = 0$ and $G(\infty) = 0$. The so-called memory function M is nothing but the time derivative of G

$$M(t) = \frac{d}{dt}G(t). \qquad (3.118)$$

The relaxation function G can be measured in a step-strain experiment. To that end, the material is instantaneously deformed at $t=0$ and this deformation is kept fixed for $t>0$. Meanwhile, the stress τ is measured. The deformation is applied as fast as possible as in practice, instantaneous actions are nearly impossible. A step strain applied at $t=0$ corresponds to

$$\mathbf{B}(0,t) = \mathbf{B}_0 H(t),\tag{3.119}$$

with the Heaviside function $H(t)$ defined by

$$H(t) = \begin{cases} 0, & t < 0 \\ 1, & t \geq 0 \end{cases}.\tag{3.120}$$

The derivative of $H(t)$ is the delta function $\delta(t)$, which has the characteristic property

$$\int f(t)\delta(t)\,dt = f(0)\tag{3.121}$$

for any smooth function $f(t)$ and for any integration interval that includes the time origin. The application of Equation 3.55 then yields

$$2\mathbf{D}(t) = \mathbf{B}_0\delta(t).\tag{3.122}$$

The substitution of this into Equation 3.115 gives

$$\tau(t) = \mathbf{B}_0 G(t).\tag{3.123}$$

Measuring τ implies measuring G except for multiplicative constants determined by the components of $\mathbf{B}_0$.

Example 3.13:   Relaxation Function of the Maxwell Model

For the Maxwell model, it can be verified that

$$G(t) = G_0 e^{-t/\lambda_0}\tag{3.124}$$

with $G_0 = \eta_0/\lambda_0$.

Experimental relaxation functions cannot adequately be described with a single exponential Maxwell model. A standard generalization is to take a sum of exponentials:

$$G(t) = \sum_{i=0}^{N} G_i e^{-t/\lambda_i}.\tag{3.125}$$

The number N of exponentials and the values of the parameters G_i and λ_i are arbitrary choices, depending on the desired accuracy for representing the experimental data.

In steady flow, the deformation tensor $\mathbf{D}$ and the stress tensor $\boldsymbol{\tau}$ are constant. The substitution of $t=0$ and constant $\mathbf{D}_0$ and $\boldsymbol{\tau}_0$ values in Equation 3.115 gives

$$\boldsymbol{\tau}_0 = 2\eta_0\mathbf{D}_0, \tag{3.126}$$

an equation that represents a linear relationship between steady-state stress and strain independent of the magnitude of $\mathbf{D}_0$. In practice, this relation only holds for fairly small deformations. Accordingly, η_0 is referred to as the zero-shear viscosity. It is related to the relaxation function by

$$\eta_0 = \int_{-\infty}^{0} G(t-t')\,dt' = \int_{0}^{\infty} G(t)\,dt. \tag{3.127}$$

So, η_0 equals the area under the $G(t)$ curve.

3.6.5 COMPLIANCE FUNCTION

For the Kelvin model, Equation 3.114 suggests that in viscoelastic flow, the strain can be written as an integral over the stress history:

$$\mathbf{B}(-\infty,t) = \int_{-\infty}^{t} j(t-t')\boldsymbol{\tau}(t')\,dt' \tag{3.128}$$

The integral of the function $j(t)$ is more important than j itself. An alternative to the step-strain experiment is the step-stress or creep experiment. At $t=0$, the stress is instantaneously increased from zero to a certain value $\boldsymbol{\tau}_0$ and kept constant for $t>0$, while the strain response is measured. This corresponds to

$$\boldsymbol{\tau}(t) = \boldsymbol{\tau}_0 H(t) \tag{3.129}$$

The substitution of this into Equation 3.128 gives

$$\mathbf{B}(-\infty,t) = \mathbf{B}(0,t) = \int_{0}^{t} j(t-t')\boldsymbol{\tau}_0\,dt' \equiv J(t)\boldsymbol{\tau}_0. \tag{3.130}$$

The function $J(t)$ is called the *compliance* (Figure 3.8). In a creep experiment, the strain initially shows a fast increase, and for $t \to \infty$ convergence to a straight line is observed, given by

$$J(t) = J_{\mathrm{e}}^{0} + J_{\mathrm{e}}^{1}t. \tag{3.131}$$

The long term creep behavior is determined by two parameters, the steady-state compliance J_{e}^{0} and the slope J_{e}^{1}

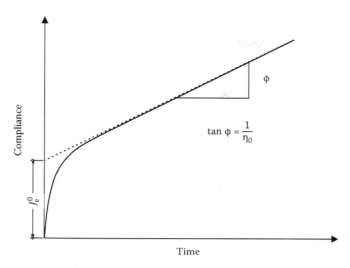

FIGURE 3.8 General experimental behavior of the compliance $J(t)$ in a step-stress experiment.

Example 3.14: Compliance in Laminar Creep Flow

Considering only the '12' components of the strain tensor **B** in Equation 3.35 gives for Equation 3.130 with $t \rightarrow \infty$

$$\gamma(0,t) = (J_e^0 + J_e^1 t)(\tau_0)_{12}. \tag{3.132}$$

Differentiation with respect to time yields

$$\dot{\gamma} = J_e^1(\tau_0)_{12}. \tag{3.133}$$

The steady-state equation (Equation 3.126) reads in this case as

$$(\tau_0)_{12} = \eta_0 \dot{\gamma}. \tag{3.134}$$

Comparison of Equations 3.133 and 3.134 yields for laminar flow

$$J_e^1 = \frac{1}{\eta_0}. \tag{3.135}$$

In principle, the compliance J contains the same information as the relaxation function G and the memory function M. However, in general, the relation between J and G is not straightforward as Example 3.15 shows.

Example 3.15:   *G* and *J* in Laminar Creep Flow

In laminar flow, a relation between $G(t)$ and $J(t)$ can be found from the long-term creep behavior. The derivation is based on the observation that the '12' components of the tensors **B** and **D** are related via time differentiation (see also Example 3.12). If the constant step stress τ_0 is applied at $t=0$, Equation 3.115 becomes for $t>0$

$$(\tau_0)_{12} = \int_0^t G(t-t')\dot{\gamma}(t')\,dt' = \int_0^t G(s)\dot{\gamma}(t-s)\,ds. \tag{3.136}$$

For large t, the system converges to a steady state with a constant shear rate $\dot{\gamma}_\infty$. From now on, two systems are compared. First, the original creep system described by Equation 3.136, and second, an imaginary, steady-state system that has a constant shear rate $\dot{\gamma}_\infty$ for all times. For the latter system, Equation 3.115 reads as

$$(\tau_0)_{12} = \int_{-\infty}^t G(t-t')\dot{\gamma}_\infty\,dt' = \dot{\gamma}_\infty \int_0^\infty G(s)\,ds. \tag{3.137}$$

For $t \rightarrow \infty$ the two expressions should be identical

$$\dot{\gamma}_\infty \int_0^\infty G(s)\,ds = \int_0^t G(s)\dot{\gamma}(t-s)\,ds. \tag{3.138}$$

Splitting up the integration interval on the left-hand side into the subintervals $[0, t]$ and $[t, \infty]$, and moving the integration over the first interval to the right-hand side, gives

$$\dot{\gamma}_\infty \int_t^\infty G(s)\,ds = \int_0^t G(s)[\dot{\gamma}(t-s) - \dot{\gamma}_\infty]\,ds \tag{3.139}$$

The next step is to integrate both sides over the positive t-axis

$$\dot{\gamma}_\infty \int_0^\infty \left[\int_t^\infty G(s)\,ds\right] dt = \int_0^\infty \left[\int_0^t G(s)[\dot{\gamma}(t-s) - \dot{\gamma}_\infty]\,ds\right] dt \tag{3.140}$$

The integrations are in the (s, t) plane over the triangular regions I (the left-hand side of Equation 3.140) and II (the right-hand side of Equation 3.140) (Figure 3.9). Interchanging the variables s and t leads to the equation

$$\dot{\gamma}_\infty \int_0^\infty \left[\int_0^s dt\right] G(s)\,ds = \int_0^\infty \left[\int_s^\infty [\dot{\gamma}(t-s) - \dot{\gamma}_\infty]\,dt\right] G(s)\,ds \tag{3.141}$$

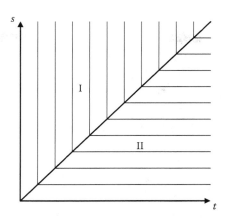

FIGURE 3.9 Sketch of the integration regions of the integrals in Equation 3.139.

The factor

$$\gamma_0 = \int_s^\infty [\dot{\gamma}(t-s) - \dot{\gamma}_\infty] \, dt = \int_0^\infty [\dot{\gamma}(t) - \dot{\gamma}_\infty] \, dt \tag{3.142}$$

is time independent.

Equation 3.141 leads to the relation

$$\frac{\gamma_0}{\dot{\gamma}_\infty} = \frac{\int_0^\infty s G(s) \, ds}{\int_0^\infty G(s) \, ds}. \tag{3.143}$$

At $t=0$, the original creep system has a vanishing strain. However, the strain of the comparing steady-state system at $t=0$ is nonvanishing; it has the value of γ_0. For the latter system, it holds at all times that

$$\gamma(0,t) = \gamma_0 + \dot{\gamma}_\infty t \tag{3.144}$$

Alternatively, the steady state can be described for all t by Equations 3.130 and 3.131

$$\gamma(0,t) = (J_e^0 + J_e^1 t)(\tau_0)_{12} \tag{3.145}$$

Comparing the latter expressions for $t=0$ and $t \to \infty$ yields

$$\gamma_0 = J_e^0 (\tau_0)_{12} \tag{3.146}$$

and

$$\dot{\gamma}_\infty = J_e^1 (\tau_0)_{12} = \frac{1}{\eta_0} (\tau_0)_{12}. \tag{3.147}$$

In view of Equations 3.127, 3.146, and 3.147, Equation 3.143 can be written in the form

$$J_e^0 = \frac{\int_0^\infty sG(s)\,ds}{\eta_0^2}. \tag{3.148}$$

Through Equations 3.135 and 3.148 one has made available the relation between the relaxation function G and the long term behavior of the creep compliance J in a creep experiment.

3.7   GENERAL OBSERVATION

The rheology of polymeric fluids is a challenging science. Describing the flow behavior and capturing the observations in a workable formalism requires advanced mathematics. Some basic approaches capture (surprisingly) a great number of macroscopic fluid-flow features but will fail to define important phenomena such as polymer melt fracture. Linear viscoelastic theories are derived with the important assumptions of small deformation and slow flows, which for most industrial processing purposes becomes unpractical. Furthermore, standard continuum mechanics neglects the nature of macromolecules. The macroscopic mathematical formalisms, that is, constitutive equations, require experimental evidence for the defining parameters from macroscopic fluid-flow observations. Irrespective of the finer details, the present introduction to basic polymer rheology should suffice to critically asses the rheological aspects of polymer melt fracture as presented in the literature. More advanced approaches to constitutive modeling and the efforts to integrate polymer architecture are addressed in Chapters 8 and 9.

REFERENCES

1. Bird, R. B. and C. F. Curtiss, Fascinating polymeric liquids. *Phys. Today*, **37**(1):36–43 (1984).
2. Keunings, R., Advances in the computer modelling of the flow of polymeric liquids. Keynote lecture. In *Eighth International Symposium on Computational Fluid Dynamics*, Bremen, Germany, September 5–10, 1999.
3. Batchelor, G. K., *An Introduction to Fluid Dynamics*. Cambridge University Press, Cambridge, U.K., 1967.
4. Bird, R. B., W. E. Stewart, and E. N. Lightfoot, *Transport Phenomena*. John Wiley & Sons Inc., New York, 1960.
5. Reynolds, O., An experimental investigation of the circumstances which determine whether the motion of water shall be direct or sinuous and the law of resistance in parallel channels. *Philos. Trans. R. Soc. Lond.*, **174**:935–982 (1883).
6. Macosko, C. W., *Rheology*. VCH Publisher, New York, 1994.
7. Ferry, J. D., *Viscoelastic Properties of Polymers*. John Wiley & Sons Inc., New York, 1980.
8. Gedde, U. W., Molar structure of crosslinked polyethylene as revealed by ^{13}C nuclear magnetic resonance and infrared spectroscopy and gel permeation chromatography. *Polymer*, **27**:269–274 (1986).

9. Mark, J. E., A. Eisenberg, W. W. Graessley, L. Mandelkern, and J. L. Koenig (Eds.), *Physical Properties of Polymers*. American Chemical Society, Washington, DC, 1984, pp. 141–150.
10. Greassley, W. W., *Polymeric Liquids and Networks: Structure and Properties*. Garland Science, New York, 2004.
11. Larson, R. G., *The Structure and Rheology of Complex Fluids*. Oxford University Press, New York, 1999.
12. Bird, R. B., R. C. Armstrong, and O. Hassager. *Dynamics of Polymeric Liquids*. Volume 1: *Fluid Mechanics*. John Wiley & Sons, New York, 1987.
13. Larson, R. G., *Constitutive Equations for Polymer Melts and Solutions*. Butterworth Publishers, Boston, MA, 1988.
14. Leonov, A. I. and A. N. Prokunin, *Nonlinear Phenomena in Flows of Viscoelastic Polymer Fluids*. Chapman & Hall, London, U.K., 1994.
15. Faber, T. E., *Fluid Dynamics for Physicists*. Cambridge University Press, Cambridge, U.K., 1995.

4 Polymer Processing

The chemist's work is often completed after polymers have been synthesized and characterized. The engineer will follow with the proper technology to handle and transform fluids, powders, and lumps into articles of interest to customers. The development of the first rubber-processing tool, the "Pickle," in 1820, was the key for rubber to become an industrial-attractive material [1]. Rubber had been known in Europe since the Spanish and Portuguese sailed the world seas several hundred years ago. However, the new, "easy-processing," tool allowed for innovative rubber applications [2]. In the following years, rubber demand rose and alternative, higher output processing techniques were needed. Eventually, roll mills, extruders, and gear pumps were developed [3]. Today, these rubber technologies have been adapted to suit the polymer-processing needs. As new polymers and new applications are developed, the drive for improved processing continues. Polymer-processing machines that allow stable polymer flow and undistorted extrudates at high rates are key attributes for success.

In this chapter, polymer melt fracture is described as it is observed in some of the most common polymer-processing techniques. Thermoforming, biaxial stretching, foaming, lamination, and compression molding are not considered. These secondary processing techniques are either used in combination with the common polymer-processing techniques or do not present melt fracture–related problems.

4.1 EXTRUSION

An essential characteristic of thermoplastic polymers is their ability to sustain a repetitive process of heating and cooling without, in principle, changing their chemical composition. The combined action of heating, mechanically kneading, and transporting polymers is called extrusion (Figure 4.1). The extruder performing this operation then forces the polymer melt into a shaping device that defines the type of extrusion. Subsequently, the shaped polymer melt is cooled into a solidified article. A typical extruder consists of a single screw inside a barrel that can be heated. The screw shape can be very complex and may serve many purposes [4,5]. Most common screws are designed to handle polymer granules (often called pellets). By rotating the screw, the granules are gradually melted and transported toward a die. The screw design ensures a homogeneous melt composition and temperature at the exit. The optimum screw geometry and extrusion processing condition (temperature, pressure) for maximizing the extruder output is defined by the viscoelastic properties of the polymer melt. For handling powders, making polymer blends, and dispersing additives homogeneously, twin-screw extruders are typically used.

Every polymer is at least twice exposed to an extrusion step. The first time, when powders, polymer solutions or melts from the polymerization reactors are formed

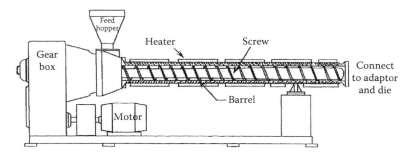

FIGURE 4.1 Schematic cross section of a typical single-screw polymer extruder. The solid polymer granules (pellets) are fed via a feed hopper into a rotating screw that is heated via external barrel heaters. The granules are gradually molten and transported forward into an adaptor and die.

into granules. The high bulk density and cylindrical or spherical shape of the granules allows for easy polymer transportation and handling. The second time is when granules arrive at converters to be shaped into an application.

4.1.1 GRANULATION

Industrial polymerization facilities mostly have a unit operation for the densification of polymer. It entails the production of granules that are safe and easy to transport and have a higher bulk density than the fluid or powder a reactor produces. The operation consists of extruding a polymer melt through a thick, disklike, perforated plate (Figure 4.2). Each perforation is a small capillary die of well-defined dimensions. The plate is exposed to a moving blade that cuts the strands emerging through the capillaries. A cooling fluid, typically water, ensures the simultaneous solidification of the thus formed granules. Depending on the viscoelastic properties of the polymer and the flow front feeding the multi-capillary disk, each capillary experiences a different throughput. In some cases and in localized areas of the die plate, this may result in polymer melt fracture or irregularly shaped granules. Depending on the severity, remediative actions need to be taken. Irregularly shaped or melt-fractured granules are more difficult to dry and cause problems during transport and further handling. At the converters, wet granules need extra drying and the often automated granules transportation and extruder feeding may get blocked or become uncontrollable.

4.1.2 FILM BLOWING

Film blowing [6] is a process where the polymer melt is forced through an annular die. The melt forms a tube that when closed can be inflated by air to form a blown tube also referred to as "bubble." The thickness of the film is controlled by the amount of inflation and the speed with which the bubble is drawn by the nip rolls (forming the bubble closure) onto the winders (Figure 4.3). Extrusion film blowing is probably the most dominant processing operation as to volume of polymers used [7,8].

FIGURE 4.2 Small-scale industrial water-cooled granulator. The picture shows a perforated disk that can be pressed against rotating cutter blades. The polymer melt extruding through the many capillary dies is cut into granules and cooled with water. (Courtesy of Reifenhäuser Gmbh, *Technical Company Literature*, Reifenhäuser Gmbh & Co. KG Maschinenfabrik, Troisdorf, Germany, 2009.)

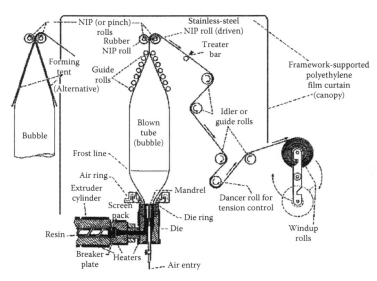

FIGURE 4.3 Schematic cross section of the front part of a blown-film extruder, the bubble, and the take-off and wind-up system. (From USI Chemicals Inc., Polyethylene *Extrusion Blown Film—An Operational Manual*, USI Chemicals Inc., Cincinatti, USA, 1970.)

The actual film thickness depends on the application and can range from a few microns to several millimeters. The quality of the produced film is defined by the thickness homogeneity along the circumference of the bubble. In turn, it defines the optical and mechanical performance of the application. Most applications require film (5–200 μm thick) of uniform transparency (no hazy regions)

FIGURE 4.4 A picture of a spiral mandrel as used in extrusion blown-film machines to achieve a homogeneous melt flow and an equal-thickness distribution in final film.

and consistent mechanical properties (thin spots lead to undesired film breakage). A uniform melt emerging out of the die ensures an even film thickness distribution. To that purpose, the core (or mandrel) of the die often has a pattern of spiral grooves (Figure 4.4).

The spirals, however, may not be optimized for all polymers alike and can work counterproductive by inducing a preferential flow instead. The result is a thickness variation of a few microns along the circumference of the film bubble that corresponds to the number of spiral inlet ports in the mandrel. At these positions, the higher film thickness appears as hazy stripes or bands (Figure 4.5). A similar feature can be induced by a spider (a mechanical system consisting of four or more bolts connecting the mandrel to the die to hold the smooth mandrel in place and centered). Precisely above the spider-arms, the film may show hazy stripes or bands. Again, a preferential higher flow rate or higher local stress lies at its origin. The observation of these stripes is related to the resolving power of the human eye for detecting rough surfaces. Rough surfaces diffuse light and give a matte, hazy appearance. Surface roughness measurements of a blown film usually indicate a higher surface roughness for the hazy (rough) stripes or bands (Figure 4.5).

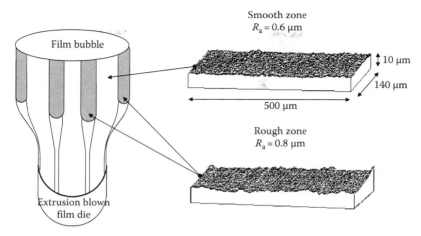

FIGURE 4.5 A schematic representation of the formation of rough bands along the circumference of the bubble of an HDPE blown film and actual roughness measurements performed in each zone. (Data from J.-P. Villemaire and J.-F. Agassant from The Dow Chemical Company.)

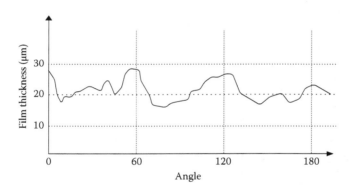

FIGURE 4.6 A schematic representation of the thickness variation along the circumference of an HDPE blown film (only 180° is shown). In this example, the average film thickness is 20 μm, and the zones showing higher thickness correspond to the position of the melt inlet ports of a spiral mandrel die configuration. The thicker zones often have a surface roughness as shown in Figure 4.5. (Data from J.-P. Villemaire and J.-F. Agassant from The Dow Chemical Company.)

A variation in the average roughness from 0.6 to 0.8 μm is sufficient to observe the phenomenon. Typically, the number of "higher-roughness" bands corresponds to the number of ports (spirals) of the particular film-blowing equipment. A higher-than-average film thickness is measured in these regions (Figure 4.6).

The stripes phenomenon is primarily observed close to the die exit and may vanish after multiaxial stretching during the bubble forming process. Present day state-of-the-art die-mandrel design technology and the rotation of the die reduces or avoids the phenomenon completely.

Closely related to these fairly broad stripes or bands is a phenomenon referred to as *die-lines* [9]. These much narrower stripes, sometimes line-like thin, may appear along the entire circumference of the film. Die-lines are not associable with preferential flow regions as described above. They are the result of either tiny defects, microscopic irregularities in the die and/or mandrel lips metal surface, or deposits from waxes, additives or cross-linked polymer accumulated over time at the die exit [10]. This, at a microscopic scale, very locally induced change in flow profile seems sufficient for a die-line to occur. They can either be the indentations of the surface, forming a line during continuous extrusion of the part or actually very local surface distortion, in terms of "line"-like stripes.

Surface roughness may also appear gradually along the entire circumference of the melt next to the die exit (Figure 4.7). It happens that the surface roughness may be found on either side of the film. The degree of surface roughness on each side depends on the die/mandrel geometry, the viscoelastic properties of the polymer, and the processing flow rate or die exit stress. These surface distortions generally become more severe at higher flow rates and limit the blown-film output. Higher flow rates eventually lead to bubble breaks or a bubble imbalance, that is, an instability.

The above-described type of surface distortion should be differentiated from an overall hazy or matte film that is induced by crystallization when processing semicrystalline polymer (e.g., PE, PP). In such cases, the crystallization kinetics of the polymer is at the origin. The bubble cooling and overall processing conditions define the time window within which the macromolecules can fold and orient to form spherulites, that is, crystals [11]. Larger crystals will diffuse light more and produce more hazy films. This type is typically referred to as *internal haze*. The overall film haziness is also determined by an external-haze component that can be associated as the combination of flow-induced surface distortions and the irregular stacking of crystals close to the surface. Distinguishing the two contributors is difficult. Depending on the overall processing condition, film thickness, size and orientation of the crystals near the surface, both internal and external haze can appear simultaneously. Wetting the surface allows measuring the internal haze independent from the external haze.

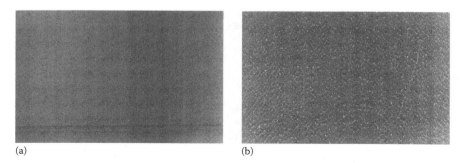

(a)   (b)

FIGURE 4.7 Photographs of two extrusion-blown films produced from the same PE. At a relatively low (a) and high (b) output the film shows a smooth and a rough surface, respectively.

For certain polymers, the bubble may not show surface distortions but becomes unstable and starts to swing in various ways. These phenomena are described as either axisymmetric periodic fluctuations in bubble diameter, fluctuations in frost line height and tension, or as a helical motion of the bubble [12]. Their origin is complex and is often associated with the melt strength of the polymer and not with melt fracture. Melt strength is related to the viscoelastic character of the polymer melt or more specifically its elongational properties. Polymers with high melt strength tend to give rise to very stable bubbles. At higher flow rates, the bubble rather ruptures than turning unstable with high haul-off speeds when maintaining or reducing the film thickness. The rupture phenomenon is a brittle failure of the film. Polymers with low melt strength tend to show unstable bubbles.

4.1.3 FILM AND SHEET CASTING

Polymer melts are extruded through a slit die to prepare cast films or sheets. Similar types of melt fracture occur as in the extrusion-blown film process. Cast films usually have a thickness ranging from the micron to the millimeter scale. Cast sheets have a thickness of several millimeters. Typically, they are designed for further, secondary processing operations like thermoforming and lamination. Uniform film thickness is critical, and requires considerable effort in polymer melt feed block and die design. Surface distortions can appear as localized bands or may span the entire breadth of the film or sheet similar to extrusion-blown film. At higher flow rates, the size of the stripes or bands may stay the same but the severity of the surface distortions increases (Figure 4.8).

Die-lines may equally be observed in this process. Also here, they originate from microscopic irregularities in or at the metal surface of the die lips. In certain cases, the die lips may be seen to be visibly damaged; in other cases a deposit from waxes, additives or cross-linked polymer accumulated over time, initiates a die-line [10]. In the latter case, cleaning the die will remove the die-lines. For much of the same reasons, die-lines appear in all extrusion involving dies generating free surfaces: film blowing, film and sheet casting, blow molding, pipe and profile extrusion, wire coating, and fiber spinning. For most applications, die-lines are not perceived as a major issue. In most instances, they either vanish after subsequent cooling and squeezing the film or sheet between the chill-roll and high gloss take-up winder, or after stretching the film, or because of printing or embossing operations. Consequently they do not hamper the optical or mechanical performance of the final application.

In a number of applications (geomembranes, construction film), a severe form of surface distortions is created intentionally. At higher flow rates, these surface distortions appear on either or both sides and span the entire width of the film or sheet depending on the feed block–die configuration.

In addition to surface distortions, the cast film or sheet extrusion process may induce an oscillating, wavy geometry in the melt flow direction (Figure 4.9). The surface remains glossy indicating the absence of a visible roughness. The waviness exists coherently on both sides of the film or sheet.

Another film-casting phenomenon is edge waviness. The melt emerging out of a slit die normally shows a constant neck-in (the film width is narrower than the slit die

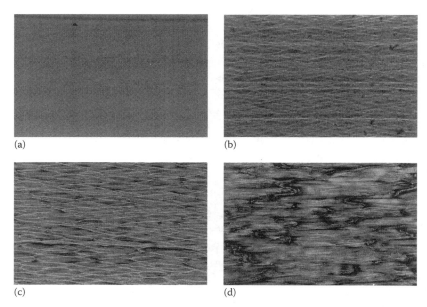

(a)   (b)

(c)   (d)

FIGURE 4.8 Illustration of the gradual development of surface distortions on LLDPE tapes at relatively low flow rates. At relatively low flow rates (a), localized hardly visible "die-lines" develop. At higher flow rates (b, c, d), die-lines become more severe and eventually result into very irregular surface distortion patterns across the entire extrudate surface. (Courtesy of L. Kale from The Dow Chemical Company.)

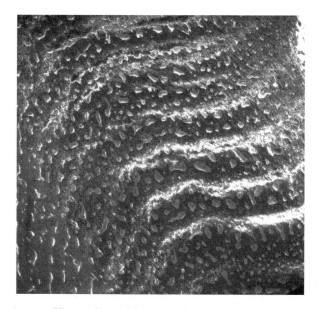

FIGURE 4.9 A wavy PP-cast film arising from flow instabilities. The droplet-like shapes are related to polymer additives rather than to the flow instability.

width). For certain processing conditions and polymers classes, the neck-in becomes unstable and starts oscillating in a direction perpendicular to the melt flow [13,14]. It creates films with uneven width and thickness, and often requires a processing rate reduction. The phenomenon is very similar, and related in origin, to what is known as draw resonance during fiber spinning. The draw resonance phenomenon is well understood and quantitative predictions can be made as to its onset [15–18]. Edge waviness (similar as bubble instability in film blowing, or draw resonance in fiber spinning) is not referred to as a melt fracture phenomenon and not further discussed in this book.

A special case of film casting exists where a molten film (a "web") is deposited on a substrate instead of a chill roll, with the intention of coating the substrate, for example, paper, cardboard, cloth, and other polymer films or sheets. Extrusion coating, as this type of film casting is referred to, can give rise to draw resonance and edge waviness. Surface distortions are rarely seen and not an issue in the final application.

4.1.4 EXTRUSION BLOW MOLDING

In a blow molding operation [19], the melt is extruded through a die in a continuous or discontinuous fashion to form a parison. The parison is subsequently shaped by inflation into a mold to produce a hollow article (e.g., bottles, drums, surf boards). For the discontinuous blow-molding operation, the extrusion is continuous but the melt accumulates into a vessel and is pushed subsequently either by the screw or a plunger through the die to form a parison. The die/mandrel configuration can be a very complex piece of mechanical engineering. During the parison forming, the die and/or mandrel move relative to one another to create a variable gap. This process is designed to adjust the thickness along the parison length so that the wall thickness is uniform after inflation. The die gap variation is made possible through converging or diverging die geometry (Figure 4.10). The most advanced blow molding technologies use flexible dies to induce ovalized and asymmetric parison geometries.

The thickness of the parison is also defined by the polymer extrudate swell. Extrudate swell indicates that the extrudate thickness is larger than the die gap [20]. The phenomenon occurs immediately after the die and depends on the viscoelastic properties of the polymer melt and the processing conditions (e.g., die geometry, flow rate, and temperature) [21–26].

During a typical blow-molding operation, the most readily observable surface distortions are die-lines. Mild forms of die-lines may vanish during the parison inflation but more severe forms make the blow-molded part useless (Figure 4.11). They originate at the die lips and for similar reasons as discussed in extrusion-blown film, and cast film and sheet extrusion. At higher flow rates, the number of die-lines tends to increase and eventually become wide bands having a high-frequency, small-amplitude surface distortion. In some cases, surface distortions only appear in narrow sections along the circumference of the parison. The transition from smooth to distorted surfaces is very sharp. In view of the variable die gap during the parison formation, it is possible to observe sharp surface distortion transitions along its

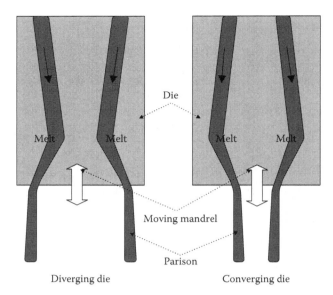

Diverging die   Converging die

FIGURE 4.10 Schematic representation of a polymer melt flowing through and exiting a diverging and a converging die/mandrel configuration. The die or the mandrel can move up and down during the parison formation and create a variable gap to shape its lengthwise thickness.

FIGURE 4.11 Die-lines on the surface of a blow-molded PE bottle. A change in die gap, due to the relative motion of the die/mandrel configuration, has caused a sudden clearly visible increase in the area of surface distortions.

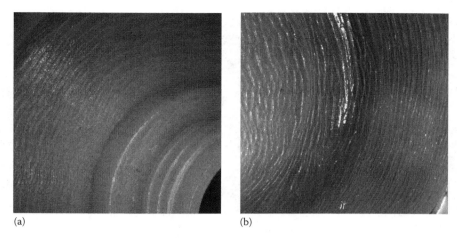

(a)   (b)

FIGURE 4.12 A blow-molded PE bottle fabricated using a converging die shows surface distortions on the inside (b) of the bottle while the outside (a) surface remains smooth (difficult to see in view of the bottle's translucent nature).

circumference. A specific position in the die/mandrel movement during the parison forming process can be associated with this transition. In other cases, the entire parison shows surface distortions. Converging dies tend to produce the surface distortions on the inside of the parison (Figure 4.12), and diverging dies produce it on the outside.

4.1.5 WIRE COATING

The wire-coating extrusion technique requires stable flows at very high flow rates. The operating principle consists of pulling a wire, cable or optical fiber through a die while it is coated with a polymer at speeds of up to 2500 m/min. The surface finish of the coated wires is expected to be smooth and glossy. The polymer may start to show a variety of defects. The surface distortions can range from localized die-lines to matte and low-amplitude high-frequency surface roughness (Figure 4.13). The degree of surface roughness varies depending on the flow rate, the type of cooling (water, air), and the polymer nature. The latter implies a formulation of multiple polymers, additives for various purposes (e.g., antioxidants, fillers, plasticizers, flame retardants, carbon black), and reagents to cross-link the polymer and enhance the mechanical performance of the wire coating. Occasionally, the wire coating is smooth and glossy but has a low-amplitude very-low-frequency wavy character. It results in a barely visible but measurable periodic thickness variation along the length of the wire coating. The phenomenon is not always visible during the actual processing operation and may only appear after a while on the final coated wire. In the latter case, nonuniform stress relaxation as a consequence of inhomogeneous mixing of formulation components or variable cross-link densities are possible causes and not melt fracture.

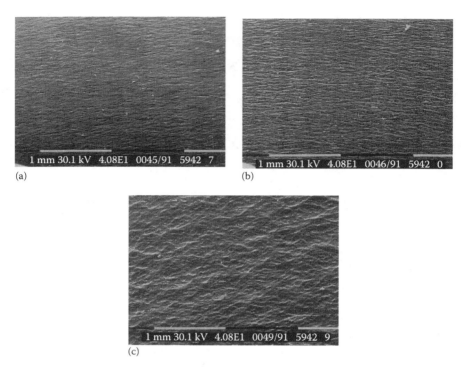

(a)   (b)

(c)

FIGURE 4.13 SEM of surface distortions of an LDPE wire coating. At low flow rates, the high-frequency small-amplitude surface distortion intensity may be affected by the cooling method (a = air cooling; b = water cooling). At higher flow rates, the surface distortions become more severe (c = air cooling). On a macroscopic scale, sample a appears smooth while sample b has a matte finish. Sample c feels rough.

4.1.6 PIPE AND PROFILE

Pipe and geometrically more complex profiles can show melt fracture in terms of surface distortions (including die-lines, Figure 4.14) and low-amplitude, low-frequency wavy patterns. The features may occur at the inside or outside of the profile. Relatively high output requirements combined with particular viscoelastic polymer properties may induce the effects. The use of formulated polymer systems complicates the understanding of melt fracture in terms of individual polymer components.

The major issue, beside aesthetics, is the failure of the part to meet strict dimensional standards and tolerances (Figure 4.15). In addition, for some polymers, very fine, hairline cracks can appear in the struts. Extruding multiwalled PC sheet may give rise to these so-called micro-crazes (Figure 4.16).

4.1.7 FIBER SPINNING

In a fiber-spinning operation, a polymer melt is extruded through a die plate containing many holes. The resulting extrudates are stretched into very thin filaments

(a)   (b)

FIGURE 4.14 Observable die-lines on the outside surface of a PE pipe (a) and surface distortions on the inside (b). In the latter picture, it is also possible to observe the changeover from a PE grade generating a smooth surface (back) to one inducing surface distortions under the same processing conditions (front). (Courtesy of S. Patterson from The Dow Chemical Company.)

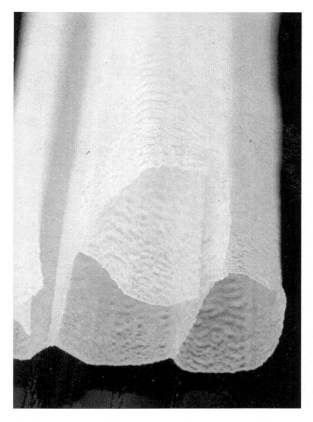

FIGURE 4.15 Profile extrusion of thermoplastic elastomers can induce severe surface distortions making the part unsuitable for use.

FIGURE 4.16   Extrusion of a multiwall PC sheet can induce local areas of surface distortion often referred to as "crazes" that diminish the optical properties. (Courtesy of M. Vreys from The Dow Chemical Company.)

with a diameter in the order of microns [17]. The number of holes varies from a few to thousands. Each hole has a very small aspect ratio (length/diameter ≪ 10). The individual filaments may show melt fracture during processing, but due to their very small size this is difficult to observe and usually not reported as a problem. The dominating fiber spinning irregularities are not referred to as melt fracture but are better known as draw resonance. Draw resonance corresponds to a regular and sustained periodic variation in the drawn filament diameter. Its origin is associated with the nonlinear dynamics of the process, being a space and time dependent disturbance around the main flow component [18]. Fiber instabilities appear when the pulling speed is increased or the pulling distance from the die is reduced. These effects have also been observed for purely viscous fluids (glucose syrup). For viscoelastic polymer melts, the periodicity reduces and eventually gives rise to fiber breaks when the elasticity of the material increases [27]. The limiting step during fiber spinning is the filament breakup. Filament breakage occurs because of the growth of surface perturbations or because of cohesive fracture [15].

Fiber instabilities as well as bubble instabilities, co-extrusion layer-instabilities, and edge waviness are studied using stability analysis techniques. Stability analysis is a procedure for determining whether a solution of the conservation equations corresponding to steady operation (i.e. a solution where all derivatives with respect to time are set to zero in the equations) can be maintained when disturbances enter the system. The analytical and numerical methods used in a stability analysis are treated in specialized texts [28]. There are three classes of analysis: those which establish conditions under which a process is absolutely unstable to any disturbance, no matter how small; those which determine the effect of small but finite disturbances near conditions corresponding to absolute instability; and those which establish conditions under which a process is absolutely stable regardless of the magnitude of the upset. Linear stability to infinitesimal disturbances is studied by obtaining the set of linear partial differential equations that describe the transient behavior of the process near the steady state [15]. Again, these types of instability are not classified as melt fracture.

4.1.8 CO-EXTRUSION

All the above melt-processing techniques can be operated in a co-extrusion mode. It implies the simultaneous extrusion of two or more polymer classes through a single die to form a multilayer polymer film, wire coating, hollow object, and profile or fiber. In addition to the above described melt fracture phenomena, irregularities can appear at the interface between polymer layers. At normal flow rates, the polymer interfaces within the die and the final article are smooth and parallel. It may happen that a nonuniform layer thickness is produced due to the surging in output rates and poor melt temperature uniformity within the individual polymer streams. In other cases, with polymers of substantially different viscosities, the lower viscosity polymer will tend to encapsulate the higher viscosity polymer. Layer thickness variation and in the extreme case layer breakup can result [18,29–31]. As flow rates increase, the interface between layers can develop waviness (Figure 4.17). Low-amplitude waviness may develop at low flow rates, but may neither be visible nor interfere with the functionality of the application. At higher flow rates, a high-amplitude wave can develop perpendicular to the flow direction. The wave crests can get carried forward and converted into a fold. Multiple folding can result in an extremely distorted interface. Some of these interfacial phenomena strongly resemble the features associated with melt fracture.

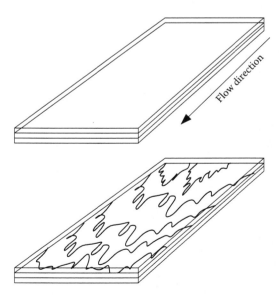

FIGURE 4.17 Interfacial instability development during three-layer co-extrusion at increasing flow rates (top to bottom). The interfacial instability may become very severe as to break up the middle layer.

4.2 INJECTION MOLDING

After extrusion, injection molding is probably the second-most widely used processing technique [32]. Injection molding combines extrusion and molding into one operation. The extrusion step is designed to melt the polymer and to transport it to the mold. The molten polymer is subsequently injected into the mold via a single movement of the extruder screw or a plunger (torpedo). The melt never forms a free surface, since it always flows in a contained space. The mold is always kept at a lower temperature than the melt. During injection, a solidified polymer layer forms at the mold walls while the melt keeps flowing, creating a fountain-like flow effect (Figure 4.18). An irregular flow of molten polymer into the mold may give rise to a flow front that bounces off one mold surface onto the other and creates localized surface distortions (Figure 4.19).

In general, different types of surface distortions are observed on injection-molded parts. The descriptive terminology tends to confuse the issue. Terms such as "smears," "splays," "scars," and "blushes" [33], or "gate blush," "tiger striping," and "shark skinning" [34], or "step defects" and "chevrons" [35] are used. They often describe a visible pattern on the molded article of variable surface roughness ranging from loss of gloss to Schallamach waves (a pattern of detachment bands when rubbers are dragged over a substrate) [36]. A specific pattern known as "tiger stripes" usually starts at the injection point and radiates out over the molded part in a periodic fashion (Figure 4.19). The same pattern is repeated on the other side of the part but out of phase by 180°. The drive to injection molding of thinner parts has revived the interest in finding solutions to avoid these patterns [37,38]. Bogaerds et al. in Hatzikiriakos and Migler's book [18] provide a detailed investigation combining experimental observations of "tiger stripes" with numerical simulations. Their findings suggest the presence of a swirling instability near the fountain flow surface that can be stabilized by materials with an increased strain hardening as proposed by Chang [35].

The presence of a rough surface either in local patches or over the entire part can also be induced either by a rough mold surface (usually an intentional effect), by spherulites near the surface upon crystallization of semicrystalline polymers (similar to effects in film), by rubber particles when processing high impact PS

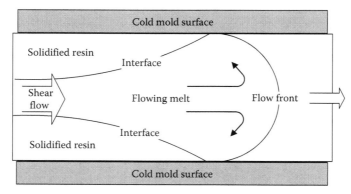

FIGURE 4.18 Schematic representation of a fountain flow during the injection of a hot, molten polymer into a cold mold.

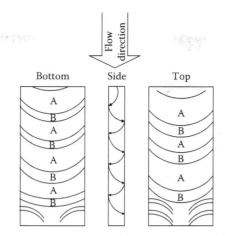

FIGURE 4.19   Schematic representation of a periodic surface defect occurring during injection molding of thin parts (A=glossy surface, B=rough surface). (Adapted from Chang, M.C.O., On the study of surface defects in the injection moulding of rubber modified thermoplastics, in *SPE ANTEC*, San Francisco, CA, 1994, pp. 360–367.)

(HIPS) or acrylonitrile-butadiene-styrene (ABS) polymers, or blends of immiscible polymers. The latter types of surface distortions are not referred to as melt fracture.

4.3   ROTATIONAL MOLDING

The production of hollow parts is done typically via blow molding, injection molding, or combinations thereof. However, for very large (e.g., 1000 L water tanks) or geometrically complex and seamless containers (e.g., fuel tanks), these processing techniques have limitations and rotational molding is the preferred technology [39,40]. For this purpose, polymer granules are grinded into a powder that is poured into a mold of the desired shape. The mold is heated while continuously being rotated along three axes to homogeneously distribute the melting powder. After a sufficiently long period of time, the heating and tumbling action will have fused the powder into a thick coating, sticking at the mold wall. After cooling, the mold is opened and the part is inspected for defects and overall performance. Sometimes, irregular grooves and pitting, resembling the skin texture of an orange, is observed, mostly at the inside of the container. This "orange peel" or "orange skin" surface distortion is not a polymer melt fracture phenomenon as induced by a strong extrusion driven flow. The origin is related to the combined effect of particle fusion and macromolecular crystallization kinetics within the boundary condition of the rotational molding operating conditions [41]. Pragmatically, the latter can be associated with local density differences of the powder particles.

4.4   CALENDERING

In 1836, well before the use of single-screw extruders became popular, calendering, as a process technology, was developed to coat cloth and leather with rubber [42].

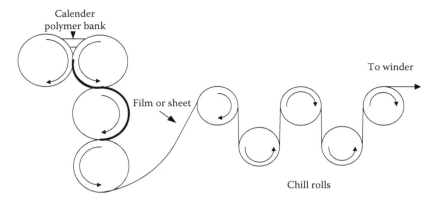

FIGURE 4.20   Typical calender rolls arrangements. The behavior of the polymer "bank" in between the first two calender rolls determines the quality of the final film or sheet.

A calender is conceptually a very simple device consisting out of three, four or five rotating cylinders (rolls) (Figure 4.20). The technology is mainly used to process PVC and rubbers into films and sheets. Most of the plastics in use for calendering are formulated systems.

In a calender, an amount of polymer is fed between two heated rolls that turn with a different velocity in opposite directions. The gap between the rolls is selected to form a sheet that sticks to one of the rolls and to keep an amount of polymer (the bank) in front of the gap. The velocity difference induces a strong shearing action between the different layers in the melt while a mixing action takes place in the transverse direction along the rolls by moving material from one roll to the other. The gap between the rolls sizes the thickness of the final film or sheet. Three types of surface distortions are observed. Loss of gloss occurs when the film on the side not in contact with the roll gets a hazy appearance. Chevrons occur when V-shaped defects, which have a higher film thickness than average, are formed in a regular fashion over its entire width. "Rockets" occur when air bubbles create spots with an elongated tail randomly distributed over the film width [43]. The latter defect can be associated with the mixing action of the calender and the shape of the bank but is not a melt fracture phenomenon.

4.5   GENERAL OBSERVATION

In the industrial practice, all polymer-processing techniques are operated at the highest possible flow rates or part production rates. At these extremes, the first show of any phenomenon that intervenes with the formation of a quality film, sheet, or part is sufficient to take corrective measures. The operation condition or polymer is modified. The term "melt fracture" is used when the extrudate shows signs of distortion. In the case of localized surface distortions (die-lines, stripes), or mild surface roughness, processing may continue. The distortions may vanish or be considered acceptable during subsequent post-extrusion treatment (stretching, molding, embossing). Sometimes, surface distortions are accepted as a useful or aesthetic attribute in a number of applications. In most instances, however, surface distortions affect the optical appearance of the extrudate and may cause additional quality issues during handling or printing. It is also noted that surface distortions originate at and are very sensitive to die defects and any kind of die deposit.

Alternatively, the extrudate can have an oscillating, wavy appearance in the flow direction. This is also a melt fracture feature where the entire volume of the extrudate is distorted. In practice, each of these phenomena occurs while the flow is considered to be stable (constant extrusion rates and pressure). There are, however, circumstances where polymers produce unstable melt flows. Typically, the extruder pressure and output fluctuates, the bubble becomes unstable, or draw resonance is observed. Further processing is impossible because the emerging melt becomes difficult to shape. The extrudate may or may not show surface distortions, and the distinction in terms of melt fracture and melt flow instabilities becomes vague. Similar features can occur at polymer–polymer interfaces during co-extrusion.

A polymer that is extruded using different processing techniques does not always show melt fracture at the same flow rates. The processing conditions can be more favorable in one technique over the other. Accordingly the industrial definition of polymer melt fracture is often linked to certain polymers in combination with a specific processing technique. The terminology used is often colorfully descriptive but not necessarily clear on the causality.

Finally, some finished article distortions often show melt fracture–like features. However, their origin is usually related to inhomogeneous crystallization, to formulation phase changes and inhomogeneous mixing, or to the relaxation of residual stresses.

REFERENCES

1. Hancock, T., *Personal Narrative of the Origin and Progress of the Caoutchouc or India Rubber Manufacture in England*. Longman, Brown, Green, Longman's and Roberts, London, U.K., 1857.
2. Burke, J., *Connections*. Little, Brown and Company, Boston, MA, 1995.
3. Tadmor, Z., L. N. Valsamis, J. C. Yang, P. S. Mehta, O. Duran, and J. C. Hinchcliffe, The corotating disk plastic processor. *Polym. Eng. Rev.*, **3**:29–62 (1983).
4. Fenner, R. T., *Principles of Polymer Processing*. Chemical Publishing Co., New York, 1980.
5. Rauwendaal, C., *Polymer Extrusion*, 4th edn. Carl Hanser Verlag, Munich, Germany, 2001.
6. Butler, T. I. and E. W. Veazey, *Film Extrusion Manual. Process, Materials, Properties*. TAPPI Press, Atlanta, GA, 1992.
7. *Encyclopedia '97*. Modern plastics (1996).
8. *Plastics in Perspective*, 3rd edn. Association of Plastics Manufacturers Europe (1995) (www.plasticseurope.org).
9. Ding, F. and A. J. Giacomin, Die lines in plastics extrusion: Film blowing experiments and numerical simulations. *Polym. Eng. Sci.*, **44**(10):1811–1827 (2004).
10. Gander, J. D. and A. J. Giacomin, Review of die lip buildup in plastics extrusion. *Polym. Eng. Sci.*, **37**:1113–1126 (1997).
11. Pucci, M. S. and R. N. Shroff, Correlation of blown film optical properties with resin properties. *Polym. Eng. Sci.*, **26**:569–575 (1986).

12. Minoshima, W. and J. L. White, Instability phenomena in tubular film, and melt spinning of rheologically characterized high density, low density and linear low density polyethylenes. *J. Non-Newtonian Fluid Mech.*, **19**:275–302 (1986).

13. Silagy, D., Y. Demay, and J.-F. Agassant, Study of the stability of the film casting process. *Polym. Eng. Sci.*, **36**:2614–2625 (1996).

14. Silagy, D., Y. Demay, and J.-F. Agassant, Study of the linear stability of the drawing of a Newtonian film. *C. R. Acad. Sci. Ser. II Mec. Phys. Chim. Astron.* **322**:283–289 (1996).

15. Petrie, C. J. S. and M. M. Denn, Instabilities in polymer processing. *AIChE J.*, **22**:209–236 (1976).

16. Larson, R. G., Instabilities in viscoelastic flows. *Rheol. Acta*, **31**:213–263 (1992).

17. Ziabicki, A., *Fundamentals of Fibre Formation*. John Wiley, London, U.K., 1976.

18. Bogaerds, A. C. B., G. W. M. Peters, and F. P. T. Baaijens, Tiger stripes: Instabilities in injection molding. In *Polymer Processing Instabilities*, Hatzikiriakos, S. G. and K. B. Migler (Eds.). Marcel Dekker, New York, 2005.

19. Rosato, D. V. and D. V. Rosato, *Blow Moulding Handbook*. Hanser, Munich, Germany, 1989.

20. Bird, R. B., R. C. Armstrong, and O. Hassager, *Dynamics of Polymeric Liquids. Vol. 1: Fluid Mechanics*. John Wiley & Sons, New York, 1987.

21. Dealy, J. M. and K. F. Wissbrun, *Melt rheology and its role in plastics processing*. Van Nostrand Reinhold, New York, 1990.

22. Koopmans, R. J., Die swell-molecular structure model for linear polyethylene. *J. Polym. Sci. A Polym. Chem.*, **26**:1157–1164 (1988).

23. Koopmans, R. J. and D. Porter, Predicting the dynamics of extrudate swell from shear viscosity experiments. In *Theoretical Applied Rheology, Proceedings of the 11th International Congress on Rheology*, Keunings, R. and P. Moldenaers (Eds.), Brussels, Belgium, 1992, pp. 372–373.

24. Koopmans, R. J., Extrudate swell of high-density polyethylene. Part III. Extrusion blow moulding die geometry effects. *Polym. Eng. Sci.*, **32**:1755–1764 (1992).

25. Koopmans, R. J., Extrudate swell of high-density polyethylene. Part II. Time dependency and effects of cooling and sagging. *Polym. Eng. Sci.*, **32**:1750–1754 (1992).

26. Koopmans, R. J., Extrudate swell of high-density polyethylene. Part I. Aspects of molecular structure and rheological characterization methods. *Polym. Eng. Sci.*, **32**:1741–1749 (1992).

27. Chang, J. C., An experimental and theoretical study of continuous filament drawing. PhD thesis, University of Delaware, 1980.

28. Chandrasekhar, S., *Hydrodynamic and Hydromagnetic Stability*. Oxford University Press, Oxford, U.K., 1961.

29. Khomani, B., Interfacial stability and deformation of two stratified power-law fluids in plane Poiseuille flow. Part I. Stability analysis. *J. Non-Newtonian Fluid Mech.*, **36**:289–303 (1990).

30. Khomani, B. and M. M. Ranjbaran, Experimental studies of interfacial instabilities in multilayer pressure driven flow of polymeric melts. *Rheol. Acta*, **36**:345–366 (1997).

31. Boyer, R. F. and H. F. Mark (Eds.), *Selected Papers of Turner Alfrey*. Marcel Dekker, New York, 1986, pp. 523–559.

32. Rubin, I. I., *Injection Moulding. Theory and Practice*. John Wiley & Sons, New York, 1972.

33. Ballman, R. L., R. L. Kruse, and W. P. Taggart, Surface fracture in injection moulding of filled polymers. *Polym. Eng. Sci.*, **10**:154–158 (1970).

34. Hobbs, S. Y., The development of flow instabilities during the injection moulding of multicomponent resins. *Polym. Eng. Sci.*, **32**:1489–1494 (1996).

35. Chang, M. C. O., On the study of surface defects in the injection moulding of rubber modified thermoplastics. In *SPE ANTEC*, San Francisco, CA, 1994, pp. 360–367.

36. Schallamach, A., How does rubber slide? *Wear*, **17**:301–312 (1971).

37. Hamada, H. and H. Tsunasawa, Correlation between flow mark and internal structure of thin PC/ABS blend injection mouldings. *J. Appl. Polym. Sci.*, **60**:353–362 (1996).

38. Heuzey, M. C., J. M. Dealy, D. M. Gao, and A. Garcia-Rejon, The occurrence of flow marks during injection moulding of linear polyethylene. *Int. Polym. Proc.*, **12**:403–411 (1997).

39. Strong, A. B., *Plastics: Materials and Processing*, 2nd edn. Prentice Hall Inc., Upper Saddle River, NJ, 2000.

40. Crawford, R. J. and J. L. Haber, *Rotational Moulding Technology*. Plastics Design Library/William Andrews Publishing, New York, 2002.

41. Soos Takacs, E., M. Emani, D. D'Agostin, and J. Vlachopoulos, Study of orange peel phenomena in rotational moulding. In *SPE ANTEC*, Cincinnati, OH, 2007, pp. 2722–2726.

42. Schidrowitz, P. and T. R. Dawson, *History of the Rubber Industry*. Heffer, Cambridge, U.K., 1952.

43. Agassant, J. F., P. Avenas, J.-P. Sergent, B. Vergnes, and M. Vincent, *La mise en forme des matieres plastiques*. Technique & Documentation, Paris, 1996.

44. Reifenhäuser Gmbh, *Technical Company Literature*. Reifenhäuser Gmbh & Co. KG Maschinenfabrick, Troisdorf, Germany.

45. USI Chemicals International Inc., Polyethylene *Extrusion Blown Films—An Operational Manual*, USI Chemicals International Inc., 1970.

5 Melt Fracture Experiments

The term *melt fracture* was coined by John Tordella in 1956 [1] to describe a critical stress phenomenon. For PE, acrylic resins, nylon, and PTFE, rough extrudates and extrudates with shapes that did not conform to the cross sections of the openings of extrusion dies were observed. He considered that these polymer melt extrudates show small cracks caused by melt fracturing that initiates and propagates in die inlets, at and above a critical pressure. Flow visualization experiments may have inspired the fracture terminology. Tordella's conjecture was that during flow, a substantial elastic strain is imposed. The inability of the polymer melt to deform elastically resulted in the fracture of the melt [2,3]. Polymer melt fracture, or melt fracture for short, has been used ever since to describe collectively all kinds of extrudate distortions. The subsequent visual inspection of extrudates has led to the introduction of many descriptive terms for distortions such as haze, matte, loss of gloss, orange peel or orange skin, sharkskin, pick hammered, wavy, screw thread, helical, pulsating, slip-stick, spurt, gross, or chaotic. Alternatively, terms have been defined that associate the phenomena with a perceived origin. Structural turbulence and elastic turbulence [4,5] are notable examples. The latter nomenclature is of doubtful appropriateness, since turbulence is based on the dissipation of energy in eddies, which does not occur in polymer melts [3,6]. Overall, semantics have definitely not contributed to the clarity of the issue and some controversy may remain with any definition. Certainly, the common use of melt fracture is insufficient to capture all phenomena, and fracture may not always be related to its origin. As a more neutral and general terminology, the terms "melt flow instability" [6,7] or "extrudate distortion" are used. The former focuses on the origin, whereas the latter defines the final result. In this chapter, the term "melt fracture" is used loosely to refer to any form of extrudate distortion. Surface and volume distortions of the extrudate are considered as two distinctive types of melt fracture (Table 5.1).

Most melt fracture researchers focus their efforts on flows through capillary or slit dies. Surface distortions are observed as small-amplitude high-frequency undulations or ripples superposed on a uniform extrudate core. They may be periodic or aperiodic, vary in amplitude and/or frequency, and/or spiral around the uniform core. Small-amplitude high-frequency distortions are commonly referred to as sharkskin. Small-amplitude low-frequency distortions are often described as ripples. In a number of cases, the surface distortion takes the form of a period fracturing. Extrudates with surface distortions emerge out of the die in a straight fashion. In contrast, volume distortions differ from surface distortions primarily because they occur throughout the extrudate. They may have a wave-like appearance usually of a helical kind. The helix can be extremely regular, may have a reproducible periodicity,

TABLE 5.1

Common Types of Melt Fracture for Viscoelastic Polymer Melts

Type	Surface Distortions		Volume Distortions		
Common descriptor	Sharkskin		Screw thread	Helical	Chaotic
Key-feature	Small-amplitude high-frequency roughness	Undulations, wave length 1/5–1/10 of strand diameter	Single or double thread	Wave-like distortions in the order of strand diameter	Incoherent, gross
Origin	Die exit	Die exit	Die exit	Die entry	Die entry

and often has a smooth surface. In other cases, the extrudate becomes incoherent and breaks up in lumps. This is sometimes referred to as gross or chaotic melt fracture. Extrudates showing volume distortions emerge out of the die in a swirling fashion. In a number of cases, volume-distorted extrudates can also have a super-imposed, distorted surface.

For certain polymers, the extrudate emerges out of the die in a pulsating fashion. This phenomenon is known as *spurt*. These extrudates show a periodic succession of no distortion, surface distortion, and volume distortions.

5.1   CONSTANT-PRESSURE AND CONSTANT-RATE EXPERIMENTS

Melt fracture studies are guided by the theory of laminar shear flow in capillary rheometers. The key assumptions of that theory are (1) a fully developed steady shear flow under isothermal conditions, (2) neglect of transients during the startup of flow, (3) a zero-velocity at the die wall, (4) a chemically stable and incompressible polymer melt, and (5) a homogeneous material and temperature field [8–12].

Most experimental devices deviate from these idealizations and require well-known corrections to obtain the true shear rate and stress at the wall [8,9,12]. Capillary rheometers are operated either with constant rate or with constant pressure (Figure 5.1).

In a constant-rate rheometer, the pressure is measured in the barrel at a certain distance from the capillary entry, or from a force-measuring device driving the piston. The pressure or force is monitored over time, and is considered to have reached a steady state when it becomes invariant with time. That value is assumed to represent the total pressure drop P over the die (Figure 5.2). The sensitivity of the pressure transducer in use proves critical for detailed melt fracture measurements. Small pressure fluctuations may not be detected because of low signal-to-noise ratio or because the transducer is not located in the right place.

The apparent wall shear stress τ_a can be calculated taking into account the experimental geometry of the capillary die:

$$\tau_a = \frac{RP}{2L} \qquad (5.1)$$

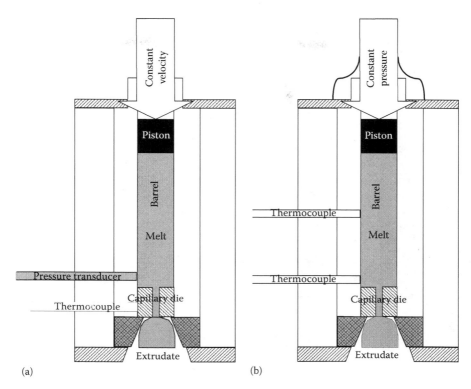

FIGURE 5.1 Schematics of (a) piston-driven constant-(flow) rate and (b) gas-driven constant-pressure capillary rheometer.

For long dies (e.g., $L > 60$ mm and die diameter $D = 2$, $R = 1$ mm), the entrance (P_{ent}) and exit (P_{ex}) pressure drops (Figure 5.2) can be neglected and a fully developed laminar flow may be assumed to exist. At higher shear rates, however, such dies may suffer from shear heating, and temperature control is critical.

For short dies (e.g., $L < 60$ mm length and die diameter $D = 2$, $R = 1$ mm), a Bagley correction [13] is required to compensate for the excessively high pressure found in the barrel relative to that in the capillary. Bagley defined an effective capillary length $L + eR$ greater than the actual length L. The Bagley procedure allows calculating the true wall shear stress τ_t from the apparent wall shear stress for capillary dies:

$$\tau_t = \frac{RP}{2(L + eR)} = \frac{R(P - P_{ent})}{2L} \tag{5.2}$$

where
  e is the Bagley correction coefficient
  P is the measured pressure

The Bagley correction factor e is a dimensionless number, the so-called *end correction*. The graphs of P versus $L/2R$ at constant shear rates ideally yield straight

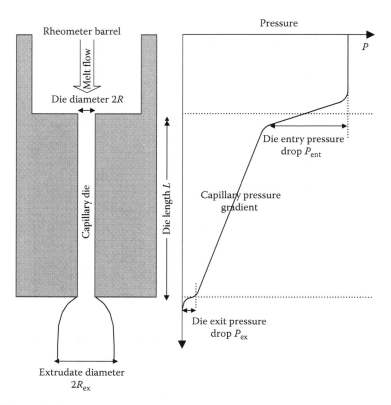

FIGURE 5.2 Schematics of the pressure profile in a capillary rheometer.

lines (Figure 5.3). The $L/2R$ axis intercept gives the end correction e. Alternatively, a pressure correction can be obtained for each shear rate by determining the intercept with the P-axis that defines the entrance pressure drop P_{ent}. The use of a zero-length ($L/2R = 0$) die [14] also allows to define P_{ent} at different shear rates. Philippoff and Gaskins [15] associated the end correction e with an elastic potential energy in steady flow. They considered the end correction e as the sum of a viscous correction n_c (the Couette term) and an elastic correction γ_R (the recoverable (elastic) shear strain):

$$e = n_c + \frac{1}{2}\gamma_R. \tag{5.3}$$

The Couette term is often set to either 1 [16] or 1.75 as proposed by Bagley [17]. Bagley [18] showed that the elastic and viscous terms could be measured by assuming Hooke's law in shear:

$$\tau_t = G\gamma_R \tag{5.4}$$

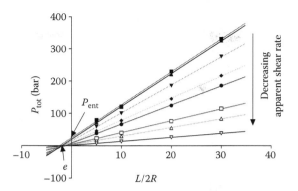

FIGURE 5.3 Typical dependencies of P on $L/2R$ at constant shear rates for estimating e by the Bagley method for an HDPE. The lines represent least-squares regression fits to the data (symbols).

where G is the elastic (shear) modulus (see Equation 3.86). Equations 5.3 and 5.4 indicate that the plots of e versus the true wall shear stress τ_t allow the viscous correction and the elastic shear modulus to be determined. Bagley found these plots to be linear for both branched (LDPE) and linear PE (HDPE) [18].

To convert the apparent wall shear rate $\dot{\gamma}_a$ to the true wall shear rate, a Rabinowitsch correction [8,19] is required. The piston speed determines the volumetric throughput ($\equiv$ output Q_i or Q) and allows calculating the apparent wall shear rate:

$$\dot{\gamma}_a = \frac{4Q_i}{\pi R^3}. \tag{5.5}$$

Equation 5.5 for a capillary die should be compared to Equation 3.102 for a slit die after substitution of $n = W = 1$. The Rabinowitsch correction adapts the apparent wall shear rate by correcting the Poiseuille flow for the non-parabolic velocity field in the capillary. The true wall shear rate for capillary dies becomes

$$\dot{\gamma}_t = \frac{4Q_i}{\pi R^3}\left(\frac{3n+1}{4n}\right) \tag{5.6}$$

with

$$n = \frac{d\log \tau_t}{d\log \dot{\gamma}_a} \tag{5.7}$$

n being the slope of a log–log graph.

Alternatively, constant-pressure rheometers are used. The melt is forced through a capillary by way of a piston or nitrogen (or any inert) gas driving a metal ball. The same basic assumptions about polymer melt flow apply as for constant-rate rheometers. Imposing a constant pressure does not instantaneously establish a steady-state flow situation. However, compared to constant-rate rheometers, the steady state is reached much quicker because under constant-rate conditions, the finite melt compressibility in the barrel plays an important role. At constant pressure, the steady state is reached much slower when the rheometer barrel is attached to a smaller diameter die [20].

For both types of rheometers, the measurements are made at discrete rate/pressure intervals to obtain a pressure (P)–output (Q_i) flow curve. Equations 5.1 and 5.5 can then be used to calculate the apparent wall shear rate and apparent wall shear stress. For each experimental setup, it can be expected that some of the laminar shear flow assumptions do not hold when melt fracture appears. Therefore, the validity as well as the usefulness of any data correction must be a point of discussion in research on the origin and onset of melt fracture. Especially, important points of discussion are related to a non-zero fluid velocity at the die wall, the isotropic homogeneity, and the compressibility of the polymer melt. In this context, it should be emphasized that the flow curve represents a global, macroscopic relationship between the imposed pressure and measured output (or vice versa). It implies that the flow curve is characteristic for the entire extrusion system (a specific polymer, barrel and capillary geometry combination) and is not an "autograph" of the polymer melt. Moreover, it is not equal (but can be related) to the local shear stress–shear rate curve as expressed by a constitutive equation (see Chapter 8). Preferably, flow curves should be reported as pressure P_{tot} (or apparent wall shear stress) versus output (Q_i) (or apparent wall shear rate), or vice versa depending on the selection of the independent variable. Therefore, the careful reporting and interpretation of experimental data is critical when relating flow curves to melt fracture phenomena. The remark made by Boudreaux and Cuculo [21] in 1977 is still valid today: "On the basis of the literature reviewed, there does not seem to be a definitive statement regarding the use of flow curves as an indicator for extrudate distortions."

In the 1970s, it became generally accepted in the literature that the experimental flow curves have two characteristic shapes: continuous or discontinuous (Figure 5.4) [21]. Each of the two flow curves in Figure 5.4 has been linked to two distinct patterns of streamlines at the capillary die entrance (Figure 5.5). The two flow curves and flow patterns are generally associated with linear (HDPE) and branched PE (LDPE), respectively. However, the LDPE-type polymer architecture is seemingly not essential for obtaining a continuous flow curve. Linear polymers such as PS and PP also yield continuous flow curves. The flow curve association with HDPE and LDPE seems to be a purely historical coincidence.

5.1.1 Discontinuous Flow Curves

A typical discontinuous flow curve (Figure 5.4a) is characterized by two partially overlapping flow regimes sometimes referred to as Branches I and II (at low- and high-Q regimes respectively). Such a flow curve is found for most linear flexible polymers (HDPE [22–25], LLDPE [26–28], PB [29–32], PI [33], TFE-THP copolymers [2], high molecular mass linear silicones [34–37], SIS [38], clay [39], and certain compounds of PVC and EPDM [40]).

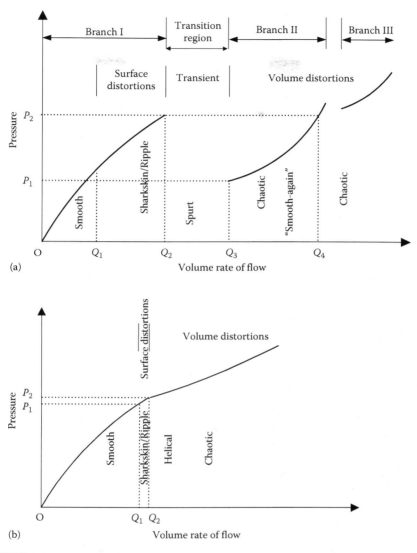

FIGURE 5.4 A schematic representation on logarithmic axes of typical flow curves for controlled-rate or controlled-pressure capillary experiments. Either (a) discontinuous or (b) continuous flow curves are measured.

In the transition region between the branches the pressure oscillates. The pressure oscillation amplitude can vary in the range of 10×10^5 to 200×10^5 Pa depending on the polymer and the experimental setup [25,41]. The discontinuous flow curve may be subdivided into at least five regimes that are associated with typical extrudate distortions (Figure 5.4a). For the selected regimes, the pressure transients for reaching the steady state are illustrated in Figure 5.6.

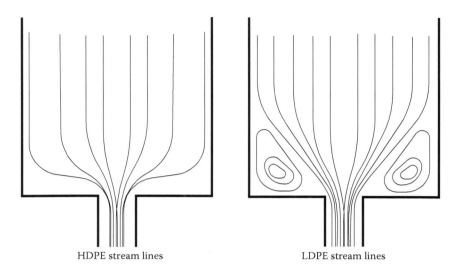

HDPE stream lines   LDPE stream lines

FIGURE 5.5 Schematics of typical stream lines pattern for HDPE and LDPE in flow regimes where melt fracture is observed.

Regime 1 (O to Q_1) (Figure 5.6a): The "normal" steady shear regime gives smooth extrudates, which, for most (uncompounded noncrystalline) polymers, will be transparent and glossy. For gradually higher outputs, the extrudate's appearance remains the same but the pressure transient will be reduced and the steady-state situation will be reached faster. The precise definition of point (P_1, Q_1) as corresponding to the onset point (P_1, Q_3) of Branch II is a matter of debate. For the time being, the only justification is the arbitrary schematic representation of the discontinuous flow curve in Figure 5.4a and some statistical support as examined in detail by Shaw when reviewing the literature findings relating to this theme [42].

Regime 2 (Q_1 to Q_2) (Figure 5.6a): The surface distortion regime is identified by the extrudate's loss of transparency, loss of glossiness, and matte appearance. It is often referred to as the "sharkskin regime" in view of its rough feel resembling the skin of a shark. The extrudate distortion takes the form of a fine-scale, low-amplitude, high-frequency surface roughness. The visual observation is very subjective and strongly dependent on the resolving power of the human eye. Below a certain surface roughness threshold of about 0.8 μm, the human eye may perceive the surface to be smooth, although other detection methods can still define surface distortions [43].

Usually, surface distortions are not associated with high-frequency pressure oscillations. Direct pressure measurements carried out in capillary dies [23] do not reveal clear oscillations. However, the noise on the measured pressure signals is difficult to interpret [44]. Such findings merely indicate that either the pressure oscillations (if present) are too small to be distinguished from the random noise of the pressure transducer in use, or the measurements are not performed in the right place in the die to observe small pressure oscillations [23]. For HDPE, Kometani et al. [45] could not observe any pressure or flow pattern fluctuation near the wall despite the observation of a surface distorted extrudate. A peculiar phenomenon was observed

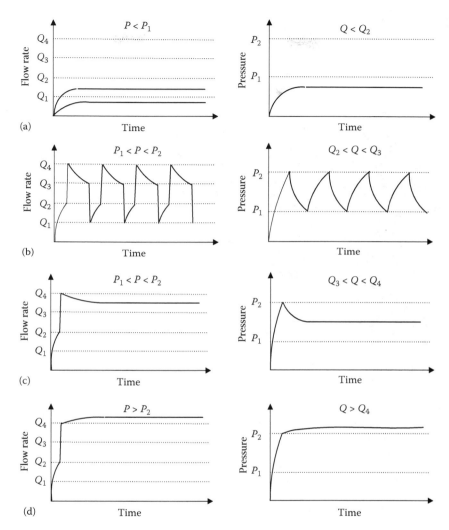

FIGURE 5.6 Schematics of flow rate transients and the associated pressure transients during the approach to steady state for the flow rate regimes indicated in Figure 5.4a.

for a metallocene catalyzed ethylene–propylene copolymer by Fernandez et al. [46]. Above an apparent wall shear stress of 0.23 MPa, they observed a corkscrew shaped extrudate split into two corkscrew strands.

Regime 3 (Q_2 to Q_3) (Figure 5.6b): The "spurt," "stick-slip," or oscillating pressure regime covers the transition between Branches I and II in the flow curve. The extrudate emerges out of the die in periodic bursts. The periodic emerging is reflected in a relatively slow pressure buildup followed by a sudden pressure drop. How slow and sudden depends on the imposed Q in between Q_2 and Q_3; it can be fast up and slow down.

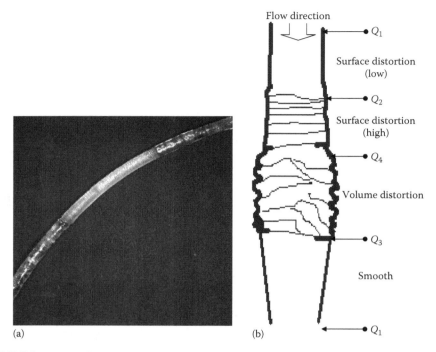

(a)   (b)

FIGURE 5.7 At flow rates in the transient regime, a periodic succession of four types of extrudate distortions is typically observed. Each type can be associated with flow rates of the hysteresis loop going from Q_1 through Q_4 through Q_1 (Figure 5.4a).

The amplitude of the pressure oscillation remains the same but the frequency increases when the amount of polymer melt in the reservoir above the capillary is emptied [25,47]. The extrudate distortion reflects the characteristics of a transient hysteresis loop. The typically observed surface and volume distortions at flow rates on Branches I and II, respectively, are separated by a smooth length of extrudate characteristic for the transition region (Figure 5.7).

The length of each type of distortion is related to the flow rate characteristics of the hysteresis loop when changing from Q_1 through Q_4 through Q_1 [25]. After the hysteresis loop has been completed, the process starts anew as long as the polymer melt is forced with constant rate through the capillary die. The smooth extrudate strand zones may be very short and sometimes hardly visible as the jump from Q_2 to Q_4 and Q_3 to Q_1 is very rapid.

In a constant-pressure measurement mode, no pressure oscillations are observed. Imposing a pressure sequence from low to high or from high to low to determine the flow curve can lead to different types of extrudate distortions. When the pressure is gradually increased from low to high in stages, a sudden break in the flow curve arises at point (P_2, Q_2) (Figure 5.4a). Any additional, very small pressure increase will immediately find a steady flow point on Branch II, and the flow rate increases dramatically as reported, for example, by Bagley et al. [48], El Kissi and

Piau [49], and Vinogradov et al. [33]. The extrudate appearance may vary with changing polymer architecture, but a common finding is that surface distortions disappear and volume distortions appear for pressures up from the point (P_2, Q_2). In many cases, the extrudate appears smooth for pressures higher than point (P_1, Q_1) and relatively close to point (P_2, Q_2) [24,48]. Near the end of Branch I, the extrudate can appear wavy [48] and is described as rippled by Tordella [3]. When starting the experiment at pressures above point (P_2, Q_4), lowering the pressure produces a break in the flow curve at point (P_1, Q_3). Further lowering of the pressure gives steady flow points on Branch I. Bagley et al. [48] report extrudates to be rough for HDPE. Similar results are obtained by Durand [25] for an HDPE. Consequently, for these materials, constant-pressure experiments indicate the coexistence of two steady flow regimes. The polymer will select either Branch I or II depending on the preceding shear history. Durand's [25] comparison between constant-pressure and constant-rate experiments demonstrates the overlap of both flow curves (Figure 5.8).

Regime 4 $(Q_3$ to $Q_4)$ (Figure 5.6c): Often referred to as the "gross-melt fracture" regime, gives in most cases helical, spiraling distortions that affect the entire volume of the extrudate. The observation of the extrudate close to the die often shows a swirling motion. In contrast, all the previous regimes show the melt to emerge straight downward. For a number of polymers, HDPE [50,51], LLDPE [52], metallocene catalyzed PE [53], and PTFE and tetrafluoroethylene-hexafluoropropylene copolymers [3], "smooth-again" extrudates are found. This so-called super-extrusion regime has the potential to allow for high-speed processing [54]. However, not all polymers seem to give rise to this phenomenon and the actual operational stability may be questionable. The pressure transient shows an initial overshoot before

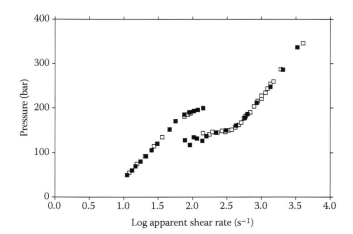

FIGURE 5.8   Flow curves obtained from constant-pressure (□) and constant-rate (■) measurements for an HDPE using a capillary rheometer equipped with a die of dimensions $L = 21.6\,mm$, $2R = 1.3\,mm$, and entry angle 90° at 160°C. (From Durand, V., Ecoulements et instabilité des polyéthylènes haute densité, PhD thesis, Ecole des Mines de Paris (CEMEF), Sophia Antipolis, France, 1993.)

reaching its steady-state value. The size of the overshoot lies between P_1 and P_2 and reduces to disappear at (P_2, Q_4).

Regime 5 (*above* Q_4) (Figure 5.6d): The chaotic regime is characterized by all kinds of strange volume distortions of the extrudate. It is difficult to define specific patterns because the extrudates are grossly distorted, and sometimes look like glued together polymer lumps or highly-stretched spiraling ribbons. The extrudates seem to sputter out of the die and usually twist in all directions. The pressure reaches a quasi-stable value. Similar to the onset flow rate of surface distortions Q_2, Q_4 is not uniquely defined. It is not necessarily associated with the end pressure of Branch I. For HDPE and LLDPE, an additional relaxation oscillation regime defined by a second transition region and a Branch III have been observed [27,28,36]. The associated pressure oscillations are not of the same intensity, if observed at all. The flow rate corresponds to very high apparent shear rates (>10,000 s^{-1}). The extrudates are all volume distorted making it very difficult to discern a regular pattern.

5.1.2 CONTINUOUS FLOW CURVES

A continuous flow curve is measured for LDPE [1], very low density metallocene catalyzed PE [55], ethylene vinylacetate copolymers [56], PP [57–60], PS [61,62], PVDF [3], PMMA [61], PA [61], branched silicones [34,36], linear and star-branched SBR [63], and PVC [64] (Figure 5.4b). No distinction is found between constant-pressure and constant-rate experiments. For most polymers, at gradually increasing flow rates, the emerging polymer melt is first smooth and transparent, then some form of undulation or waviness (ripple) starts to occur. Eventually, this turns into helical distortions until a totally chaotic, grossly distorted extrudate appears [3]. Surface distortions as defined by fine-scale, low-amplitude, high-frequency roughness seem to appear as a loss of gloss (matte), only in a very brief output range [3] that is followed immediately by more severe volume distortions of the extrudate. For the mentioned polymers, surface distortion may exhibit some unusual forms. The most commonly seen is the screw thread, a single or double entwined helix around a stable central core. The extrudates emerge straight out of the die until helical distortions start to appear. Then the emerging polymer melt starts to swirl. Although the flow curves are continuous, it is still possible to observe an effect that strongly resembles "spurt." For LDPE [65,66] and high-energy irradiation (electron beam) modified PP [66,67], a periodic phenomenon is observed showing a succession of volume-distorted, smooth, and volume-distorted regions in the extrudate (Figure 5.9). At gradually higher flow rates, the smooth region becomes shorter until it completely disappears and a totally volume-distorted extrudate remains. Eventually, at even higher flow rates, the extrudate sputters out of the die and twists in all directions. The spurt-like extrudate is typically not linked to a transition region in the flow curve. For all experimental flow settings that yield this feature, no reference is made to pressure oscillations except in the case of modified PP. In this specific case, the pressure oscillations are very small (<10^6 Pa) and difficult to detect with standard pressure transducers. Only when using fast pressure measuring transducers can these signals be picked up [67]. A similar spurt phenomenon can be observed for PS although the pressure relaxation oscillations are significantly

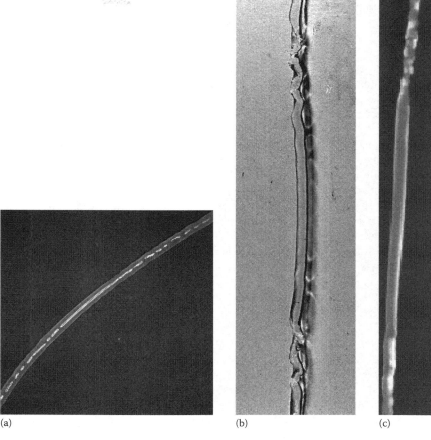

(a) (b) (c)

FIGURE 5.9 A periodic, spurt-like extrudate distortion for an LDPE (a), PS (b), and branched PP (c). For each material, no surface distortions appear but a succession of volume distorted, smooth, and volume distorted regions appear (flow direction top to bottom). Although resembling "spurt," the flow curve of LDPE and branched PP is continuous, and very small (<10^6 Pa) pressure relaxation oscillations are detected, only when using fast pressure-measuring transducers in the commercial constant-rate capillary rheometers. PS has a much larger pressure fluctuation and a discontinuous flow curve. The smooth middle part corresponds to the pressure build-up phase.

larger. The overall surface is smooth but regularly twisted when the built-up pressure relaxes.

The experimental findings of discontinuous and continuous flow curves in combination with the observation of a variety of extrudate distortions stimulated researchers to use more elaborate experimental techniques for investigating melt fracture.

The aim of most research revolved around a number of issues: How is melt fracture initiated? Where is melt fracture initiated? Can a critical value be defined that would determine the initiation of melt fracture in terms of measurable quantities? What is the role of elasticity when processing viscoelastic polymer melts?

5.2 FLOW VISUALIZATION

A number of techniques are applied to visualize the flow characteristics of polymer melts. Initially, experimental findings indicated that extrudate distortions are the result of flow instabilities originating at the entrance of the die. As a result, many experiments are directed toward observing the flow in the reservoir and at the die entrance.

The flow pattern is visualized by following the motion of particles either naturally accompanying the melt (air bubbles, dust, and gel particles), or placed in the melt (carbon black, carborundum particles, iron filings, poppy seeds, chopped glass fibers). In addition to merely observing the path of the particles and determining the streamlines, the velocity of the particles in the melt can be measured. Simple optical particle tracking detection systems or more sophisticated laser Doppler velocimetry (LDV) equipment can be used [68–70]. The latter technique allows only point-wise velocity measurements and the determination of the entire flow profile is a time consuming exercise. Another flow visualization method consists of injecting or placing dyes or other pigments in the melt and to observe the developing melt flow patterns. It is a fairly straightforward technique that still requires skilled experimentation. An altogether different way of flow visualization involves the viewing of flowing melts in between crossed polars. This flow birefringence technique shows a pattern of lines (fringes) corresponding to positions where the stresses are equal.

Besides the die entry, flow visualization is performed in the die as well as at the die exit. These experiments, however, are more difficult to analyze. The dimensions are much smaller and the flow rates much higher. Especially when flow instabilities occur, it is difficult to resolve the particle positions, the colors or the stress fringes. The use of high intensity laser technology may resolve these issues somewhat. Some of the more relevant flow visualization work in relation to melt fracture is done by Benbow and Lamb [71], Tordella [2], Vinogradov et al. [72], Vinogradov and Malkin [73], Bartos and Homolek [74], Galt and Maxwell [75], Maxwell and Galt [76], Den Otter [34,35,77], White [78], Checker et al. [79], Ma et al. [80], El Kissi [81], Münstedt et al. [70], Hertel and Münstedt [82], Robert et al. [83,84], Combeaud et al. [85], Piau et al. [29], Martyn et al. [86,87], Gough et al. [88,89], and Gough and Coates [90]. Measurements at the exit region have been carried out by Gogos and Maxwell [91], Checker et al. [79], Barone and Wang [92–95], Migler et al. [96,97], and Kometani et al. [45].

All investigators used windows based on glass, quartz, or acrylic materials for optical transparency (the latter only for solutions or amorphous polymers that are processable at low temperatures). Most flow visualization experiments are performed using slit dies in an attempt to create a quasi (2D) flow situation. The high polymer melt temperatures and pressures put certain mechanical constraints on the experimental setup making the flow conditions not necessarily ideal [98]. Martyn et al. [86,87]

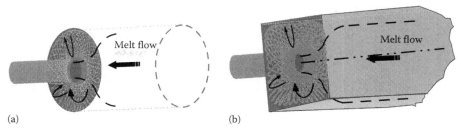

(a)   (b)

FIGURE 5.10 Schematic showing ideal natural flow of a polymer melt in (a) an axisymmetric and (b) pseudoaxisymmetric die entry geometry. (Courtesy of M. Martyn, T. Gough, and P. D. Coates from The Dow Chemical Company.)

investigated the entrance flow for axisymmetric contractions (Figure 5.10). Rusch [69] developed an experimental setup for full 3D velocity profile measurements through particle tracking before and in a capillary die for an LDPE melt. Die entry vortices were identified although no attention was paid to extrudate distortions in that study.

In contrast, flow curve measurements and extrudate observations are mostly done using capillary rheometers. When melt flow instabilities occur (e.g., vortex formation), a one-to-one correspondence between slit and capillary die experimental results may be suspect. The flow in the die develops a full three-dimensional (3D) character. Visualization experiments by Chiba and Nakamura [99] with dilute polyacrylamide solutions in a circular entry die demonstrate the 3D character of flow instabilities originating at the die entry (Figure 5.10a). The Coates group in Bradford (UK) developed an experimental setup to visualize a 3D vortex development of an LDPE resin in axisymmetric contraction flows (Figure 5.11) [86–90].

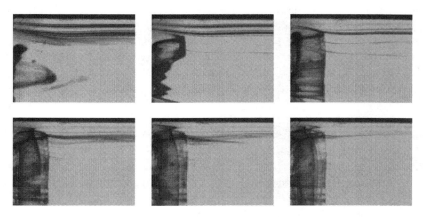

FIGURE 5.11 Flow (from right to left) of an LDPE melt through an axisymmetric contraction (Figure 5.10b), showing the development of a vortex before the die entry. Carbon black is injected at the side of the flow (top) and is gradually smeared over a layer that rolls up making entry vortex visible. The video coverage of the sequences captured in these six still pictures here indicates a motion from top to bottom and bottom to top indicating the full 3D nature of the vortex. (Courtesy of M. Martyn, T. Gough, and P. D. Coates from The Dow Chemical Company.)

5.2.1 PARTICLE TRACKING

Tordella [2,3] and Tordella and Wilkens [100] observed the flow of LDPE by follow-ing the motion of hard and soft tracer particles made by mixing carbon black with high- and low-viscosity PE. With hard tracer particles, at flow rates below those for the onset of extrudate distortion, they observed a radial flow with the particles at the center of the reservoir above the die inlet moving faster than those near the edges of the reservoir. At higher flow rates, the particles converged towards the ori-fice whereas those in the corners of the reservoir appeared to be trapped. When extrudate distortions could be observed, he noticed a see-saw movement between the particles at the center and those at the corner of the reservoir. With the soft particles, he saw conical streamlines surrounded by circulating stagnant regions in the corners, at low flow rates. At even higher flow rates, the center streamline broke or "fractured" and allowed material from the stagnant regions to flow through the orifice (Figure 5.12). When the stagnant region was depleted, the center streamlines reformed and new stagnant regions developed. The distorted extrudate was com-posed of alternate material from the central streamlines and the stagnant material in the corners. Similar observation had been made by Clegg [101]. The findings were later confirmed by many other investigators providing consensus that for LDPE, melt fracture is initiated at the die entrance.

In contrast, HDPE and polymers characterized by a discontinuous flow curve have not been observed to show any dead spaces or swirling motion of the flow lines (Figure 5.5) [49,50,91].

For the polymer solutions of polyacrylamide and hydroxymethyl-cellulose, Nakamura et al. [102] observed oscillating flows similar to those reported for LDPE. Some 10 years earlier, the same had been reported by Nguyen [103], Cable [104], and Cable and Boger [105] for polyacrylamide in glucose and water solutions. Many of their flow visualization results are reproduced in a book by Boger and Walters [106]. With tapered entry dies, the flow behavior remains essentially the same except that there are no or very small stagnant areas at high flow rates. It appears that the visco-elastic fluid shapes its own tapering when forming die entry vortices.

The direct recording of the individual motion of randomly distributed small particles in polymer melts to determine their velocity profile was first reported by Maxwell and Galt [75,76]. These authors observed the motion of irregularly shaped carborundum particles of size 8 μm for several LDPE melts in a 6 mm capillary tube, and slit dies of varying dimensions. Three distinct flow regions were identified. Close to the wall, particles moved first with one velocity than with another, one of which could be a zero-value. This stop-and-go flow was called stick-slip and associ-ated with melt elasticity. In the central part of the flow field, all particles moved with nearly the same velocity. In between these two regions, a viscoelastic laminar shear flow velocity pattern was noted. The time-averaged velocity profiles were very stable for the apparent wall shear stress ranging from 14 to 27 kPa. No reference was made to melt fracture in these papers. Den Otter et al. [107] using a similar velocity mea-surement approach worked under stable and "unstable" conditions of an LDPE melt flow, and observed a zero-velocity at the wall in all cases in contrast to the previous authors. Subsequent investigations [34] on HDPE, LDPE, and linear and branched

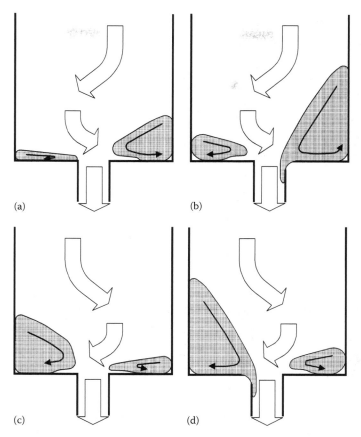

FIGURE 5.12 Schematic representation of an oscillating flow of material from the funnel and the stagnant areas from the reservoir in the die. (Adapted from Nakamura, K. et al., *J. Textile Machinery Jpn.*, 36, 49, 1987.)

PDMS using dust or gel particles as tracer material led to a similar conclusion. Wall slip, being smaller than 1 mm/s, was considered not important for these materials and was therefore assumed to be equal to zero. Less firm conclusions were made for ethylene–propylene rubber velocity measurements. Close to the wall, experiments indicated a velocity tending to zero at normal processing temperature, while a slip velocity was found at room temperature [77]. The size of the tracer particles in relation to its sensitivity for detection in combination with the quality of the optical setup prevented measurements at the wall. Bartos and Holomek [74] studied the flow of PB in a capillary die using spherical aluminum tracer particles with size of less then 10 μm. In the laminar flow regime, a zero wall velocity was reported although some experiments implied a slip velocity. Extrusion of pure polymer through capillaries coated with tracer particles showed the majority of the coating to be smeared onto the surface of the extrudate. Similarly, the stress development over time in a piston-driven constant-rate capillary rheometer showed an irregular oscillation around a mean value, which grows at increasing flow rates. In these flow regimes (O to Q_2) (Figure 5.4a), showing no disturbance in laminar flow and velocity profile, the extrudate surfaces were either smooth or distorted. The surface distortions were concluded to originate at the die exit in accordance to Clegg [101], Benbow and Lamb [71], and Bagley et al. [108]. In the transition regime (Q_2 to Q_3), the periodic pulsation of the flow rate was related to a melt velocity change over the entire cross section of the capillary at the same time and same value. It indicated the presence of a slippage or "a reaction of the material to a deforming stress which can be described as slip" [74].

At flow rates related to volume distortions, a significant velocity scatter at any point in the capillary was noted as well as slip. El Kissi [81] measured stable velocity profiles at the entrance and exit of the die and did not observe any macroscopic slip, in accordance with Bartos and Holomek [74] for low flow rates below the spurt regime. Piau et al. [29], also using LDV for PB in steel slit dies, showed fully developed and stable parabolic velocity profiles. No polymer slippage at the die wall was suggested even in the regime where surface distortions were observable. The overall flow curve did not show any spurt regime. For PTFE dies and the same PB under identical experimental conditions, the flow curve showed a spurt regime. At low flow pressure, a fairly regular velocity profile was observed while in the spurt regime a jerky, translational motion of the particles could be noted. At higher flow pressures, a plug flow was observed indicating wall slip.

Münstedt et al. [70] used LDV to connect observed pressure oscillations during spurt to oscillations in wall velocity for an HDPE material. They also observed that significant slip occurs at Branch I of the flow curve, before the onset of oscillations. No slip was found for an LDPE material at similar conditions. Robert et al. [83] confirmed these results for another HDPE, although they found negligible slip for Branch I of the flow curve. Schwetz et al. [109] and Hertel and Münstedt [82] measured 3D velocity profiles in contraction flows for branched polymers such as LDPE and showed clear relations between secondary flow patterns and strain hardening, that is, elongational characteristics. Combeaud et al. [85] used LDV to measure velocity profiles of PS in contraction flow. They observe velocity oscillations corresponding to volumetric distortions on the corresponding extrudates.

Archer et al. [110] followed with a microscope the motion of 1.5 μm spherical glass particles dispersed in a PS solution, placed between the glass plates of a plane Couette cell. Step shear experiments of 0.5–5 strain units showed that the particles slip within a few micrometers from each wall.

Migler et al. [96,97] used optical velocimetry and microscopy to observe flow near the die exit for LLDPE with and without polymer processing aid (PPA). They used a capillary sapphire tube connected to a twin-screw extruder. They related surface distortions to the specific kinematics near the exit, and showed that "sharkskin" can also occur in the presence of a slip-creating coating (PPA).

5.2.2 FLOW BIREFRINGENCE

The flow birefringence technique is used by Tordella [2], Ballenger and White [58], Oyanagi [111], Brizitsky et al. [112], Checker et al. [79], Arda and Mackley [113], Arai [114], Beaufils et al. [43], Piau et al. [29], Robert et al. [84], Combeaud

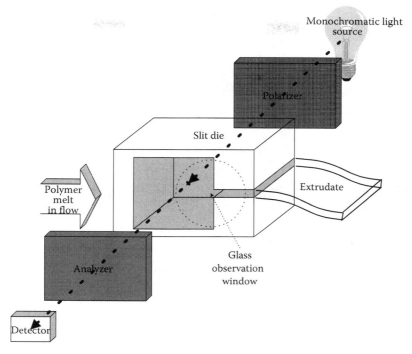

FIGURE 5.13 Schematic flow birefringence setup.

et al. [85], and Barone and Wang [92,95] to study melt flow instabilities of different polymers. The experimental setup generally consists of a monochromatic light source (e.g., a LASER) illuminating slit die geometry placed between crossed polars (Figure 5.13).

The flow birefringence pattern that develops under steady-state conditions is analyzed. Images shown in Figure 5.14 represent iso-stress lines (also referred to as stress fringes and iso-chromes) for a specific flow situation. A high fringe concentration is observed at the die entry and exit corners (the so-called corner singularities).

Tordella [2] observed flow birefringence pattern instability for LDPE and HDPE. For LDPE, the results supported the tracer particle observation. Melt fracture is initiated as a flow instability at the die entry. Die entry initiated melt fracture decreases in magnitude with decreasing die entry angle. For HDPE, the first departure from laminar flow occurred within the capillary at the die wall and not at the die entrance. The breaks of iso-chromes and their grainy structure was interpreted as an elongational rupture (the limit of elastic deformation is reached) and associated with the first surface roughness appearance on the extrudate. In the spurt regime, a large increase in birefringence at the die entry occurred, which decreased with distance in the die. It was considered to be an indication of slip or sharp yielding within the capillary similar to that observed for rubber. All flow birefringence results convinced

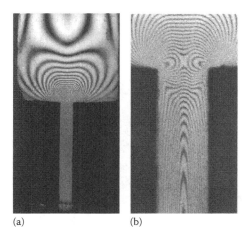

(a)   (b)

FIGURE 5.14 Flow birefringence of an HDPE melt flowing through a slit die (15:1 contraction) (a) in a regime where surface distortions occur. A detail (b) of the die entrance and exit is shown. No pulsation or vortex formation can be observed. (Data from J.-P. Villemaire and J. F. Agassant from The Dow Chemical Company.)

Tordella [2] that melt flow instabilities and the resulting melt fracture are related to the elastic strain behavior of the polymer in accordance to considerations by Philippoff et al. [115].

On studying HDPE polymers, Oyanagi [111] observed that the number of fringes increased with increasing output. The development of a high-frequency fringe pulsation was noted in the reservoir at the die entry, at the point the spurt regime started. Some 15 years later, Sornberger et al. [116] reported similar findings.

In a number of cases, however, at the die entry and at very high flow rate, it was possible to observe the formation of so-called die lip vortices. If they occurred, the extrudates showed volume distortions [106].

Flow birefringence studies in the die during the occurrence of extrudate surface distortions have been done by Vinogradov et al. [72], Beaufils et al. [43], and Sornberger et al. [116]. They all agree that the birefringence method shows no flow instability in the die at the onset of extrudate surface distortions (Figure 5.14).

The flow birefringence of a high molar weight nearly uniform PS is reported by Kamath [117] (Figure 5.15). No vortices were observed in the entry region but highly irregular flow lines that suggested a swirling motion. This propagated into a "disordered" exit flow pattern. The experimental setup did not allow observing extrudate distortions as the exit region was a reservoir filled with polymer melt.

Combeaud et al. [85] reported combined flow birefringence and LDV measurements on a commercial linear PS. They observed velocity and stress oscillations in the entrance region, which correspond with volumetric distortions in the extrudate. Their observations indicated a transient periodic motion related to a gradual change over from helical extrudates to sudden discharges yielding strongly irregularly deformed extrudates. Carefully considering the entry region, they noted that from time to time the vortex suddenly discharges into the main flow and destabilizes

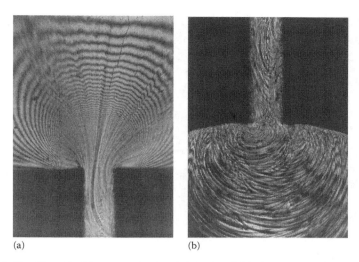

(a)   (b)

FIGURE 5.15 Flow birefringence of a high-MM ($M_w = 510\,\text{kg/mol}$), nearly uniform polystyrene. (Data from M. Mackley from The Dow Chemical Company.)

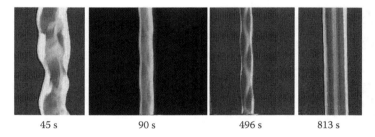

45 s   90 s   496 s   813 s

FIGURE 5.16 Extrudate distortions (from right to left) of an LDPE melt through a pseudo-axisymmetric contraction (Figure 5.10b) show a transient character. During startup, over a period of >13 min, the extrudate appearance does change from very distorted to undistorted. (Courtesy of M. Martyn, T. Gough, and P. D. Coates from The Dow Chemical Company.)

it. It was also concluded that transient vortex destabilization exists, but cannot be an explanation for the onset of volume distortion, as low angle convergents also show periodic volume distortions without an observable vortex. Martyn et al. [86,87] likewise observed a transient volume distortion for LDPE (Figure 5.16). It was noted that the use of pigmented centerline indicated its oscillation being out of phase with the periodic shape of the extrudate (see 90 and 496 s images).

Robert et al. [84] performed flow birefringence on two HDPE materials in the die land of a slit die. They reported stress oscillations that correspond to pressure oscillations and the stick-slip phenomenon. At the exit singularities, Piau et al. [27] found for PB a pulsation of fringes, with a frequency that equals the low-amplitude high-frequency surface distortion. The onset of a surface distortion and its increased severity at higher flow rate does not perturb the flow upstream of the die exit. This was in line with the findings of Sornberger et al. [116] who performed similar experiments for LLDPE. Barone and Wang [92] examined PB melts at the slit die exit. The flow birefringence signal was observed to oscillate periodically near the die exit in tune with a period identical to the extrudate surface distortion. In contrast, a steady behavior is observed in the die inland region. Additional experiments [95] with an ethanol-coated die removed the presence of oscillations and surface distortions.

5.2.3 STACKED COLORS

Another flow visualization technique is to stack alternately different colored disks of the same polymer. The pigmentation of the polymer should remain sufficiently small in order not to influence the melt rheology. The technique was applied by Benbow and Lamb [71], Metzger and Hamilton [118], Hürlimann and Knappe [119], Oyanagi [111], White [78], Bergem [120], Uhland [23], Ma et al. [80], Vos et al. [121], and Sombatsompop and Wood [122]. Alternatively, as done by Lyngaae-Jorgensen and Marcher [123] or Benbow and Lamb [71], color markings can be injected in the die during extrusion to follow the flow pattern.

For steady laminar flows, a cut in the flow direction of the extrudate showed that their composition consists of concentric alternately colored cylinders. It demonstrated the flow to be perfectly axisymmetric and laminar in the die [108]. In the spurt regime of *trans*-1,5-polypentene, Bergem [120] showed for surface distortions that resemble a screw thread that the discontinuities of the black lines correspond to a rupturing of an outside layer of the extrudate near the die exit. He also carried out an experiment where first HDPE filled with 0.5% carbon black is extruded through a corrugated capillary die, followed by a removal of the black HDPE out of the reservoir. The extrusion was repeated with uncolored HDPE at a constant rate in the middle of the spurt regime. The experiment showed a continuous layer of black polymer at the die wall. This suggested that the polymer did not slip at the die surface. Bergem explained that a slip was not necessary for the occurrence of a flow curve discontinuity. It could be explained by a lubrification mechanism of a thin surface layer of a disentangled polymer in low-viscosity state. Although physically different in nature, both mechanisms would give rise to the same macroscopic observations. Similar experiments were repeated by Barone and Wang [93] for a PB and a metallocene catalyzed HDPE. The barrel was half filled with polymer containing, respectively, 10% and 0.5% carbon black with uncolored polymer on top. The lower half of the transparent slit die was coated with fluoro-polymer. Upon starting the piston at a constant rate, an extrudate with a black skin and natural core is formed. The black is observed to stick to the uncoated die wall indicating a wall adhesion versus a slipping flow regime in the coated part. Such observations are associated with local slip while uncolored regions are associated with local stick.

At higher flow rates, volume distortions appear and the extrudate is no longer composed of concentric alternately colored cylinders. In this regime, Metzger and Hamilton [118] observed strong layer exchange that resulted in a loss of layer axisymmetry even though the extrudate exterior was smooth. For HDPE, Oyanagi [111] obtained results resembling those of Bergem [120] and Benbow and Lamb [71]. Starting from steady conditions, at gradually increasing flow rates, the laminar flow pattern is lost and becomes unsymmetrical. These situations give rise to wave forms in which the wave pitches ranged from small and limited to the outer-layer (surface distortion) to rough and eventually to deeper pitches that extended to the centre of the extrudate (volume distortion). When volume distortions appeared, the color markings were found alternately left and right to the axis of the extrudate.

For LDPE, Hürlimann and Knappe [119], starting from stacked alternated black and natural layers, showed clearly the vortex formation before the die exit. At 150°C, the onset of volume distortions was associated with a critical extensional stress being independent of extension or extension rate, indicating the critical importance of the centerline extensional flow characteristics. These findings substantiate the later findings of Combeaud et al. [85] and the theoretical considerations of McKinley et al. [124]. The centerline instability as observed by Martyn et al. [86,87] and shown in Figure 5.16 can be associated with the Hürlimann and Knappe [119] findings of exceeding an extensional stress.

Cogswell [125,126] provided an analysis of surface distortions by co-extruding similar materials but with different colors at the core and the surface layer. Cracks were formed in the skin layer only. Cogswell concluded that surface distortions correspond to a tearing of the exterior extrudate layer at the die exit. More than 22 years later Barone et al. reconfirmed these results [127]. A local die exit stress oscillation induces the periodic surface distortions.

Related to these experiments, interesting flow visualization studies are reported into the origin of layer thickness variation in multi-layer coextrusion by Dooley and Hughes [128]. They showed for two differently colored but identical PS layers, a gradual rearrangement of the interface during flow in a square channel. Later, they associated the secondary flow phenomenon to the effects of first and second normal stress differences by comparing experiments with nonlinear viscoelastic flow simulations [129–131]. The theoretical and experimental studies of interfacial instabilities in multilayer pressure-driven flows by Khomani [132], and Khomani and Ranjbaran [133] also indicated the importance of first and second normal stress differences, implying the importance of extensional flow characteristics. In two (A-B)- and three-layer (A-B-A) flows using PP, HDPE, and LLDPE, interfacial instabilities are shown reminiscent of surface distortions. The authors explained that a thin, less-viscous layer adjacent to the wall can stabilize long-wave disturbances in the core layer, while short- and intermediate-wavelength disturbances are stabilized with a more elastic fluid as the core component.

5.3 CRITICAL NUMBERS

For the predictive purposes of polymer melt fracture onset, it is useful to define critical numbers. Experimental support for any one does not constitute a physical explanation. Moreover, no mechanism is implied by the observation that melt fracture seems to always occur at a critical number value [6,134]. There is an accumulating evidence of the presence of characteristic points as indicated by a change of slope or a discontinuity in the flow curves. Still, the accurate determination of such a critical point and its relation to melt fracture has some inherent weaknesses. First, a flow curve is made up out of a set of discrete points, each independently measured under the assumption of laminar steady flow. Second, the measured pressures or flow rates are global, averaged values characteristic for the entire extrusion system and not for the polymer alone. Third, a transient process exists before a steady-state flow is obtained that depends on the entire extrusion system. Fourth, the melt fracture onset is usually defined for conditions where the visual detection of a distorted extrudate is possible, and this does not need to correspond with the actual onset of melt fracture. The latter point is clearly illustrated by Beaufils [43,135]. For an HDPE and an LLDPE, they demonstrated by measuring the extrudate surface topology with a roughness measuring device that the surface is distorted well before a visual observation can be performed. Another detection challenge is mentioned in the paper by White [78] stating that "many early researchers missed the spirals (ripple-like surface distortions) because they apparently had passed over the flow rates where spiraling extrudates formed when changing the flow rate in measuring continuous flow curves."

5.3.1 REYNOLDS NUMBER

The first critical number that comes to mind is the Reynolds number, Re, in analogy with turbulent flow of Newtonian fluids. For capillary dies it is defined as

$$Re = \frac{v \rho R}{\eta} \tag{5.8}$$

where
  v is the mean velocity of flow
  ρ is the density
  R is the capillary radius
  η is the fluid viscosity

The typical criterion for the onset of turbulence is an Re value of 1000–2000. The high viscosity of polymer melts immediately indicates that melt fracture appears at Reynolds numbers that are much below the onset of turbulent flow [3,6].

5.3.2 WEISSENBERG AND DEBORAH NUMBER

The influence of viscoelastic forces on a flow can be characterized by the dimensionless Weissenberg number, We. We is the product of shear rate $\dot{\gamma}$ (a measure of the velocity gradient and defined in Equation 3.30) and a characteristic relaxation time λ

$$We = \lambda \dot{\gamma}. \tag{5.9}$$

The Weissenberg number also represents the ratio of the elastic and viscous forces in the fluid. In flows that are time dependent, it is possible to define a Deborah number, De, which is the ratio of a characteristic relaxation time of the fluid and a characteristic time of flow t_f

$$De = \frac{\lambda}{t_f}. \qquad (5.10)$$

The characteristic time t_f can be a residence time, the time of a flow transient, or the time during which a particle experiences a velocity gradient passing through a die contraction. A Deborah number exceeding unity indicates that the flow is dominantly elastic. The Deborah number is often proportional to the Weissenberg number with the proportionality constant depending on dimensionless geometric or operating parameters of the flow. The use of these dimensionless parameters is inspired by the belief that melt flow instabilities are related to the elastic characteristics of the polymer [134]. In that context, the Weissenberg number should be a useful defining parameter of the elastic energy contained in the flowing melt. White [136] formulated We as the ratio of the first normal stress difference and the shear stress for conditions of uniform simple shear equal to those in a cone-plate rheometer. McIntire [137] examined the use of the Weissenberg number for determining the onset of melt fracture using classical linear hydrodynamic stability analysis but presented no firm conclusions. The belief that acceleration effects at the die entry are critical for the onset of melt fracture led others [138,139] to use the Deborah number. The Deborah number represents a ratio of timescales implying a fluid-like behavior for small numbers and a solid-like behavior for large numbers. The correlation of We and De to melt flow instabilities was examined by Barnett for various polymers [140]. Some correlation was found with the onset of melt fracture but still a number of exceptions are noted such as for various PPs. McKinley et al. [124] provide a theoretical overview for simple unidirectional and more complex 2D flow geometries. For a purely elastic instability, they indicate a coupling between the curvature of fluid streamlines and the elastic normal stresses that gives rise to a tension along the streamline. They propose a dimensionless criterion that must be exceeded for the onset of purely elastic instabilities to occur in planar contractions (e.g., flows through a capillary die):

$$\left[\frac{\lambda_1 U}{R} \frac{\tau_{11}}{\eta_0 \dot{\gamma}} \right]^{1/2} \geq M_{crit} \qquad (5.11)$$

where

λ_1 is the fluid's relaxation time
U is the characteristic stream wise fluid velocity
R is a characteristic radius of curvature of the streamline
τ_{11} is the tensile stress in the "1," i.e., flow direction
η_0 is the zero-shear viscosity of the fluid
$\dot{\gamma}$ is a characteristic value of the local deformation rate

The McKinley number bears some relationship to the Deborah and Weissenberg number. The experimental data by Hürlimann and Knappe [119] for LDPE would provide a valid test case.

5.3.3 RECOVERABLE STRAIN

A more direct measure of polymer elasticity is the recoverable strain γ_R as defined in
Equation 5.4. When the recoverable strain reaches a value between 1 and 10, extrudate distortion has been observed [3,6,7,21]. The modulus G in Equation 5.4 can be
approximated by the ratio of the zero-shear viscosity and a characteristic relaxation
time λ (see Equation 3.106):

$$G \approx \frac{\eta_0}{\lambda}. \tag{5.12}$$

Combining Equations 5.4 and 5.12 with Newton's law $\tau = \eta_0 \dot{\gamma}$, the recoverable strain
can be expressed as the product of shear rate and a characteristic relaxation time λ:

$$\gamma_R = \frac{\tau}{G} \approx \frac{\eta_0 \dot{\gamma} \lambda}{\eta_0} \approx \lambda \dot{\gamma}. \tag{5.13}$$

Hence, the recoverable strain is roughly equal to the Weissenberg number.

Spencer and Dillon [62] found for PS that the product of the critical shear stress
τ_c for onset of melt flow instabilities and the weight-average MM M_w gives a constant
number of about 3×10^{11}. This number is related to the recoverable strain as defined
by Bagley [18] using an equation derived by Wall [141]:

$$G = \frac{\rho R T}{M_w}, \tag{5.14}$$

where
  R is the universal gas constant
  T is the absolute temperature

Accordingly, the equation for recoverable strain can be rewritten as

$$\gamma_R \rho R T = M_w \tau_c \approx 3 \times 10^{11}. \tag{5.15}$$

Vlachopoulos and Alam [142] investigated this relationship for PS, PP, LDPE, and
HDPE and corroborated the results for PP of Vinogradov et al. [143] that this number is only a rough approximation. They proposed a related number that took into
account the broader MMD of commercial polymers by including the higher MM
averages:

$$\gamma_R \approx 2.65 \left[\frac{M_z M_{z+1}}{M_w^2} \right]. \tag{5.16}$$

Vinogradov et al. [143] defined a recoverable strain of 3.7 and 4.3 for surface
and volume distortions, respectively. Pomar et al. [144] found for LLDPE and

LLDPE-octadecane solutions, a recoverable strain of 1.73 for the onset of surface distortions. Boger and Williams [145] showed that Equation 5.16 could equally well be represented in terms of the apparent power-law index n:

$$\gamma_R \approx \frac{3n+1}{2n}. \tag{5.17}$$

Cogswell et al. [146] obtained a recoverable strain of about unity for a silicone polymer (PDMS). All these data suggest that no absolute universal number exists that is valid for defining or predicting melt fracture for all polymers. In addition, as Petrie and Denn [6] remarked: "A successful criterion has no implications as far as the mechanism is concerned, ..., it does not in any way contradict our claim that melt elasticity is an essential part of any mechanism."

5.3.4 CRITICAL STRESS

Nearly, all investigators report the presence of one or more critical stress values to indicate the onset of melt fracture. The numbers found are very much associated with the nature of the polymer and related either to surface or volume distortions. The most unambiguous critical stress value related to melt fracture is defined by the flow curve discontinuity. The critical stress τ_c, defined by the end point of Branch I, is generally accepted to indicate the onset of melt fracture. For the continuous flow curves, a change of slope may be identified (Figure 5.4b) [66,67]. Often, a second critical stress value τ_c^* is reported and associated with the onset of surface distortions [42]. For PS, Spencer and Dillon [62] associated τ_c with the onset of more or less regular spirals (ripples) similar to the later findings of Vinogradov et al. [143] for PP, a material that generates a continuous flow curve. Lupton and Regester [22] identify τ_c with a matte extrudate appearance. The determination of τ_c^* at stresses lower than τ_c has caused much debate in relation to LLDPE melt fracture, that is, onset of surface distortions. In the case of discontinuous flow curves, an obvious τ_c^* choice would be the stress associated with the onset of Branch II. However, no experimental evidence seems to support such a claim. In most cases, extrudate surface distortions always appear at lower stresses than can be expected from the onset stress of Branch II [25,35,135]. In contrast, for constant-pressure experiments with HDPE, Bagley et al. [108] observed surface distortion well above the onset point of Branch II. Alternatively, the change of slope is forwarded as a measure of τ_c^* [116]. For an essentially continuous flow curve presenting data on either a linear or logarithmic scale may sometimes be helpful to define that critical point. Table 5.2 summarizes most of the τ_c^* values reported in the literature. In general, 10^5 Pa seems to be a consensus number for τ_c^*.

Table 5.3 summarizes the critical stress values τ_c associated with the onset of the spurt regime or extrudate volume distortions as reported in the literature. Again, a wide range ($1 \times 10^5 - 5 \times 10^5$ Pa) of critical stress values is found. This significant scatter may in part be related to imprecise melt fracture observations, failure to report the precise type of extrudate distortion, or not mentioning whether apparent

TABLE 5.2
Critical Wall Shear Stress τ_c^* Associated with the Onset of Surface Distortions as Reported by a Number of Authors

Authors	τ_c^* (MPa)	Polymer	T (°C)
Tordella [2]	0.15	HDPE	150
Zhu [31]	0.10–0.15	Star PB	25
Den Otter [34]	0.05	PDMS	20
Herranen and Savolainen [52]	0.35	LLDPE	237 and 267
Bartos [57]	0.08	PP	200
	0.091	PP	220
	0.096	PP	240
	0.1	PP	260
	0.0695	LDPE	125
	0.076	LDPE	150
	0.089	LDPE	175
	0.105	LDPE	225
	0.125	PS	200
	0.4	PMMA	200
Benbow and Lamb [71]	0.09–0.15	LDPE	130–200
Bartos and Holomek [74]	0.04–0.1	PB	Ambient
Martyn et al. [87]	0.57	LDPE	180
Oyanagi [111]	0.2–0.35	HDPE	130–230
Beaufils et al. [135]	0.06 and 0.1	LLDPE	145–205
Vinogradov et al. [143]	0.03–0.097	PP	180–240
Pomar et al. [144]	0.1–0.2	LLDPE	135
Ballenger et al. [147]	0.14–0.17	HDPE	190
Ramamurthy [148,150]	0.14	LLDPE	160 and 260
Utracki and Gendron [149]	0.25	LLDPE	190
Kalika and Denn [151]	0.26	LLDPE	215
Lim and Schowalter [152]	0.1	PB	25
Hatzikiriakos [153]	0.09	HDPE	180
Kurtz [154]	0.14	LLDPE	190
Tzoganakis et al. [155]	0.22	LLDPE	215–230
El Kissi and Piau [156]	0.17	LLDPE	190
Hatzikiriakos et al. [157]	0.18	LLDPE	200
Rosenbaum et al. [158]	0.18	TFE-HFE	300–350
Naguib and Park [159]	0.12	Linear PP	180–210
Naguib and Park [159]	0.07–0.09	LCB PP	180–210
Santanach Carreras et al. [160]	0.17–0.21	SEBS	170–230
Vega et al. [161]	0.20	mPE	190
Delgadillo et al. [162]	0.15–0.19	LLDPE	150–190
	0.16–0.19	LLDPE/LDPE blends (<50% LD)	
	No surface distortions	LLDPE/LDPE blends (>50% LD)	

TABLE 5.2 (continued)
Critical Wall Shear Stress τ_c^* Associated with the Onset of Surface
Distortions as Reported by a Number of Authors

Authors	τ_c^* (MPa)	Polymer	T (°C)
Doerpinghaus and Baird [163]	0.16	LLDPE	150
	0.16	mLLDPE	150
	0.17	mBLDPE	150
Gendron et al. [164]	0.161–0.206	LLDPE	200

or true wall shear stress numbers are used. It should also be noted that reported values are based on the implicit assumptions of laminar flow and only an average stress across the experimental setup. Therefore, the values in Tables 5.2 and 5.3 can only be considered as indicative. It should also be remembered that the flow curve is a representation of the complete experimental setup. It does not represent a locally initiated phenomenon well. At best, it indicates system instability averaged over multiple polymer architecture, flow geometry, and condition variables.

Cogswell [126] defines the onset point for surface distortions not only in terms of true wall shear stress, but also via a critical elongation rate. An approximate expression for the critical elongation rate $\dot{\varepsilon}$ (at the die wall) is given by Equation 5.18 for a capillary die of radius R and a melt flow velocity change Δv when exiting the die:

$$\dot{\varepsilon} = \frac{10\Delta v}{R} \tag{5.18}$$

For a silicone polymer (PDMS) Cogswell obtained values between 3 and $5\,s^{-1}$ depending on the die radius. Kurtz [170] determines this to be $10\,s^{-1}$ for LLDPE based on the calculation of the exit acceleration, assuming a zero velocity at the die wall. In a somewhat similar fashion, Hürliman and Knappe [119] defined a critical extensional stress of $1.1–1.2\,MPa$ in the die inlet as the onset point of melt flow instabilities, for LDPE at 150°C.

Migler et al. [96,97] used their observations on surface distortions of LLDPE to propose a new kinematic variable that unites results with and without processing aids. The reconfiguration rate, a combination of strain and strain rate, is proposed. The universal applicability of this parameter to different materials remains to be studied.

The identification of critical stress numbers implies a sudden onset point of surface or volume distortions. For fine-scale, low-amplitude high-frequency surface distortions, visual observation of the extrudate alone is insufficient, but it remains the most direct and common operational procedure [47,150,151]. The difficulty is the subjectivity of the observation. Over the years, very little has been done to quantify surface distortions, let alone volume distortions. Surface analysis, however, is not new and many techniques are available. Whitehouse [171] reviews some of the

TABLE 5.3
Critical Wall Shear Stress τ_c Associated with the Onset of the Spurt Regime or Volume Distortions as Reported by a Number of Authors

Authors	τ_c (MPa)	Polymer	T (°C)
Lupton and Regester [22]	0.32	HDPE	190
Uhland [23]	0.165	HDPE	180
Drda and Wang [24]	0.25–0.337	HDPE	160–200
Durand [25]	0.235–0.25	HDPE	160
Vinogradov et al. [33]	0.36	PB	22
	No spurt–0.15	PI	22
Zhu [31]	0.35	Linear and star PB	25
Ferri and Canetti [38]	0.37	SIS	140–220
El Kissi and Piau [49]	0.06	PDMS	23
Kazatchkov et al. [60]	0.12–0.15	PP	200–260
Spencer and Dillon [62]	0.05–1.4	PS	175–250
Münstedt et al. [70]	0.187	HDPE	170
Oyanagi [111]	0.45–0.5	HDPE	130–230
Hürlimann and Knappe [119][a]	1.1–1.2	LDPE	150
Vlachopoulos and Alam [142]	0.073–0.21	PS	170–230
	0.10–0.14	PP	200
	0.10–0.14	LDPE	150
	0.14–0.26	HDPE	150
Vinogradov et al. [143]	0.6–0.25	PP	180–240
Ramamurthy [148,150]	0.435	HDPE	220
	0.435	LLDPE	220
Utracki and Gendron [149]	0.175–0.414	HDPE	190
	0.33	LLDPE	190
Kalika and Denn [151]	0.39–0.43	LLDPE	215
El Kissi and Piau [156]	0.25	LLDPE	190
Doerpinghaus and Baird [163]	0.32 (spurt)	LLDPE	150
	0.41 (spurt)	mLLDPE	150
	0.44 (volume)	mBLDPE	150
	0.09 (volume)	LDPE	150
Gendron et al. [164]	0.385–0.425	LLDPE	200
Ui and Moi [165]	0.2–0.3	HDPE	150–240
Blyler and Hart [166]	0.2	HDPE	160
Utracki and Dumoulin [167]	0.185–0.3	HDPE	
Atwood and Schowalter [168]	0.38	HDPE	225
Becker et al. [169]	0.25	HDPE	160

[a] Extensional stress.

important types of instrumentation for measuring surface texture. Essentially, two parallel branches of instrumentation are available to obtain more quantitative measures: one mimics the eye and the other follows the tactile example of the nail. These two, however, measure different things. The optical instrumentation looks for lateral structure, namely, the spacing and detail in the plane of the surface, whereas the stylus instrumentation examines heights in the plane perpendicular to the surface.

5.4   MELT FRACTURE OBSERVATION

The quantitative description of the molten distorted strand emerging out of a die is a delicate operation. It requires a direct, online measurement tool or, as is usually the case, the examination off-line of solidified strands. The way these strands are cooled can be expected to significantly influence the extrudate distortions. A slow cooling allows for a relaxation of the distortions and a potential reduction of their severity. A fast cooling solidifies the extrudate almost immediately and allows for more accuracy in determining the dimensions of the surface distortions. Volumetric distortions however may be enhanced due to inhomogeneous shrinkage, in particular for semicrystalline polymers. Furthermore, cutting a strand of a sizeable length and holding it to cool can induce undesired elongation due to sagging. Most authors provide only limited details on their sampling procedure [150,151,172,173]. Howells and Benbow [4] magnify their photographs seven times to have a better visualization of the topological details, but do not specify how the samples were prepared. El Kissi and Piau [49] and El Kissi et al. [174] study melt fracture of PDMS. At room temperature, PDMS is in the molten state, which makes it very difficult to handle, if not impossible. Consequently, photographic means were used to study the melt fracture. The same procedure is used to study LLDPE [156] and PB [29]. The evidence for a gradual increase in the severity of melt fracture for the latter polymers appears to be less convincing because of the smaller size of the initial surface distortions.

5.4.1   Microscopy

Optical microscopy [144,170] or scanning electron microscopy [154,175,176] provides more precise information. However, the accurate measurement of the frequency and amplitude of the surface topology remains a difficult task. Rudin et al. [177] used magnifying glasses to follow the evolution of surface distortions for strands that are slowly cooled in air. Kurtz [170], Karbashewski et al. [178], and Pomar et al. [144] used optical microscopy to study surface distortions. The magnification of the surface allows measuring the wavelength of the surface distortions. Only Kurtz and Piau et al. provide quantitative information. For PB, Piau et al. [29] defined a frequency between 0.01 and 1000 Hz. For LLDPE, Kurtz [170] defined an interval between 10 and 500 Hz. Via indirect illumination of the samples, he also determined the amplitude to be between 40 and 200 µm.

Scanning electron microscopy allowed even more detailed observation of the samples. The depth of vision of the pictures is much better. The magnification is usually similar to that of optical microscopy (20–40 times) and essentially defined by the width of the strand area that can be examined.

Clegg [101] and Venet [41] have studied the longitudinal cuts of the deformed extrudate using optical microscopy. Such an approach allows excellent visualization of the regular nature of the surface distortions and a more accurate definition of their amplitude.

5.4.2 PROFILOMETRY

Profilometry is well known for the analysis of coatings and metal surfaces. The stylus method essentially uses a calliper consisting of two arms: one touching a reference surface and the other the surface under consideration. The arm toward the test piece ends with a diamond stylus with a tip dimension such that it can probe the detailed geometry of the surface [179]. By probing the surface in a fixed coordinate framework, a 3D picture of the surface topology can be generated. Subsequently, the data can be analyzed by defining characteristic numbers that quantify the surface roughness. The average roughness (R_a) and total roughness (R_t) are defined as

$$R_a = \frac{1}{n} \sum_i |Z_i - Z_a| \tag{5.19}$$

and

$$R_t = Z_{max} - Z_{min} \tag{5.20}$$

with

$$Z_a = \frac{1}{n} \sum_1^n Z_i \tag{5.21}$$

where
  Z_i is the perpendicular distance between the highest and lowest point of successive peaks and troughs measured by the stylus
  Z_a is the average value
  Z_{max} is the highest value
  Z_{min} is the lowest measured perpendicular distance over the total scanned area
    [180]

The measurements are easy to perform and highly reproducible. However, the dimensions of the stylus need to be smaller than the distortions on the surface for reliable results (Figure 5.17).

Sornberger et al. [116] and Beaufils and coworkers [43,135] were the first to apply this technique to polymers. They characterized the surfaces of HDPE and LLDPE extrudates emerging from a slit die. Beaufils was able to precisely define

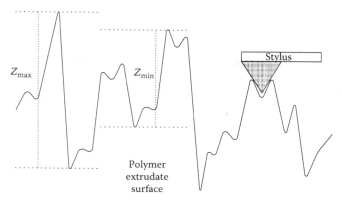

FIGURE 5.17 Profilometry is a stylus method used to determine the surface roughness of extrudates. The stylus tip dimension needs to penetrate the detailed geometry of the extrudate surface to present accurate roughness measures.

the threshold for the onset of surface distortions (Figure 5.18). Even at very low flow rates, the visually smooth extrudate surface had a measurable degree of roughness with values averaging between 0.1 and 0.3 µm. (Polished metal surface can have average roughness values of about 0.05 µm.) At this stage, the roughness is considered to be unorganized, but evolves and gradually gets organized having higher average roughness values until it becomes visible when the extrudate turns matte. This happens for an average roughness of about 0.8–1 µm. These quantitative investigations reveal that the amplitude and wavelength increase with flow rate ranging, respectively, from 0.1 to 10 µm (based on average roughness) and 160 to 310 µm. The total roughness R_t may be 5–10 times higher than R_a. For $R_a > 25$ µm the resolving power of the stylus reached its limits.

Tong and Firdaus [181] applied profilometry to film surface distortions. The roughness is measured on both sides of the film at once and the stylus could distort the surface topology due to the very thin nature of the film.

Venet [41] combined several off-line surface roughness techniques (optical microscopy, profilometry, and observation of extrudate cross sections with optical microscopy) for LLDPE and metallocene-catalyzed PE. The observation of extrudate cross-sections had also been applied by Clegg [101] and Bergem [120]. Venet observed that these three techniques are in very good agreement and allow following accurately the development of surface distortions with increasing flow rate.

5.4.3 INDIRECT METHODS

Several attempts have been made to develop techniques for online melt fracture analysis. Ajji et al. [182] put a Göttfert Rheotens® [10,183] device at 70 cm below a capillary rheometer to measure on line, but not continuously, the appearance of surface distortions. The strand is cooled solid and pulled between the wheels of the Göttfert Rheotens® device (Figure 5.19). The measured force signal is analyzed in relation to the various flow rates. At the onset of surface distortions, a significant change in

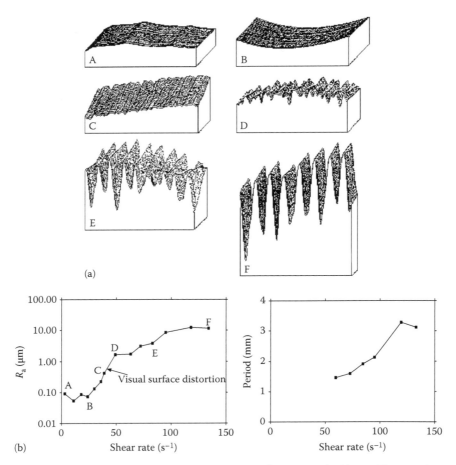

FIGURE 5.18 Dependency of surface distortions on flow rate probed by profilometry for an LLDPE. A slit die is used at 205°C. The measurements are 2.5 by 125 μm (A, B, C, D, E) and 30 by 800 μm for F. (From Beaufils, P. et al., *Int. Polym. Process.*, 4(2), 78, 1989. Copyright Carl Hanser Verlag, München. With permission.)

the base signal is observed. This approach, however, does not allow quantifying the precise surface topology of the extrudates.

Kale and Hager [184] developed a method to scan the surface of a polymer tape produced with a small extruder moving at rates of up to 400 mm/s. The tape is illuminated and the reflected light is detected using fiber optics. Applying fast-Fourier analysis techniques on the reflected light signal, a measure for the surface roughness of the tape can be defined. This technique allows a continuous operation and does not mechanically interact with the extrudate (Figure 5.20).

Tzoganakis et al. [155] and Le Gall et al. [185] performed an image analysis of the pictures of surface distortions to define a fractal dimension. This led to a relative metric for defining the degree of surface distortion.

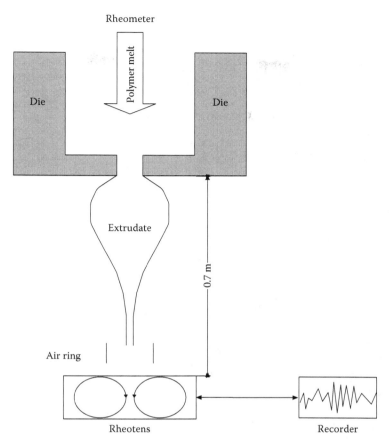

FIGURE 5.19   Schematic representation of a method to probe the onset of surface distortions with a Göttfert Rheotens® rheometer. (After Ajji, A. et al., *Polym. Eng. Sci.*, 33(23), 1524, 1993.)

An ultrasonic technique has been used to define the onset and degree of surface distortions. Herranen and Savolainen [52] used this technique on line when extruding LLDPE and LLDPE/LDPE blends through capillary dies. Gendron et al. [164] revisited the technique for the study of three different melt flow rate LLDPE. The polymer flows through a die that is equipped with two aligned steel buffer rods on which piezoelectric transducers are mounted: one for producing longitudinal waves and one for detecting the echo. A well-defined acoustic pulse is generated that travels to the polymer–steel interface where part of it is transmitted into the polymer sample. The transmitted pulse is reflected back and forth between the two steel/polymer interfaces (Figure 5.21). This produces a series of echo signals that are detected and analyzed using fast-Fourier transforms. Deviations from steady-state slow flow, ultrasound analysis patterns are correlated to the off-line observations of surface distortions.

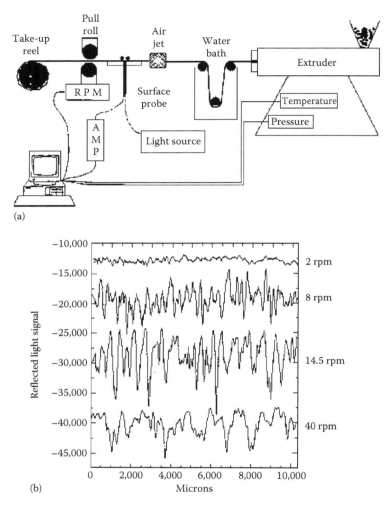

(a)

(b)

FIGURE 5.20 Schematics of an indirect online method to probe surface distortions. The light reflected from the surface of a tape moving at rates of up to 400 mm/s is detected and analyzed. An example of the reflection signal for a smooth, matte, and rough surfaces are shown. (Courtesy of L. Kale from The Dow Chemical Company.)

5.4.4 MELT FRACTURE QUANTIFICATION

Irrespective of the observation methodology, all the literature findings clearly demonstrate a gradual development of surface distortions, and in particular, when considering fine-scale, low-amplitude high-frequency distortions. At first, the surface is smooth and transparent. At higher flow rates, the usually transparent and smooth strands may develop a haze and lose their glossy features. This is accompanied by the formation of small, dispersed cavities that grow and organize into valleys and ridges along the transverse direction of the strand. Eventually

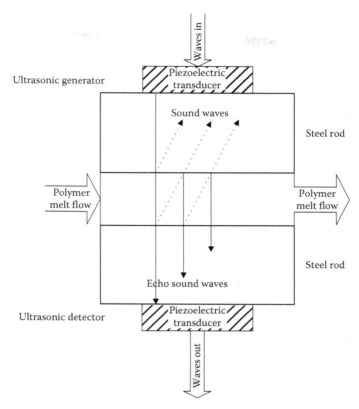

FIGURE 5.21 Schematic representation of the ultrasonic setup for characterizing extrusion instabilities. (After Gendron, R. et al., Ultrasonic characterization of extrusion instabilities, in *SPE ANTEC*, Toronto, Canada, 55(Vol. 2), 1997, 2254–2256.)

they become very regular and periodic. For some polymers, the ridges' amplitude becomes so high, that they tumble over much to the analogy of the crest of sea waves [41] (Figure 5.22).

Kurtz [186] showed that the surface roughness amplitude increases linearly with wall shear stress. Tordella [3] estimated the size of roughness as 1/5–1/10 of the overall diameter of the extrudate. Benbow and Lamb [71] defined a series of parallel crack-like structures as having a depth varying from 50 μm to half the extrudate thickness and a wavelength similar to the depth of the fissures. The surface distortion frequency and amplitude increase as flow rate rises [43,135,154,187] (Figures 5.23 and 5.24).

The dependence of the surface distortion amplitude on shear stress as measured by Venet and coworkers [41,188] showed a sigmoidal shape (Figure 5.24). In agreement with Beaufils and coworkers [43,135], surface distortions are present (<0.8 μm) before they are detected visually. An initial weak rise in the amplitude followed by a faster rise at higher shear stress is typical (Figure 5.24). For different

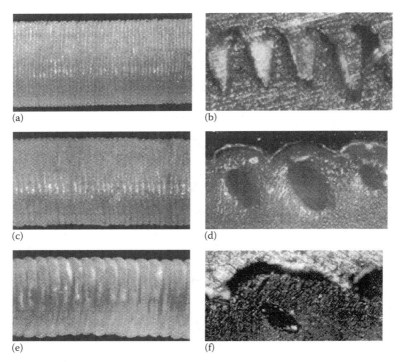

(a) (b) (c) (d) (e) (f)

FIGURE 5.22 Cross-section of an extrudate showing the collapse of intense surface ridges. The constant-rate measurements are made at 190°C using an $L/2R=0$, $2R=1.39$, and entry angle 90°. (Courtesy of C. Venet from The Dow Chemical Company.)

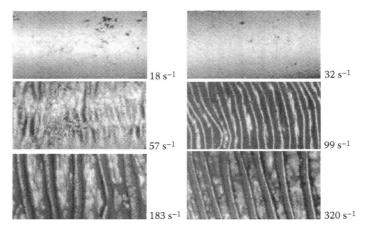

18 s^{-1} 32 s^{-1} 57 s^{-1} 99 s^{-1} 183 s^{-1} 320 s^{-1}

FIGURE 5.23 Optical microscopy of surface distortions observable at selected flow rates for an LLDPE at 190°C with $L/2R=0$, $2R=1.39$, and entry angle 90°. (Courtesy of C. Venet from The Dow Chemical Company.)

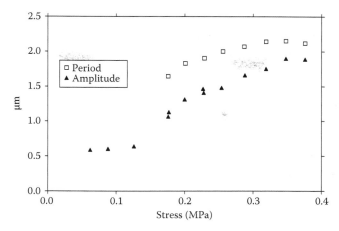

FIGURE 5.24 The period and amplitude of the surface distortions as measured by pro-filometry for the samples shown in Figure 5.23. (From Venet, C., Propriétés d'écoulement et défauts de surface de résins polyéthylènes, PhD thesis, Ecole des Mines de Paris (CEMEF), Sophia Antipolis, France, 1996.)

PEs (LLDPE, high energy irradiated LLDPE and a very low density metallocene-catalyzed PE), Venet and coworkers [41,188] find critical stress τ_c^* values ranging from 0.08 to 0.12 MPa.

The periodic organization seems to begin at stress levels between 0.13 and 0.17 MPa in agreement with observations by Beaufils and coworkers [43,135]. The gradual surface distortion development as expressed in average roughness or amplitude on linear axes is not linearly proportional to shear stress (or shear rate). Only in a limited shear stress regime can a linear relationship, as proposed by Kurtz [187], be considered as a valid approximation. In the linear regime, the shear stress (or shear rate) range approximately varies between 0.1 and 0.2 MPa (or 100–200 s^{-1}) [43,188]. All the above findings indicate that a sigmoidal transition takes place with τ_c^* anywhere between 0.1 and 0.3 MPa. Therefore, the reported τ_c^* numbers in the literature are indicative only.

In contrast to surface distortions, very few papers [85,119] are available on the quantitative character of volume distortions. Volume distortions are usually far too irregular to find a single measurable quantity.

5.5 CHANGE OF SLOPE

Sometimes, Branch I of the discontinuous flow curve has a distinctive slope change for $\tau = \tau_c^*$. In the case of LLDPE, the change of slope is observed, and related to the onset of surface distortions by Kurtz [170], Ramamurthy [150], Kalika and Denn [151], Hatzikiriakos and Dealy [47], and Rosenbaum et al. [158]. In contrast, Lim and Schowalter [152] for narrow PB, and El Kissi and Piau [156] for LLDPE observed a smooth, continuous evolution of the slope for an LLDPE. Beaufils et al. [43] observed a change of slope at the visual onset point (>0.1 MPa) but not at the real onset point (<0.1 MPa) for surface distortions. The critical stress hypothesis and the change of slope spurred a lot of speculation as to its origin. For many polymers, and without the aid of extrudate observations, a slope change in the flow curve may be difficult to spot [64,65,151]. Shaw [42] performed a comprehensive statistical analysis of data collected from many papers that claim a change of slope. He concluded that a general criterion based on a change of slope cannot be obtained.

5.6   WALL SLIP

Two contrasting boundary conditions for flowing macroscopic fluids are "stick" and "slip" at the die wall. For the stick-boundary condition, the fluid is at rest at the die wall. Slip refers to the failure of the fluid to adhere to the wall and allows a finite tangential fluid velocity at the wall. Many investigators associated slip at the wall with surface and volume distortions. Most of the experimental evidence and argumentation is indirect. Den Otter [34] even argued that slip at the wall using particle tracking techniques can never be proven, because a spherical particle of radius r will have a speed at the wall varying between 0 and $2\dot{\gamma}r$, with $\dot{\gamma}$ the wall shear rate. Furthermore, the wall stick (no slip) condition is an almost universally accepted boundary condition in both Newtonian and non-Newtonian fluid mechanics. Numerous flows have been successfully modeled for Newtonian fluids [189]. Irrespective of the boundary condition physics, several macroscopic methods can be applied to infer a finite wall-slip velocity. However, the slip-determination experiments must be carried out with utmost care as pointed out by Mackay and Henson [190].

5.6.1   THE MOONEY METHOD

The most frequently applied procedure to calculate a wall slip velocity is the Mooney method [191]. The apparent wall shear rate $\dot{\gamma}_a$ is corrected for the presence of finite slip velocity v_s according to

$$\dot{\gamma}_{a,s} = \dot{\gamma}_a + \frac{4v_s(\tau_t)}{R} \tag{5.22}$$

with R the capillary die radius.

The Mooney method assumes that v_s is a unique function of the true wall shear stress τ_t with v_s and P constant along the length of the capillary. Under those experimental conditions, v_s can be obtained by measuring the apparent wall shear rate as a the function of the capillary die radius at constant wall shear stresses. At different wall shear stress values for a plot of $\dot{\gamma}_a$ versus $1/R$, the slope of the straight lines obtained is $4v_s$ (Figure 5.25), according to Equation 5.22.

Lupton and Regester [22] used the Mooney method to determine the slip velocity for HDPE. They showed that v_s increases with apparent shear stress and has a discontinuity corresponding to the discontinuous flow curve typical for HDPE. The apparent shear rate increase at the discontinuity (onset spurt regime) was for 80% accounted for by the increased v_s. The remaining 20% was considered to be associated with viscous heating. In the spurt regime, Lupton and Regester [22], and Ui et al. [165], also for HDPE, found slip velocities of 50 mm/s. For PB, using direct

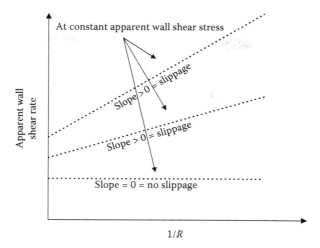

FIGURE 5.25 Schematics of a Mooney plot for determining the slip velocity in capillary dies. The straight lines are obtained at constant shear stress. A slope exceeding zero indicates a finite fluid velocity at the die wall.

velocity measurements, Bartos and Holomek [74] found values of 2 mm/s at the onset of spurt. Ramamurthy [148] calculated for LLDPE slip velocities using the apparent wall shear stress to make the Mooney plot. He considered the onset of extrudate surface distortion to be caused by a nonzero wall velocity. For a stress value of 0.1 MPa (below the observable onset of surface distortions reported at 0.14 MPa), a slip velocity of 1 mm/s was measured. The surface distortion onset was taken to coincide with the sudden change of slope in the flow curve.

A correction of the flow curve for slip velocities and using true shear stresses (including the Bagley correction) made the change of slope disappear [150] (Figure 5.26). The slip velocity was found to be proportional to the true wall shear stress following a power-law relation with an exponent close to 2. For the same LLDPE, Kalika and Denn [151] applied these observations to deduce the slip velocity by extrapolating the flow curve at low apparent wall shear rates to the regime where surface defects were expected. The apparent wall shear stress values, however, were not corrected for the pressure entry effect at fairly high $L/2R$ ratios (>100). The slip velocity was considered to be overestimated, and should be between 10% and 40% of the reported 10 mm/s, a value obtained at a stress of 0.35 MPa [41]. Kalika and Denn found power-law dependence of v_s on stress but with an exponent of 6 [151]. Hatzikiriakos and Dealy [47,153] formalized a steady-state slip model following the same power-law approach between v_s and τ_t:

$$v_s = a\tau_t^m, \tag{5.23}$$

where
  a is a temperature-dependent slip coefficient
  m is a temperature-independent exponent

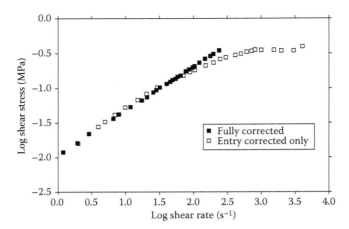

FIGURE 5.26 Schematic representation of the effect of a slip-velocity correction on the flow curve of an LLDPE as reported by Ramamurthy [150]. The correction makes the slope change disappear. (From Ramamurthy, A.V., *Adv. Polym. Technol.*, 6, 489, 1986, Figure 7. With permission from Wiley.)

The values for m were between 3 and 3.4 while v_s ranged from 1 to 10 mm/s for an HDPE using a sliding plate rheometer [192]. Further experiments on blends of two HDPE with a constant-rate capillary rheometer gave m values between 2.6 and 4.1 [47]. The difference between capillary and sliding plate rheometer was related to the effect of pressure on v_s. Kalika and Denn [151] and Hill et al. [193] supported these results. Stewart [194] tried to reconcile all these experimental finding for capillary dies by proposing a slip model based on the concept of activation rate theory.

Vinogradov and Ivanova [195], on studying ethylene–propylene elastomers with capillary rheometers argued that the wall slip must increase considerably as the melt approaches the die exit. In addition, they showed that high-capillary die pressures suppressed extrudate distortions, possibly by decreasing the slip velocity [5]. White et al. [196] found for styrene-butadiene rubber that the slip velocity decreased with pressure. The Mooney method was modified to take into account a slip velocity v_s-pressure dependence. The pressure effect on the viscosity was considered to be negligible. Tzoganakis et al. [155] used the same experimental method as Hatzikiriakos and Dealy [47] and included the pressure dependence to calculate the slip velocity for an LLDPE. Similar to the results of Ramamurthy [148], they calculated slip in the die at 0.12 MPa before the onset of surface distortions became visible at 0.22 MPa. Hatzikiriakos [197] also used this modified Mooney method to predict the position in the die of the onset point for slip to occur at different flow rates and found that the maximum slip velocity lies at the die exit.

In capillary rheometer experiments, Crawford et al. [198] observed no slippage for PDMS gum below 60 kPa and a sudden jump in slip velocity at 80 kPa. Kamerkar and Edwards [199] for PP, LDPE, and HDPE measured slip flow in straight-walled and semihyperbolically converging capillary dies. The latter die showed much less slip flow as caused by adding 2% stearic acid than for a standard capillary die.

5.6.2 The Laun Method

A considerably less labor-intensive method for measuring slip velocities was developed by Laun [200]. Instead of using independent measurements with capillary dies of different radii, a method based on the squeezing of a disk of polymer melt between parallel plates was proposed. The principles of squeezing flow of polymers have long been established [201–205]. Squeezing creates a radial and axial shear rate profile. The maximum shear rate $\dot{\gamma}_R$ is reached at the rim of the plate with radius R. For squeezing a sample of thickness H, at a speed $\dot{H}$, with a squeezing force F, the rim shear rate $\dot{\gamma}_R$ and the rim shear stress τ_R can be obtained from

$$\dot{\gamma}_R = \frac{3(-\dot{H})R}{H^2}, \tag{5.24}$$

$$\tau_R = \frac{2HF}{\pi R^3}. \tag{5.25}$$

The general relation between the force and the speed, including a slip velocity v_s that is only proportional to the local shear stress τ,

$$F = \frac{3\pi\eta_0(-\dot{H})R^4}{2H^3}\left[\frac{1}{1+6\eta_0\tau/H} + \frac{2H^2}{R^2}\right]. \tag{5.26}$$

The square brackets contain three contributions representing shear, wall slip ($6\eta_0\tau/H$) with η_0 the Newtonian viscosity, and the equibiaxial elongation ($2H^2/R^2$). For $H \ll R$ and a small slip velocity, the equibiaxial contribution can be neglected. Substitution of F and $\dot{H}$ from Equation 5.26 into Equations 5.24 and 5.25 yields the relation

$$\dot{\gamma}_R = \frac{\tau_R}{\eta_0} + \frac{6v_s}{H}. \tag{5.27}$$

When the rim shear stress is kept constant during the squeeze test by controlling the force such that

$$F(H) = \frac{F_0 H_0}{H}, \tag{5.28}$$

where F_0 and H_0 are the initial force and sample thickness respectively, then a Mooney type of plot can be obtained of $\dot{\gamma}_R$ versus $1/H$. The slope of the obtained straight line allows determining the slip velocity (Figure 5.27). For viscoelastic polymers, a steady state has to be reached before slip velocities can be measured.

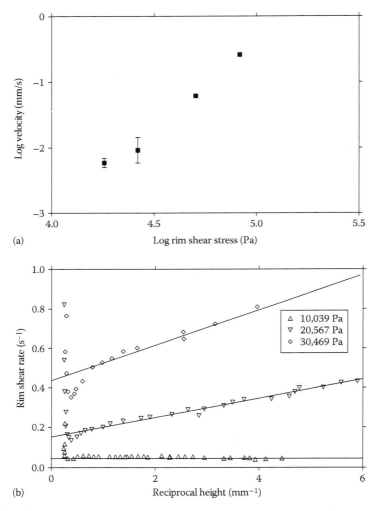

(a)

(b)

FIGURE 5.27 A Laun plot for measuring the slip velocity of polymer melts using a squeezing flow. The slope of the straight lines obtained for an LDPE at different stress levels defines the slip velocity (a). The associated slip velocity as a function of rim shear stress is shown as well (b). (Courtesy of M. Laun from The Dow Chemical Company.)

The Laun plot is considered to be more sensitive than the Mooney plot in the sense that lower slip velocities can be measured (below 1 mm/s) (Figure 5.27). Thus far, this method has not been used to relate the slip velocity to melt fracture phenomena. However, the reported results suggest that slip starts to occur at very low stresses and well below those associated with a visual onset of melt fracture. Such findings would be in line with results for LLDPE from Ramamurthy [148] and Tzoganakis et al. [155].

5.6.3 OTHER METHODS

Besides the various indirect approaches of particle tracking, flow birefringence, and the Mooney and Laun methods, additional methods have been developed in attempts to measure slip velocities directly.

Atwood [206] developed a hot-film probe method. The principle of this hot-film velocimetry method consists of placing a thin platinum film in contact with a flowing medium and measuring heat transfer changes between the probe and the fluid. As the flow conditions change, the heat transfer from the hot sensor to the medium varies. Slip enhances the heat transfer. The result is a change in voltage in the electronics of the probe, which is adjusted to keep the film at a constant temperature. The eventual measurements have to be interpreted with care as the method is very sensitive to calibration, viscous dissipation in the melt, and the distortion and noise of the signal [168,206]. For HDPE, no slip is measured until a critical stress of 0.38 MPa, which corresponds to the spurt regime onset. Similar experiments for PB by Lim and Schowalter [152] yielded convincing results that wall slip occurs at the onset of the spurt regime.

Migler et al. [207] were the first to directly measure the flow velocity of a high-MM polymer within the first 100 nm from a solid wall. The experiments make use of an optical technique that combines two methods: evanescent-wave-induced fluorescence [208] and fringe pattern fluorescence recovery after photo bleaching [209]. A small drop of fluorescently labeled polymer melt is sandwiched between two planar substrates with a refractive index exceeding that of the polymer. The top substrates are separated a distance D, and can move at a variable velocity to create a shearing action. Interference fringes are produced by splitting a laser beam into two beams and recombining them at the bottom surface. If the incidence angle of the laser beam is greater than a critical angle, total internal reflection occurs, and energy still penetrates into the polymer in the form of an evanescent wave with an exponential decay perpendicular to the substrate surface. This energy, confined within a penetration depth d, can excite the labeled polymer. In the crossing region of the two laser beams, the mixing of the evanescent waves creates an interference pattern. A device modulates the phase shift between the beams at a selected frequency, creating a sinusoidal oscillation of the fringe pattern with certain amplitude. The experimental sequence consists of two steps. First, a bleaching light pulse is shown on the sample with neither shear nor modulation of the fringe position. A fraction of the fluorescent polymer is photo-bleached in the bright fringes. Second, immediately thereafter, the shear and phase modulation are turned on to the fluorescence. Because the fluorescence intensity is the product of the bleaching and reading intensity, a region of thickness $d/2$ is probed. This procedure allows determining the occurrence of slip during a shearing action.

For the PDMS, sample slip velocities have been measured as low as 0.01 up to 1000 μm/s in a shear rate range of 0.02–40 s^{-1}. A clear transition is observed between a regime of weak and strong slip (Figure 5.28). The shape of the curve is sigmoidal and corresponds closely to previous surface distortion observations. However, Migler et al. [207] did not present melt fracture observations there.

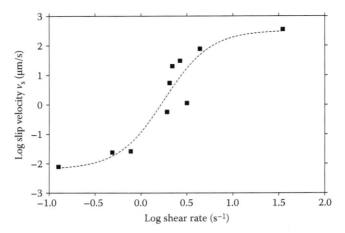

FIGURE 5.28 The relation between slip velocity and true wall shear rate as determined by Migler et al. [207] has a sigmoidal character in a double logarithmic plot. (From Migler, K. B. et al., *Phys. Rev. Lett.*, 70, 287, 1993. Copyright 1993 by the American Physical Society. With permission.)

The Migler et al. technique probes at scales close to those of the molecular dimensions. The data provided experimental support for the theoretical considerations of Brochard and de Gennes [210]. Their conjecture is that high-MM polymer melts flowing between smooth, non-adsorbing surfaces should always slip at any shear rate. Polymer chains anchored to the wall will strongly reduce slippage. Leger et al. [211] provide additional support for the validity of the Migler et al. experiments and the Brochard–de Gennes theory.

For an HDPE, the LDV results of Münstedt et al. [70] allowed, although indirectly, to calculate slip velocities from the velocity profiles inside a die for Branch I flow regimes (Figure 5.4a). In addition, a plug-like strong slip flow was measured on Branch II.

A high-resolution differential interference contrast microscope in combination with particle tracking and LDV was used by Mizunuma and Takagi [212] to observe a shear flow having a negative pressure gradient at the die exit as generated by a wedge-shaped slider in a plane Couette flow cell. Linear PDMS showed an alternating slip and stick flow corresponding to a "valley" and a "ridge," respectively.

An extensive study on PB, using a sliding plate rheometer and a pressure-controlled capillary rheometer is reported by Park et al. [213]. A generated curve of shear stress versus slip velocity showed a maximum and minimum, which was associated to the occurrence of spurt flow in capillaries.

5.7 COMPRESSIBILITY

Polymer melts are usually considered to be incompressible fluids. The assumption is correct at relatively low pressure and for flow conditions where no melt fracture occurs. For most polymers, the melt compressibility is a very small number and for PE, it is of the order of 10^{-9} Pa^{-1} [25,214]. However, compressibility plays a role during constant-rate capillary experiments [20]. When pressure oscillations appear, the melt emerging out of the die periodically slows down, although the plunger continues to move at a constant rate. The amount of melt supplied to the system by the moving piston is not equal to the amount exiting the die at each instant during the experiment. Accordingly, melt compressibility in the barrel is a necessary but not-sufficient condition for pressure oscillation defects to occur and not its cause [25,33,179,215]. At constant temperature, the thermodynamic melt compressibility χ is defined as

$$\chi = -\frac{1}{V}\left(\frac{dV}{dP}\right)_T,$$ (5.29)

where
   V is the melt volume
   P is the pressure

As spurt is a time-dependent phenomenon, Equation 5.29 can be rewritten to include the pressure–time dependency $P(t)$ and the melt density $\rho(t)$:

$$\frac{1}{\rho}\frac{d\rho(t)}{dt} = \chi\frac{dP(t)}{dt}.$$ (5.30)

Lupton and Regester [22], Vinogradov et al. [33], Weil [215,216], Okubo and Hori [217], and Molenaar and Koopmans [218] made use of compressibility to describe the spurt regime pressure oscillations accurately. In Chapter 9, this approach will be worked out in detail. Hatzikiriakos and Dealy [214] described a method to infer the melt compressibility from a single start-up pressure transient during a constant piston speed capillary experiment. They modeled the pressure dependency of melt density as

$$\rho(P) = \rho_{atm}(1+\chi P).$$ (5.31)

Equation 5.31 is used in the modified Mooney method [47] to express the dependency of wall slip on pressure. The use of the start-up pressure transient illustrates the importance of compressibility in the kinematics of transient regimes.

5.8   GENERAL OBSERVATION

In the literature, a considerable amount of experimental work is available on melt fracture. Nearly all experiments deal with the macroscopic behavior of polymer and consider the melt as a continuum. The many reports, however, still leave a lot of questions unanswered. In particular, the relation of melt fracture to polymer architecture requires further elucidation. In addition, determining and understanding the interaction between polymer melt and die wall seems macroscopically understood but remains challenging at the molecular level. All surfaces have a specific roughness that will impact the wall–molecular interaction physics and the local deformation behavior of the molecule. Furthermore, the development of mathematical models that quantify melt fracture phenomena is needed for dealing effectively with practical problems.

REFERENCES

1. Tordella, J. P., Melt fracture—Extrudate roughness in plastic extrusion. *SPE J.*, **February**: 36–40 (1956).
2. Tordella, J. P., Unstable flow of molten polymers: A second site of melt fracture. *J. Appl. Polym. Sci.*, **7**: 215–229 (1963).
3. Tordella, J. P., Unstable flow of molten polymers. In *Rheology* (F. R. Erich, Ed.). Academic Press, New York, 1969, pp. 57–92.
4. Howells, R. E. and J. J. Benbow, Flow defects in polymer melts. *Trans. Plast. Ind.*, **30**: 240–253 (1962).
5. Vinogradov, G. V. and L. I. Ivanova, Wall slippage and elastic turbulence of polymers in the rubbery state. *Rheol. Acta*, **7**(3):243–254 (1968).
6. Petrie, C. J. S. and M. M. Denn, Instabilities in polymer processing. *AIChE J.*, **22**(2):209–236 (1976).
7. Larson, R. G., Instabilities in viscoelastic flows. *Rheol. Acta*, **31**:213–263 (1992).
8. Bird, R. B., W. E. Stewart, and E. N. Lightfoot, *Transport Phenomena*. John Wiley & Sons Inc., New York, 1960.
9. Bird, R. B., R. C. Armstrong, and O. Hassager, *Dynamics of Polymeric Liquids*. Volume 1: *Fluid Mechanics*. John Wiley & Sons, New York, 1987.
10. Dealy, J. M., *Rheometers for Molten Plastics*. Van Nostrand Reinhold, New York, 1982.
11. Dealy, J. M. and K. F. Wissbrun, *Melt Rheology and Its Role in Plastics Processing*. Van Nostrand Reinhold, New York, 1990.
12. Brydson, J. A., *Flow Properties of Polymer Melts*. Iliffe Books, London, U.K., 1970.
13. Bagley, E. B., End corrections in capillary flow of polyethylene. *J. Appl. Phys.*, **28**(5):624–627 (1957).
14. Westover, R. F. and B. Maxwell, Flow behavior and turbulence in polyethylene. *J. Soc. Plast. Eng.*, **13**:27–36 (1957).
15. Philippoff, W. and F. H. Gaskins, The capillary experiment in rheology. *Trans. Soc. Rheol.*, **2**(1):263–284 (1958).
16. Leblanc, J. L., Factors affecting the extrudate swell and melt fracture phenomena of rubber compounds. *Rubber Chem. Technol.*, **54**(5):905–929 (1981).
17. Bagley, E. B., The separation of elastic and viscous effects in polymer flow. *Trans. Soc. Rheol.*, **5**(1):355–368 (1961).
18. Bagley, E. B., Hooke's law in shear and polymer melt fracture. *J. Appl. Phys.*, **31**:1126–1127 (1960).
19. Rabinowitsch, B., Über die Viskosität und Elastizität von Solen. *Z. Phys. Chem. A*, **149**:1–26 (1929).
20. Pearson, J. R. A., *Mechanics of Polymer Processing*. Elsevier, New York, 1985.
21. Boudreaux, E. J. and J. A. Cuculo, Polymer flow instability: A review and analysis. *J. Macromol. Sci.—Rev. Macromol. Chem. C*, **16**(1):39–77 (1977).
22. Lupton, J. M. and J. W. Regester, Melt flow of polyethylene at high rates. *Polym. Eng. Sci.*, **5**(4):235–245 (1965).
23. Uhland, E., Das anomale Fliessverhalten von Polyathylene hoher Dichte. *Rheol. Acta*, **18**:1–24 (1979).
24. Drda, P. P. and S. Q. Wang, Stick-slip transition at polymer melt/solid interfaces. *Phys. Rev. Lett.*, **75**(14):2698–2701 (1995).

25. Durand, V., Ecoulements et instabilité des polyéthylènes haute densité. PhD thesis, Ecole des Mines de Paris (CEMEF), Sophia Antipolis, France, 1993.

26. Wang, S. Q., P. A. Drda, and Y. W. Inn, Exploring molecular origins of sharkskin, partial slip, and slope change in flow curves of linear low density polyethylene. *J. Rheol.*, **40(5)**:875–898 (1996).

27. Piau, J.-M., N. El Kissi, F. Toussaint, and A. Mezghani, Distortions of polymer melt extrudates and their elimination using slippery surfaces. *Rheol. Acta*, **34**:40–57 (1995).

28. Li, H., H. P. Hürlimann, and J. Meissner, Two separate ranges for shear flow instabilities with pressure oscillations in capillary extrusion of HDPE and LLDPE. *Polym. Bull.*, **15**:83–88 (1986).

29. Piau, J. M., N. El Kissi, and A. Mezghani, Slip flow of polybutadiene through fluorinated dies. *J. Non-Newtonian Fluid Mech.*, **59**:11–30 (1995).

30. Wise, G. M., M. M. Denn, A. T. Bell, J. W. Mays, K. Hong, and H. Iatrou, Surface mobility and slip of polybutadiene melts in shear flow. *J. Rheol.*, **44(3)**:549–567 (2000).

31. Zhu, Z., Wall slip and extrudate instability of 4-arm star polybutadienes in capillary flow. *Rheol. Acta*, **43**:373–382 (2004).

32. Dao, T. T. and L. A. Archer, Stick-slip dynamics of entangled polymer liquids. *Langmuir*, **18**: 2616–2624 (2002).

33. Vinogradov, G. V., A. Y. Malkin, Y. G. Yanovskii, E. K. Borisenkova, B. V. Yarlykov, and G. V. Berezhnaya, Viscoelastic properties and flow of narrow distribution polybutadienes and polyisoprenes. *J. Polym. Sci.: A2*, **10**:1061–1084 (1972).

34. Den Otter, J. L., Mechanisms of melt fracture. *Plast. Polym.*, **38(135)**:155–168 (1970).

35. Den Otter, J. L., Melt fracture. *Rheol. Acta*, **10(2)**:200–207 (1971).

36. Piau, J. M. and N. El Kissi, The different capillary flow regimes of entangled polydimethylsiloxane polymers: Macroscopic slip at the wall, hysteresis and cork flow. *J. Non-Newtonian Fluid Mech.*, **37**:55–94 (1990).

37. Piau, J.-M., S. Nigen, and N. El Kissi, Effect of die entrance filtering on mitigation of upstream instability during extrusion of polymer melts. *J. Non-Newtonian Fluid Mech.*, **91**:37–57 (2000).

38. Ferri, D. and M. Canetti, Spurt and melt flow distortions of linear styrene-isoprene-styrene triblock copolymers. *J. Rheol.*, **50(5)**:611–624 (2006).

39. Kulikov, O. L. and K. Hornung, Wall detachment and high rate surface defects during extrusion of clay. *J. Non-Newtonian Fluid Mech.*, **107**:133–144 (2002).

40. Agassant, J. F., P. Avenas, J.-P. Sergent, B. Vergnes, and M. Vincent, La mise en forme des matieres plastiques. Technique & Documentation, Paris, France, 1996.

41. Venet, C., Propriétés d'écoulement et défauts de surface de résins polyéthylènes. PhD thesis, Ecole des Mines de Paris (CEMEF), Sophia Antipolis, France, 1996.

42. Shaw, M. T., Detection of multiple flow regimes in capillary flow at low shear stress. *J. Rheol.*, **51(6)**:1303–1318 (2007).

43. Beaufils, P., B. Vergnes, and J. F. Agassant, Characterization of the sharkskin defect and its development with the flow conditions. *Int. Polym. Process.*, **4(2)**:78–84 (1989).

44. Becker, J., C. Klason, J. Kubat, and P. Saha, Frequency analysis of pressure fluctuations in a single screw extruder. *Int. Polym. Process.*, **6(4)**:326–331 (1991).

45. Kometani H., H. Kitajima, T. Matsumura, T. Suga, and T. Kanai, Visualisation of flow instabilities for high density polyethylene. *Int. Polym. Proc.*, **21(1)**:32–40 (2006).

46. Fernandez, M., A. Santamaria, A. Munoz-Escalona, and L. Mendez, A striking hydrodynamic phenomenon: Split of a polymer melt in a capillary flow. *J. Rheol.*, **45(2)**:595–602 (2001).

47. Hatzikiriakos, S. G. and J. M. Dealy, Role of slip and fracture in the oscillating flow of HDPE in a capillary. *J. Rheol.*, **36(5)**:845–884 (1992).

48. Bagley, E. B., I. M. Cabott, and D. C. West, Discontinuity in the flow curve of polyethylene. *J. Appl. Phys.*, **29**:109–110 (1958).

49. El Kissi, N. and J. M. Piau, Flow of entangled polydimethylsiloxanes through capillary dies: Characterisation and modelisation of wall slip phenomena. In *Third European Rheology Conference and Golden Jubilee of the British Society of Rheology* (D. R. Oliver, Ed.). Elsevier Applied Science, London, U.K., 1990.

50. Schreiber, H. P., E. B. Bagley, and A. M. Birks, Filament distortion and die entry angle effects in polyethylene extrusion. *J. Appl. Polym. Sci.*, **4(12)**:362–363 (1960).

51. Fields, R. T. and C. F. W. Wolf, Fabrication of thermoplastic resins. U.S. Patent 2991508. E.I. du Pont de Nemours and Company (1961).

52. Herranen, M. and A. Savolainen, Correlation between melt fracture and ultrasonic velocity. *Rheol. Acta*, **23(4)**:461–464 (1984).

53. Chambon, F. and J.-M. C. Dekoninck, Method for processing polyolefins at high shear rates. Exxon Chemical Patents Inc., US Patent 5320798A, 1994.

54. Dennison, M. T., Flow instability in polymer melts: A review. *J. Plast. Inst.*, **35(120)**:803–808 (1967).

55. Hatzikiriakos, S. G., I. B. Kazatchkov, and D. Vlassopoulos, Interfacial phenomena in the capillary extrusion of metallocene polyethylenes. *J. Rheol.*, **41(6)**:1299–1316 (1997).

56. Fujiki, T., M. Uemura, and Y. Kosaka, Flow properties of molten ethylene-vinyl acetate copolymer and melt fracture. *J. Appl. Polym. Sci.*, **12(2)**:267–279 (1968).

57. Bartos, O., Fracture of polymer melts at high shear stress. *J. Appl. Phys.*, **35(9)**:2767–2775 (1964).

58. Ballenger, T. F. and J. L. White, The development of the velocity field in polymer melts in a reservoir approaching a capillary die. *J. Appl. Polym. Sci.*, **15**:1949 (1971).

59. Vinogradov, G. V. and N. V. Prozorovskaya, Rheology of polymers. Viscous properties of polypropylene melt. *Rheol. Acta*, **3(3)**:156–163 (1964).

60. Kazatchkov, I. B., S. G. Hatzikiriakos, and C. W. Stewart, Extrudate distortion in the capillary/slit extrusion of a molten polypropylene. *Polym. Eng. Sci.*, **35(23)**:1864–1871 (1995).

61. Ballenger, T. F. and J. L. White, Experimental study of flow patterns in polymer fluids in the reservoir of a capillary rheometer. *Chem. Eng. Sci.*, **25**:1191–1195 (1970).

62. Spencer, R. S. and R. E. Dillon, The viscous flow of molten polystyrene II. *J. Colloid Sci.*, **4(3)**:241–255 (1949).

63. Goutille, Y., J.-C. Majesté, J.-F. Tassin, and J. Guillet, Molecular structure and gross melt fracture triggering. *J. Non-Newtonian Fluid Mech.*, **111**:175–198 (2003).

64. Clegg, P. L., Flow in various extrusion processes. *Br. Plast.*, **39**:96 (1966).

65. Perez-Gonzalez, J., L. Perez-Trejo, L. de Vargas, and O. Manero, Inlet instabilities in the capillary flow of polyethylene melts. *Rheol. Acta*, **36**:677–685 (1997).

66. Den Doelder, J., R. Koopmans, M. Dees, and M. Mangnus, Pressure oscillations and periodic extrudate distortions of long-chain branched polyolefins. *J. Rheol.*, **49(1)**:113–126 (2005).

67. Sammler, R. L., R. J. Koopmans, M. Mangnus, and C. Bosnyak, On the melt fracture of polypropylene. In *SPE ANTEC*, Atlanta, GA, 1, 1998, p. 300.

68. Meissner, J., Polymer melt flow measurements by laser doppler velocimetry. *Polym. Testing*, **3**:291–301 (1983).

69. Rusch C., Particle Tracking zur Analyse der Strömung einer Polyethyleneschmeltze vor und in einer Kapillardüse. PhD thesis, ETH, Zürich, Switzerland, 2001.

70. Münstedt, H., M. Schmidt, and E. Wassner, Stick and slip phenomena during extrusion of polyethylene melts as investigated by laser-Doppler velocimetry. *J. Rheol.*, **44(2)**:413–427 (2000).

71. Benbow, J. J. and P. Lamb, New aspects of melt fracture. In *SPE ANTEC*, Los Angeles, CA, 3, 1963, pp. 7–17.

72. Vinogradov, G. V., N. I. Insarova, B. B. Boiko, and E. K. Borisenkova, Critical regimes of shear in linear polymers. *Polym. Eng. Sci.*, **12(5)**:323–334 (1972).
73. Vinogradov, G. V. and A. Y. Malkin, *Rheology of Polymers*. Springer Verlag, Berlin, Germany, 1980.
74. Bartos, O. and J. Holomek, Unstable flow of amorphous polymers through capillaries. I. Velocity profiles of polymer having discontinuous flow curve. *Polym. Eng. Sci.*, **11(4)**:324–334 (1971).
75. Galt, J. and B. Maxwell, Velocity profiles for polyethylene melts. *Mod. Plast.* **42(December)**:115–189 (1964).
76. Maxwell, B. and J. C. Galt, Velocity profiles for polyethylene melt in tubes. *J. Polym. Sci.*, **62(174)**:50–53 (1962).
77. Den Otter, J. L., Rheological measurements on two uncrosslinked, unfilled synthetic rubbers. *Rheol. Acta*, **14**:329–336 (1975).
78. White, J. L., Critique on flow patterns in polymer fluids at the entrance of a die and instabilities leading to extrudate distortion. *Appl. Polym. Symp.*, **20**:155–174 (1973).
79. Checker, N., M. R. Mackley, and D. W. Mead, On the flow of molten polymer into, within and out of ducts. *Philos. Trans. R. Soc. Lond.*, **308(A)**:451–477 (1983).
80. Ma, C.-Y., J. L. White, F. C. Weissert, and K. Min, Flow patterns in carbon black filled polyethylene at the entrance to the die. *J. Non-Newtonian Fluid Mech.*, **17**:275–287 (1985).
81. El Kissi, N., Stabilité des écoulements de polymères fondus. PhD thesis, Institut National Polytechnique de Grenoble, Grenoble, France, 1989.
82. Hertel, D. and H. Münstedt, Dependence of the secondary flow of a low-density polyethylene on processing parameters as investigated by laser-Doppler velocimetry. *J. Non-Newtonian Fluid Mech.*, **153**:73–81 (2008).
83. Robert, L., Y. Demay, and B. Vergnes, Stick-slip of high density polyethylene in a transparent slit die investigated by laser Doppler velocimetry. *Rheol. Acta*, **43**:89–98 (2004).
84. Robert, L., B. Vergnes, and Y. Demay, Flow birefringence study of the stick–slip instability during extrusion of high-density polyethylenes. *J. Non-Newtonian Fluid Mech.*, **112**:27–42 (2003).
85. Combeaud, C., B. Vergnes, A. Merten, D. Hertel, and H. Münstedt, Volume defects during extrusion of polystyrene investigated by flow induced birefringence and laser-Doppler velocimetry. *J. Non-Newtonian Fluid Mech.*, **145**:69–77 (2007).
86. Martyn M. T., R. Spares, T. Gough, and P. D. Coates, Start-up instabilities in axisymmetric contraction geometries for polyolefin melts. In *SPE ANTEC*, Dallas, TX, 59, pp. 1031–1035, 2001.
87. Martyn M. T., C. Nakason, and P. D. Coates, Stress measurements for contraction flows of viscoelastic polymer melts, *J. Non-Newtonian Fluid Mech.*, **91**:123–142 (2000).
88. Gough T., R. Spares, and P. D. Coates, Towards 3-D stress and velocity measurements in polymer melt flows. In *Polymer Process Engineering 07* (P. D. Coates, Ed.). University of Bradford, Bradford, U.K., pp. 411–434, 2007.
89. Gough, T., R. Spares, and P. D. Coates, In-process measurements of full field stress birefringence and velocities in polymer melt flows. In *Polymer Process Engineering 05*, (P.D. Coates, Ed.), University of Bradford, Bradford, U.K., 2005, pp. 260–287.
90. Gough T. D. and P. D. Coates, Experimentally assessed three dimensionality of polymer melt flows through abrupt contraction dies. *AIP Conf. Proc.*, **1027(1)**:177–179 (2008).
91. Gogos, C. G. and B. Maxwell, Velocity profiles of the exit region of molten polyethylene extrudates. *Polym. Eng. Sci.*, **6(4)**:353–358 (1966).
92. Barone, J. and S.-Q. Wang, Flow birefringence study of sharkskin and stress relaxation in polybutadiene melts. *Rheol. Acta*, **38(5)**:404–414 (1999).

93. Barone, J. R. and S. Q. Wang, Adhesive wall slip on organic surfaces. *J. Non-Newtonian Fluid Mech.*, **91(1)**:31–36 (2000).

94. Barone, J. R., S. Q. Wang, J. P. S Farinha, and M. A. Winnik, Polyethylene melt adsorption and desorption during flow on high-energy surfaces: Characterization of postextrusion die wall by laser scanning confocal fluorescence microscopy. *Langmuir*, **16(17)**:7038–7043 (2000).

95. Barone, J. R. and S. Q. Wang, Rheo-optical observations of sharkskin formation in slit-die extrusion. *J. Rheol.*, **45(1)**:49–60 (2001).

96. Migler, K. B., C. Lavallée, M. P. Dillon, S. S. Woods, and C. L. Gettinger, Visualizing the elimination of sharkskin through fluoropolymer additives: Coating and polymer–polymer slippage. *J. Rheol.*, **45(2)**:565–581 (2001).

97. Migler, K. B., Y. Son, F. Qiao, and K. Flynn, Extensional deformation, cohesive failure, and boundary conditions during sharkskin melt fracture. *J. Rheol.*, **46(2)**:383–400 (2002).

98. Ottinger, H.-C., End effects in flow-birefringence experiments for polymer melts. *J. Rheol.*, **43(1)**:253–259 (1999).

99. Chiba, K. and K. Nakamura, Instabilities in a circular entry flow of dilute polymer solutions. *J. Non-Newtonian Fluid Mech.*, **73**:67–80 (1997).

100. Tordella, J. P. and J. B. Wilkens, Comments on a proposed mechanism of the departure from steady laminar flow in molten polymers. *J. Appl. Polym. Sci.*, **11(12)**:2590 (1967).

101. Clegg, P. L., The flow of molten polymer and their effect on fabrication. *Br. Plast.* **30**:535–537 (1957).

102. Nakamura, K., S. Ituaki, T. Nishimura, and A. Horikawa, Instability of polymeric flow through an abrupt contraction. *J. Textile Machinery Jpn.*, **36**:49 (1987).

103. Nguyen, T. H., The influence of elasticity on die entry flows. PhD thesis, Monash University, Clayton, Australia, 1976.

104. Cable, P. J., Laminar entry flow of viscoelastic fluids. PhD thesis, Monash University, Clayton, Australia, 1976.

105. Cable, P. J. and D. V. Boger, A comprehensive experimental investigation of tubular entry flow of viscoelastic fluids: Part III. Unstable flow. *AIChE J.*, **25(1)**:152–159 (1979).

106. Boger, D. V. and K. Walters, *Rheological Phenomena in Focus*. Elsevier, Amsterdam, the Netherlands, 1993.

107. Den Otter, J. L., J. L. S. Wales, and J. Schijf, The velocity profiles of molten polymers during laminar flow. *Rheol. Acta*, **6(3)**:205–209 (1967).

108. Bagley, E. B., S. H. Storey, and D. C. West, Post extrusion swelling of polyethylene. *J. Appl. Polym Sci.*, **7(5)**:1661–1672 (1963).

109. Schwetz, M., H. Münstedt, M. Heindl, and A. Merten, Investigations on the temperature dependence of the die entrance flow of various long-chain branched polyethylenes using laser-Doppler velocimetry. *J. Rheol.*, **46(4)**:797–815 (2002).

110. Archer, L. A., Y.-L. Chen, and R. G. Larson, Delayed slip after step strains in highly entangled polystyrene solutions. *J. Rheol.*, **39(3)**:519–525 (1995).

111. Oyanagi, Y., Irregular flow behavior of high density polyethylene. *Appl. Polym. Symp.*, **20**:123–136 (1973).

112. Brizitsky, V. I., G. V. Vinogradov, A. I. Isaev, and Y. Y. Podolsky, Polarization optical investigation of normal and shear stresses in flow of polymers. *J. Appl. Polym. Sci.*, **20**:25–40 (1976).

113. Arda, D. R. and M. R. Mackley, Sharkskin instabilities and the effect of slip from gas-assisted extrusion. *Rheol. Acta*, **44**:352–359 (2005).

114. Arai, T., Pressure variation and flow birefringence of polymer melts in flows through straight ducts with approaching channel. *Rheol. Acta*, **13(4–5)**:859–863 (1974).

115. Philippoff, W., F. H. Gaskins, and J. G. Boodryan, Flow birefringence and stress. V. Correlation of recoverable shear strains with other rheological properties of polymer solutions. *J. Appl. Phys.*, **28** (**10**):1118 (1957).

116. Sornberger, G., J. C. Quantin, R. Fajolle, B. Vergnes, and J. F. Agassant, Experimental study of the sharkskin defect in linear low density polyethylene. *J. Non-Newtonian Fluid Mech.*, **23**:123–135 (1987).

117. Kamath, V. M., The modelling of viscoelastic behaviour for mon- and polydisperse polymer melts. PhD thesis, Churchill College, Cambridge University, Cambridge, U.K., 1990.

118. Metzger, A. P. and C. W. Hamilton, The oscillating shear phenomenon in high density polyethylenes, In *SPE ANTEC*, Atlantic City, NJ, 4, 1964, pp. 107–112.

119. Hürlimann, H. P. and W. Knappe, Relation between the extensional stress of polymer melts in the die inlet and in melt fracture. *Rheol. Acta*, **11**(**3–4**):292–301 (1972).

120. Bergem, N., Visualization studies of polymer melt flow anomalies in extrusion. In *Seventh International Congress on Rheology* (C. Klason, Ed.). Goteborg, Sweden, August 1976, pp. 50–54.

121. Vos, E., H. E. H. Meijer, and G. W. M. Peters, Multilayer injection moulding. *Int. Polym. Process.*, **6**(**1**):42–50 (1991).

122. Sombatsompop, N. and A. K. Wood, Flow analysis of natural rubber in a capillary rheometer. 2: Flow patterns and entrance velocity profiles in the die. *Polym. Eng. Sci.*, **37**(**2**):281–290 (1997).

123. Lyngaae-Joergensen, J. and B. Marcher, "Spurt fracture" in capillary flow. *Chem. Eng. Commun.*, **32**(**1–5**):117–151 (1985).

124. McKinley, G. H., P. Pakdel, and A. Öztekin, Rheological and geometric scaling of purely elastic flow instabilities. *J. Non-Newtonian Fluid Mech.*, **67**:19–47 (1996).

125. Watson, J. H., The mystery of the mechanism of sharkskin. *J. Rheol.*, **43**(**1**):245–252 (1999).

126. Cogswell, F. N., Stretching flow instabilities at the exits of extrusion dies. *J. Non-Newtonian Fluid Mech.*, **2**(**1**):37–47 (1977).

127. Barone, J. R., N. Plucktaveesak, and S. Q. Wang, Interfacial molecular instability mechanism for sharkskin phenomenon in capillary extrusion of linear polyethylene. *J. Rheol.*, **42**(**4**):813–832 (1998).

128. Dooley, J. and K. Hughes, Coextrusion layer thickness variation—Effect of polymer viscoelasticity on layer uniformity. In *SPE ANTEC*, Boston, MA, 53, 1995, pp. 69–74.

129. Dooley, J. and L. Dietsche, Numerical simulation of viscoelastic polymer flow: Effect of secondary flows on multilayer coextrusion. *Plast. Eng.*, **52**(**4**):37–39 (1996).

130. Debbaut, B., T. Avalosse, J. Dooley, and K. Hughes, On the development of secondary motions in straight channels induced by the second normal stress difference: Experiment and simulation. *J. Non-Newtonian Fluid Mech.*, **69**(**2–3**):255–271 (1997).

131. Dooley J., Viscoelastic flow effects in multilayer coextrusion. PhD thesis, TU Eindhoven, Eindhoven, the Netherlands, 2002.

132. Khomani, B., Interfacial stability and deformation of two stratified power-law fluids in plane Poiseuille flow. Part I: Stability analysis. *J. Non-Newtonian Fluid Mech.*, **36**:289–303 (1990).

133. Khomani, B. and M. M. Ranjbaran, Experimental studies of interfacial instabilities in multilayer pressure driven flow of polymeric melts. *Rheol. Acta*, **36**:345–366 (1997).

134. Pearson, J. R. A., Mechanisms for melt flow instability. *Plast. Polym.*, **August**:285–291 (1969).

135. Beaufils, P., Etude des défauts d'extrusion des polyéthylènes lineaires. Approche experimentale et modélisation des écoulements. PhD thesis, Ecole des Mines de Paris (CEMEF), Sophia Antipolis, France, 1989.

136. White, J. L., Dynamics of viscoelastic fluids, melt fracture, and the rheology of fiber spinning. *J. Appl. Polym. Sci.*, **8**:2339–2357 (1964).

137. McIntire, L. V., Initiation of melt fracture. *J. Appl. Polym. Sci.*, **16(11)**:2901–2908 (1972).

138. Reiner, M., The Deborah number. *Phys. Today*, **17(1)**:62 (1964).

139. Metzner, A. B., J. L. White, and M. M. Denn, Behavior of viscoelastic materials in short-time processes. *AIChE. J.*, **62(12)**:81–92 (1966).

140. Barnett, S. M., A correlation for melt fracture. *Polym. Eng. Sci.*, **7(3)**:168–174 (1967).

141. Wall, F. T., Statistical thermodynamics of rubber. II. *J. Chem. Phys.* **10**:485–488 (1942).

142. Vlachopoulos, J. and M. Alam, Critical stress and recoverable shear for polymer melt fracture. *Polym. Eng. Sci.*, **12(3)**:184–192 (1972).

143. Vinogradov, G. V., M. L. Friedman, B. V. Yarlykov, and A. Y. Malkin, Unsteady flow of polymer melts: Polypropylene. *Rheol. Acta*, **9(3)**:323–329 (1970).

144. Pomar, G., S. J. Muller, and M. M. Denn, Extrudate distortions in linear low-density polyethylene solutions and melt. *J. Non-Newtonian Fluid Mech.*, **54(1–3)**:143–151 (1994).

145. Boger, D. V. and H. L. Williams, Predicting melt flow instability from a criterion based on the behavior of polymer solutions. *Polym. Eng. Sci.*, **12**:309 (1972).

146. Cogswell, F. N., J. G. H. Gray, and D. A. Hubbard, Rheology of a fluid whose behavior approximates to that of a Maxwell body. *Bull. Br. Soc. Rheol.*, **15(2)**:29 (1972).

147. Ballenger, T. F., I. J. Chen, J. W. Crowder, G. E. Hagler, D. C. Bogue, and J. L. White, Polymer melt flow instabilities in extrusion: Investigation of the mechanisms and material and geometric variables. *Trans. Soc. Rheol.*, **15(2)**:195–215 (1971).

148. Ramamurthy, A. V., Wall slip in viscous fluids and influence of materials of construction. *J. Rheol.*, **30(2)**:337–357 (1986).

149. Utracki, L. A. and R. Gendron, Pressure oscillation during extrusion of polyethylene. II. *J. Rheol.*, **28(5)**:601–623 (1984).

150. Ramamurthy, A. V., LLDPE rheology and blown film fabrication. *Adv. Polym. Technol.*, **6**:489–499 (1986).

151. Kalika, D. S. and M. M. Denn, Wall slip and extrudate distortion in linear low-density polyethylene. *J. Rheol.*, **31(8)**:815–834 (1987).

152. Lim, F. J. and W. R. Schowalter, Wall-slip of narrow molecular weight distribution polybutadienes. *J. Rheol.*, **33(8)**:1359–1382 (1989).

153. Hatzikiriakos, S. G., Wall slip of linear polyethylenes and its role in melt fracture. PhD thesis, McGill University, Montreal, Canada, 1991, p. 248.

154. Kurtz, S. J., The dynamics of sharkskin melt fracture in LLDPE. In *Polymer Processing Society, PPS7*, Hamilton, Canada, 1991, pp. 54–55.

155. Tzoganakis, C., B. C. Price, and S. G. Hatzikiriakos, Fractal analysis of the sharkskin phenomenon in polymer melt extrusion. *J. Rheol.*, **37(2)**:355–366 (1993).

156. El Kissi, N. and J. M. Piau, Adhesion of linear low-density polyethylene for flow regimes with sharkskin. *J. Rheol.*, **38(5)**:1447–1463 (1994).

157. Hatzikiriakos, S. G., P. Hong, W. Ho, and C. W. Stewart, The effect of Teflon coatings in polyethylene capillary extrusion. *J. Appl. Polym. Sci.*, **55(4)**:595–603 (1995).

158. Rosenbaum, E. E., S. G. Hatzikiriakos, and C. W. Stewart, Flow implications in the processing of tetrafluoroethylene/hexafluoropropylene copolymers. *Int. Polym. Process.*, **10(3)**:204–212 (1995).

159. Naguib, H. E. and C. B. Park, A study on the onset surface melt fracture of polypropylene materials with foaming additives. *J. Appl. Polym. Sci.*, **109**:3571–3577 (2008).

160. Santanach Carreras, E., N. El Kissi, and J.-M. Piau, Block copolymer extrusion distortions, exit delayed transversal primary cracks and longitudinal secondary cracks: Extrudate splitting and continuous peeling. *J. Non-Newtonian Fluid Mech.*, **131(1–3)**:1–21 (2005).

161. Vega, J. F., M. Fernandez, A. Santamaria, A. Munoz-Escalona, and P. Lafuente, Rheological criteria to characterize metallocene catalyzed polyethylenes. *Macromol. Chem. Phys.*, **200**:2257–2268 (1999).

162. Delgadillo-Velazquex, O., G. Georgiou, M. Sentmanat, and S. Hatzikiriakos, Sharkskin and oscillating melt fracture: Why in slit and capillary dies and not in annular dies? *Polym. Eng. Sci.*, **48**:405–414 (2008).

163. Doerpinghaus, P. J. and D. G. Baird, Comparison of the melt fracture behavior of metallocene and conventional polyethylenes. *Rheol. Acta*, **42**:544–556 (2003).

164. Gendron, R., L. Piche, A. Hamel, M. M. Dumoulin, and J. Tatibouet, Ultrasonic characterization of extrusion instabilities. In *SPE ANTEC*, Toronto, Canada, 55(2), 1997, pp. 2254–2256.

165. Ui, J., Y. Ishimaru, H. Murakami, N. Fukushima, and Y. Moi, Study of flow properties of polymer melts with the screw extruder. *Polym. Eng. Sci.*, **4(4)**:295–305 (1964).

166. Blyler, L. L. J. and A. C. J. Hart, Capillary flow instability of ethylene polymer melts. *Polym. Eng. Sci.*, **10(4)**:193–203 (1970).

167. Utracki, L. A. and M. M. Dumoulin, Pressure oscillations during the extrusion of polyethylene: Part 1. *Polym. Plast. Technol. Eng.*, **23**:194 (1984).

168. Atwood, B. T. and W. R. Schowalter, Measurements of slip at the wall during flow of high density polyethylene through a rectangular conduit. *Rheol. Acta*, **28**:134–146 (1989).

169. Becker, J., P. Bengtsson, C. Klason, J. Kubat, and P. Saha, Pressure oscillations during capillary extrusion of high-density polyethylene. *Int. Polym. Process.*, **6(4)**:318–325 (1991).

170. Kurtz, S. J., Die geometry solutions to sharkskin melt fracture. In *Theoretical and Applied Rheology* (B. Mena, A. Garcia-Rejon, and C. Rangel Nafaile, Eds.). UNAM Press, Mexico City, Mexico, 1984, pp. 399–407.

171. Whitehouse, D. J., Surface metrology. *Meas. Sci. Technol.*, **8**:955–972 (1997).

172. Constantin, D., Linear-low-density polyethylene melt rheology: Extensibility and extrusion defects. *Polym. Eng. Sci.*, **24(4)**:268–274 (1984).

173. De Smedt, C. and S. Nam, The processing benefits of fluoroelastomer applications in LLDPE. *Plast. Rubber Proc. Appl.*, **8**:11–16 (1987).

174. El Kissi, N., L. Leger, J. M. Piau, and A. Mezghani, Effect of surface properties on polymer melt slip and extrusion defects. *J. Non-Newtonian Fluid Mech.*, **52(2)**:249–261 (1994).

175. Lin, Y. H., Explanation for slip-stick melt fracture in terms of molecular dynamics in polymer melts. *J. Rheol.*, **29(6)**:605–637 (1985).

176. Moynihan, R. H., D. G. Baird, and R. Ramanathan, Additional observations on the surface melt fracture behavior of linear low-density polyethylene. *J. Non-Newtonian Fluid Mech.*, **36**:255–263 (1990).

177. Rudin, A., H. P. Schreiber, and D. Duchesne, Use of fluorocarbon elastomers as processing additives for polyolefins. *Polym. Plast. Technol. Eng.*, **29(3)**:199–234 (1990).

178. Karbashewski, E., A. Rudin, L. Kale, W. J. Tchir, and H. P. Schreiber, Effects of polymer structure on the onset of processing defects in LLDPE's. In *SPE ANTEC*, Montreal, Canada, 49, 1991, pp. 1378–1381.

179. Reason R. E., The measurement of surface texture. In *Modern Workshop Technology*, Part 2 (W. Baker, Ed.). Macmillan, London, U.K., 1970.

180. Quantin, J. C., Application de la rugosimetrie tridimensionelle a l'étude des surfaces. PhD thesis, Ecole des Mines de Paris (CEMEF), Sophia Antipolis, France, 1986.

181. Tong P. P. and V. Firdaus, Post die effects on sharkskin melt fracture of LLDPE. In *Polymers Laminations & Coatings Conference Proceedings*. TAPPI Press, Atlanta, GA, 1993, pp. 293–296.

182. Ajji, A., S. Varennes, H. P. Schreiber, and D. Duchesne, Flow defects in linear low density polyethylene processing: Instrumental detection and molecular weight dependence. *Polym. Eng. Sci.*, **33(23)**:1524–1531 (1993).

183. Meissner, J., Development of a universal extensional rheometer for the uniaxial extension of polymer melts. *Trans. Soc. Rheol.*, **16**:405–420 (1972).

184. Kale, L. and D. Hager, A new look at melt fracture in linear low density polyethylene. In *Polymers Laminations & Coatings Conference Proceedings*. TAPPI Press, Atlanta, GA, 1993, pp. 283–291.

185. Le Gall, F., O. Bartos, J. Davis, and P. Philip, On a classification of surface defects of extruded polymer melts. In *Polymer Processing Society, European Regional Meeting*, Stuttgart, Germany, 1995.

186. Kurtz, S. J., The control of sharkskin patterns in film extrusion. In *SPE ANTEC*, New Orleans, LA, 1993, 51, pp. 454–456.

187. Kurtz, S. J., The dynamics of sharkskin melt fracture: Effect of die geometry. In *Theoretical Applied Rheology, Proceedings of the 11th International Congress on Rheology* (P. Moldenaers and K. Roland, Ed.). Polyolefins Div. Res. Dev., Union Carbide Chemicals and Plastics Co. Inc., Bound Brook, NJ. Elsevier, Amsterdam, the Netherlands, 1992, pp. 377–379.

188. Venet, C. and B. Vergnes, Experimental characterization of sharkskin in polyethylenes. *J. Rheol.*, **41(4)**:873–892 (1997).

189. Schowalter, W. R., The behavior of complex fluids at solid boundaries. *J. Non-Newtonian Fluid Mech.*, **29**:25–36 (1988).

190. Mackay, M. E. and D. J. Henson, The effect of molecular mass and temperature on the slip of polystyrene melts at low stress levels. *J. Rheol.*, **42(6)**:1505–1517 (1998).

191. Mooney M., Explicit formula for slip and fluidity. *J. Rheol.*, **2**:210–222 (1931).

192. Giacomin, A. J., T. Samurkas, and J. M. Dealy, Novel sliding plate rheometer for molten plastics. *Polym. Eng. Sci.*, **29(8)**:499–504 (1989).

193. Hill, D. A., T. Hasegawa, and M. M. Denn, On the apparent relation between adhesive failure and melt fracture. *J. Rheol.*, **34(6)**:891–918 (1990).

194. Stewart, C. W., Wall slip in the extrusion of linear polyolefins. *J. Rheol.*, **37(3)**:499–513 (1993).

195. Vinogradov, G. V. and L. I. Ivanova, Viscous properties of polymer melts and elastomers exemplified by ethylene-propylene copolymer. *Rheol. Acta*, **6(3)**:209–222 (1967).

196. White, J. L., M. H. Han, N. Nakajima, and R. Brzoskowski, The influence of materials of construction on biconical rotor and capillary measurements of shear viscosity of rubber and its compounds and considerations of slippage. *J. Rheol.*, **35(1)**:167–189 (1991).

197. Hatzikiriakos, S. G., The onset of wall slip and sharkskin melt fracture in capillary flow. *Polym. Eng. Sci.*, **34(19)**:1441–1449 (1994).

198. Crawford, B., J. K. Watterson, P. L. Spedding, S. Raghunathan, W. Herron, and M. Proctor, Wall slippage with siloxane gum and silicon rubbers. *J. Non-Newtonian Fluid Mech.*, **129(1)**:38–45 (2005).

199. Kamerkar, P. A. and B. J. Edwards, Experimental study of slip flow in capillaries and semihyperbolically convergent dies. *Polym. Eng. Sci.*, **47(2)**:159–167 (2007).

200. Laun, H. M., Squeezing flow rheometry to determine viscosity, wall slip and yield stresses of polymer melts. In *Polymer Processing Society, PPS12*, Sorrento, Italy, May 1996.

201. Leider P. J., Squeezing flow between parallel disks. II Experimental results. *Ind. Eng. Chem. Fundam.* **13**:342–346 (1974).

202. Leider P. J. and R. Bird, Squeezing flow between parallel disks. I. Theoretical analysis. *Ind. Eng. Chem. Fundam.*, **13**:336–341 (1974).

203. Lee S. J., M. M. Denn, M. J. Chrocheet, and A. B. Metzner, Compressive flow between parallel disks. 1. Newtonian fluid with transverse viscosity gradient. *J. Non-Newtonian Fluid Mech.*, **10**:3–30 (1982).

204. Chatraei S. H., C. W. Macosko, and H. H. Winter, Lubricated squeezing flow: A new biaxial extensional rheometer. *J. Rheol.*, **25(4)**:433–443 (1981).

205. Soskey P. R. and H. H. Winter, Equibiaxial extension of two polymer melts: Polystyrene and low density polyethylene. *J. Rheol.*, **29(5)**:493–517 (1985).

206. Atwood, B. T., Wall slip and extrudate distortion of high density polyethylene. PhD thesis, Department of Chemical Engineering, Princeton University, Princeton, NJ, 1982.

207. Migler, K. B., H. Hervet, and L. Leger, Slip transition of a polymer melt under shear stress. *Phys. Rev. Lett.*, **70(3)**:287–290 (1993).

208. Ausserre, D., H. Hervet, and F. Rondelez, Depletion layers in polymer solutions: Influence of the chain persistence length. *J. Phys. Lett.*, **46**:929–934 (1985).

209. Davoust J., P. Devaux, and L. Leger, Fringe pattern photobleaching, a new method for the measurement of transport coefficients of biological macromolecules. *EMBO J.*, **1(10)**:1233–1238 (1982).

210. Brochard F. and P. G. de Gennes, Shear dependent slippage at polymer/solid interface. *Langmuir*, **8**:3033–3037 (1992).

211. Leger, L., H. Hervet, G. Massey, and E. Durliat, Wall slip in polymer melts. *J. Phys.: Condens. Matter* **9**:7719–7740 (1997).

212. Mizunuma H. and H. Takagi, Cyclic generation of wall slip at the die exit of plane Couette flow. *J. Rheol.*, **47(3)**:737–757 (2003).

213. Park H. E., S. T. Lim, F. Smillo, J. M. Dealy, and C. G. Robertson, Wall slip and spurt flow of polybutadiene. *J. Rheol.*, **52(50)**:1201–1239 (2008).

214. Hatzikiriakos, S. G. and J. M. Dealy, Start-up pressure transients in a capillary rheometer. *Polym. Eng. Sci.*, **34(6)**:493–499 (1994).

215. Weill A., About the origin of sharkskin. *Rheol. Acta*, **19(5)**:623–632 (1980).

216. Weill A., Oscillations de relaxation du polyethylene de haute densite et defauts d'extrusion. PhD thesis, Universite Louis Pasteur, Strasbourg, France, 1978.

217. Okubo S. and Y. Hori, Model analysis of oscillating flow of high density polyethylene melt. *J. Rheol.*, **24(2)**:253–257 (1980).

218. Molenaar J. and R. Koopmans, Modeling polymer melt flow instabilities. *J. Rheol.*, **38(1)**:99–109 (1994).

6 Melt Fracture Variables

The Nobel Prize laureate Pierre Gilles de Gennes [1] once stated: "Compared with the giants of quantum physics, we soft matter theorists look like the dwarfs of German folk tales. We are strongly motivated by industrial purposes. We see fundamental problems emerging from practical questions." In its historical context, the research drive to understand melt fracture has always been closely linked to technological progress in industry. The need to find more differentiated or cheaper products and more efficient synthesis methods has motivated research over the years. As a result, a wide variety of plastics has become commercially available. However, melt fracture remains a challenge for high-speed polymer processing. In order to find potential improvements, the influence of polymer architecture and/or polymer-processing conditions is studied.

6.1 POLYMER ARCHITECTURE

The melt fracture behavior of nearly all commercially significant polymers has been investigated extensively. The literature, however, is strongly biased toward PE. Not surprisingly as PE gives rise to the widest variety of melt fracture phenomena. Most other polymers show less extrudate distortion variety or are essentially melt fracture free during typical processing operation. Consequently, melt fracture is not always perceived as a problem for most polymers, even though the same polymers still have the potential for giving rise to extrudate distortions under different processing conditions.

As presented in Chapter 5, experimental extrudate distortion observations for one polymer appear difficult to generalize for all polymers. In part, this may be attributed to the difficulty of defining a common polymer architecture measure for all polymers (e.g., the entanglement MM M_e), but also as Yamane and White [2] pointed out, the fact: "that many polymers used in most investigations are poorly characterized … increases the difficulty in a universal interpretation of the results." Phrases like: "The sample A was presumably a LLDPE of industrial grade [3]" support the above statement.

Many polymers remain difficult to fully characterize. Even the latest, most advanced analytical tools offer only averaged parameters. Therefore, typical relationships between polymer architecture variables and melt fracture mostly have a qualitative character. The only certainty is that the type of melt fracture strongly depends on the polymer architecture [4,5].

At certain flow rates, the surface distortions of LLDPE are more pronounced than those of HDPE [6]. The apparent shear stress for the onset of surface distortions in LLDPE is lower and the surface distortions develop higher amplitude at increasing flow rates. In some HDPE, surface distortions may even be absent [7,8]. Karbashewski et al. [9] found that increasing the amount of short chain branched molecules in a LLDPE polymer reduces surface distortions. Modifying essentially linear PE via high-energy irradiation [10] or peroxide modification [11] can change the melt fracture appearance totally. Piau et al. [12,13], working with linear and branched PDMS, observed a higher surface distortion amplitude as well as a higher shear stress onset for the linear polymers. Differences in the monomer composition of linear polymers such as for PP, PS, and TFE-HFP copolymers yields very different flow curves and melt fracture behavior in comparison to HDPE [14–18]. In PVC, extrudate surface distortions frequently vanish at high shear rates. Cogswell [19] explained such phenomena in relation to the formation of polymer aggregates arising from strong chlorine dipole interactions.

These and other experimental findings often led to conjectures on the causes of melt fracture in relation to polymer architecture. More extensive systematic studies into the influence of repeat unit, polymer MM, MMD, branch type, branch length, and branch distribution are not available in the open literature. At best, empirical correlations are presented between the polymer architecture variables and some flow curve characteristics (e.g., critical shear stress τ_c) and the associated melt fracture observation. Blyler and Hart [20] compared the effect of ethyl and butyl short-chain branching in poly(ethylene-*co*-butene) and poly(ethylene-*co*-hexene), respectively. Vinogradov et al. [21] compared PB and PI in terms of basic material parameters and rheological behavior. Vlachopoulos and Alam [22] linked the melt fracture observations of various PP, PS, HDPE, and LDPE with τ_c and rationalized the observations in terms of the Graessley entanglement theory [23]. Sometimes, molecular theories are invoked to explain the observations. Lin [24,25] applied the Doi–Edwards theory [26–28] for PE, PI, and PS to develop a polymer architecture foundation of melt fracture. Venet [10] compared various PE of the same melt flow rate (MFR) but different density. Wang and coworkers [29,30] and Yang et al. [31] developed a molar understanding of melt-wall interfacial interaction in terms of the rheological material parameters of HDPE, LLDPE, PS, LDPE, EVA, and PP. Allal et al. [32] invoked the elasticity theory of Griffith [33] to show that the period of the surface distortion as observed for linear polymers is proportional to the amplitude and that the critical onset elongational stress is a function of the plateau modulus, M_n and M_e (Equation 6.1):

$$\tau_{ce} = \frac{1}{2} G_N^0 \frac{N_e}{\sqrt{N_0}} \tag{6.1}$$

where

τ_{ce} is the critical elongational stress for the onset of surface distortion

G_N^0 is the plateau modulus

N_e is the number of entanglements per molecule ($N_e = M_e/M_0$, with M_0 the monomer MM, and M_e the MM between entanglement points)

N_0 is the number of monomers per molecule ($N_0 = M_n/M_0$)

They also defined a critical shear stress as associated with the onset of slip flow but explicitly stating that the onset of surface distortions does not originate from slip:

$$\tau_c = \frac{9}{4\pi} G_N^0 \frac{N_e}{\sqrt{N_0}} \tag{6.2}$$

Vlachopoulos and Alam [22,34] showed for a large number of commercial polymers (PP, PS, LDPE, HDPE) that the critical shear stress associated with a visual melt fracture detection is of the order of 10^5 Pa (no distinction was made between surface or volume distortions). The critical shear stress decreased with increasing MM M_w (Figure 6.1). Other studies using different polymers arrived at similar conclusions (see Table 5.2) in relation to surface distortions. Vlachopoulos and Alam [22] derived an empirical fit between τ_c (in dynes/cm^2) and the weight average MM M_w (in g/mol) for

$$\text{Polystyrene (PS): } \tau_c = 0.796 \times 10^6 + \frac{1.164 \times 10^{11}}{M_w} \tag{6.3}$$

$$\text{Polypropylene (PP): } \tau_c = 0.892 \times 10^6 + \frac{1.435 \times 10^{11}}{M_w} \tag{6.4}$$

$$\text{High density polyethylene (HDPE): } \tau_c = 0.810 \times 10^6 + \frac{1.061 \times 10^{11}}{M_w} \tag{6.5}$$

$$\text{Low density polyethylene (LDPE): } \tau_c = 0.552 \times 10^6 + \frac{0.430 \times 10^{11}}{M_w} \tag{6.6}$$

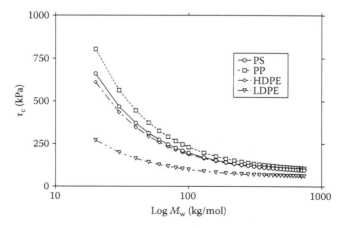

FIGURE 6.1 The influence of M_w on the critical stress for the onset of melt fracture according to the empirical fit proposed by Vlachopoulos and Alam. (Adapted from Vlachopoulos, J. and Alam, M., *Polym. Eng. Sci.*, 12(3), 184, 1972.)

Equations 6.3 through 6.6 indicate that τ_c is inversely proportional to M_w ($\tau_c = \alpha + \beta/M_w$). This functional form defines a strong τ_c increase for decreasing M_w below 100 kg/mol. Above 100 kg/mol, an increasing M_w has little effect on τ_c ($\approx 10^5$ Pa) (Figure 6.1).

The effect of MMD changes was not taken into account for these polymers. Baik and Tzoganakis [35] reported the Bagley-corrected τ_c to reduce with increasing M_w (136–348 kg/mol) and M_w/M_n for a series of peroxide-degraded PP:

$$\tau_c = -1.73306 \times 10^{-7} \times M_w + 0.16503 \tag{6.7}$$

$$\tau_c = -0.00611 \times \frac{M_w}{M_n} + 0.17202 \tag{6.8}$$

Ramamurthy [36] and Kurtz [37] for LLDPE, and Lim and Schowalter [38] for PB ($M_w = 78$–240 kg/mol), considered the wall shear stress onset of surface distortions not to be influenced by the MM. In view of the studied polymer MM, the results qualitatively agree with the findings of Vlachopoulos and Alam [22]. Piau et al. [39] suggest a strong dependency on the ratio of M_w and entanglement MM M_e [40] for the onset of surface distortion. Higher values of M_w/M_e should decrease τ_c (Figure 6.2).

For LLDPE, Ramamurthy [36] stated that only the severity and not the onset shear stress of the surface distortions (0.14 MPa) depends on the MM. Many authors

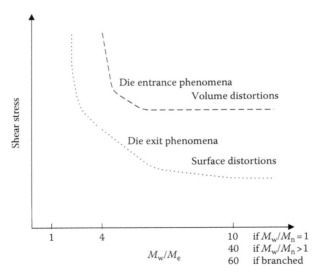

FIGURE 6.2 A schematic representation of the influence of entanglement density in terms of M_w/M_e on τ_c for the onset of surface and volume distortions as proposed by Piau et al. [39]. The scaling term M_w/M_e depends on M_w/M_n and branching. (From Piau, J.-M. et al., *Rheol. Acta*, 34, 40, 1995. With permission.)

confirm that at a given flow rate, the amplitude of the distortion increases with increasing M_w [19,41–44] while concomitantly the distortion frequency decreases [42]. Lai and Knight [45] using a constant-pressure rheometer [46], show that a very nonuniform (bimodal) LLDPE ($M_w/M_n = 15$), containing a large portion of low-MM macromolecules, does not give a low-amplitude high-frequency surface distortion (sharkskin) but a spiraled screw-thread surface distortion at 0.04 MPa. The small flow curve discontinuity that is typically observed for LLDPE of the same MFR, disappears when changing M_w/M_n from 2 to 5. Howells and Benbow [8] and Weill [7] found that for very broad MMD HDPE ($M_w/M_n > 10$), the extrudates hardly have any surface distortions. For a series of LLDPE of nearly equal M_w (90–110 kg/mol) and varying MMD ($M_w/M_n = 3.3–12.7$), Goyal et al. [47] showed a change in the melt fracture character of the extrudate. An evolution from a low-amplitude high-frequency surface distortion to a smooth but rippled extrudate was observed when MMD increased. The apparent shear rate and shear stress of melt fracture onset rise until M_w/M_n approaches 10, and then suddenly fall for broader MMD.

Venet [10] compares a set of LLDPEs with identical MFR one of which included a gamma-irradiation-induced LCB containing polymer [48]. For the branched polymer, the onset of surface distortions was shifted to higher, apparent shear rates, and the surface distortions appeared to be less pronounced. The flow curve discontinuity and the associated pressure oscillations could not be observed for the same processing conditions ($L/2R = 16$, $2R = 1.39$ mm, 190°C). Only at lower temperature (150°C), small pressure oscillations could be detected (at the die entry). The extrudate, however, does not show the characteristics of a transition region (spurt). Lai et al. [49–51] observed that very small amounts of LCB (0.01–3 LCB/1000 carbon atoms) in metallocene-catalyzed polyolefin polymers ($M_w/M_n = 1.5–2.5$) increased τ_c.

Blyler and Hart [20] and Durand [52] for a set of HDPE polymers with the same MMD, and Kurtz [37] for LLDPE, noted a discontinuous flow curve, and a translation of Branch I toward higher, apparent shear rates as M_w decreased (at identical temperature and capillary die geometry). Branch II remained nearly unchanged and was considered independent of M_w. The flow curve discontinuity was reduced, but τ_c remained unchanged (Figure 6.3). For the higher M_w polymers with the same M_w/M_n, Branch II started at lower pressures, and the pressure oscillation amplitude increased (Figure 6.4; Table 6.1).

For HDPE, Wang and Drda [53] using a constant-pressure capillary rheometer did not show pressure oscillations but a transition region onset at lower shear stress for the highest M_w ($M_w = 130$ and 316 kg/mol) at 180°C. The higher M_w HDPE showed a flow curve with a wider transition zone. Vinogradov et al. [54] showed for PB and PI that the flow curve transition region also covers a wider shear rate range with increasing M_w. In addition, at constant M_w and higher M_w/M_n, the shear rate range of the flow curve discontinuity becomes smaller again. Lin [24] suggested on theoretical grounds that the flow curve transition region can be eliminated by broadening the MMD. Experiments on PI, PS, and PE appear to support that suggestion. He further deduced qualitatively that the transition region is a general phenomenon of linear flexible polymers that will occur when M_w is high and MMD is narrow.

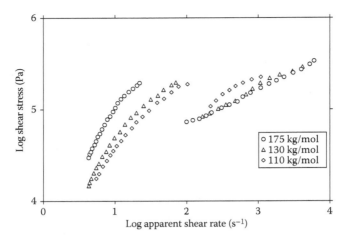

FIGURE 6.3 Dependency of the discontinuous flow curve on M_w for three HDPE homopolymers of varying M_w and $M_w/M_n = 5$. The data were measured with a constant-rate rheometer equipped with a capillary die ($L/2R = 8$, $2R = 1.3$ mm, $180°$ entry angle) at $160°C$ and corrected for entry-pressure effects. (From Durand, V., Ecoulements et instabilité des polyéthylènes haute densité, PhD thesis, Ecole des Mines de Paris (CEMEF), Sophia Antipolis, France, 1993.)

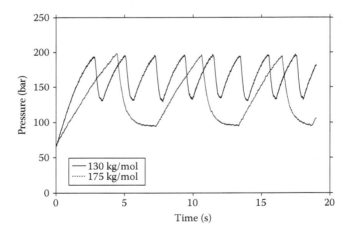

FIGURE 6.4 Pressure oscillations for two HDPE homopolymers of varying M_w (130 and 175 kg/mol) and $M_w/M_n = 5$ ($L/2R = 8$, $2R = 1.3$ mm, $180°$ entry angle, $160°C$). (From Durand, V., Ecoulements et instabilité des polyéthylènes haute densité, PhD thesis, Ecole des Mines de Paris (CEMEF), Sophia Antipolis, France, 1993.)

Fernandez et al. [55] studied 10 mPEs of varying M_w between 90 and 215 kg/mol, and relatively constant M_w/M_n of about 2.1–2.7. The onset of surface distortions varies greatly with shear rate, but is constant in terms of shear stress at a value of 0.20 MPa. This stress is independent of M_w, and holds for linear and sparsely branched samples. Similar results were observed by the same group on another set of metallocene

TABLE 6.1
Influence of Polymer Architecture on Surface and Volume Distortions at Fixed Processing Conditions

Polymer Architecture			Melt Fracture			Flow Curve	
M_w	M_w/M_n	LCB	Amplitude	Frequency	τ_c	Breadth of transition region	Pressure oscillation
↑	—	—	↑	↓	—	↑	↑
—	↑	—	↓	↑	τ_c^*↑ τ_c —	↓	↓
—	—	↑	↓	↓	↑—↓	Disappears	Very small (<5 bar)

Note: Constant L/R, entry angle, die material and temperature; —, unchanged.

samples [56]. Kazatchkov et al. [57] studied melt fracture of two series of LLDPE, varying both in M_w (50–110 kg/mol) and M_w/M_n (3.3–12.7). The effect of no uniformity was first to increase the onset rate of surface distortion due to increased shear thinning. At wide MMD, the onset rate for melt fracture decreased again due to the onset of volumetric distortions. The optimum M_w/M_n at an M_w of about 100 kg/mol was about 9. The critical rate was found to increase strongly with decreasing M_w. Kazatchkov et al. [57] do not explicitly quantify the effect of polymer architecture on critical stress rather than rate, but do claim that it follows the same trend.

Wang et al. [58] studied 14 mPE with different levels of LCB between 0.06 and 0.98 LCB/chain. They found a decrease in critical stress for surface distortions with increasing LCB level. They included a similar commercial sample and found a very different result. They confronted their results with other literature results that either point to constant stress [56] or increasing stress [10] with increasing LCB level, and conclude that the effect of LCB is complex and not simply captured by a single topological metric like LCB/chain or LCB/1000C.

Perez et al. [59] studied blends of mLLDPE with LDPE. Surface distortions are postponed to higher rate with increasing level of LDPE above 10 wt%. Delgadillo-Velazquez and Hatzikiriakos [60] studied blends of an LLDPE with different LDPEs. At low levels of LDPE (<20%), there is limited effect of onset stress and the rate of both sharkskin and spurt. The pure LDPEs show a critical stress and rate that strongly vary between the materials, depending on their polymer architecture. Blends with more than 20% LDPE do not show stick-slip oscillations. Den Doelder et al. [61] showed that the critical stress for the onset of volumetric distortions of various LDPEs decreases with increasing M_w, at similar M_w/M_n and LCB levels, while the onset stress varies from 0.07 to 0.2 MPa. A high melt strength–PP (different type of LCB) had a much lower onset stress, <0.03 MPa.

Dao and Archer [62] studied stick-slip for a series of narrow model PB with varying M_w. They found the critical onset stress to be independent from M_w. Ferri and Canetti [63] showed that the onset stress for spurt for narrow styrene–isoprene–styrene block polymers increased slightly with increasing PS content. There is little dependence on M_w. Fujiyama and Inata [64] have studied various series of PP. They concluded that the critical stress for the onset of melt fracture is independent of M_w and increases with increasing M_z/M_w. Furthermore, they studied narrow PP [65] from, for example, a vis-breaking process. The narrower samples were more prone to surface distortions. Rather than providing a model, they show that PP with M_w/M_n of 7 do not give surface distortion, whereas M_w/M_n of 2.1 and 3.1 do, at 0.09 MPa. Naguib and Park [66] studied the effect of LCB on PP surface distortions. The onset stress was 0.12 MPa for various linear PP and decreased with a slope of 0.3 MPa/ (#LCB/1000C) to a value of 0.075 MPa at an LCB level of 0.22/1000C.

6.2 POLYMER-PROCESSING VARIABLES

Extrusion of polymer melts can occur under different processing conditions. Commonly, flow rate and melt temperature as well as die geometry and die material can be changed. However, separating the influence of each independent variable on melt fracture is difficult. Most melt fracture studies using capillary rheometers focus on evaluating the effects of the ratio of capillary length L to die radius R. Other variables are capillary entrance angle, die surface finish (roughness, coatings), die material (steel, aluminum, glass, etc.), and die and/or melt temperature.

6.2.1 LENGTH–RADIUS RATIO

The critical wall shear stress for the onset of surface and volume distortions is often found to be independent of the L/R ratio. For linear PE and TFE-HFP copolymers, Tordella [67] observed an L/R independence for the onset of extrudate surface roughness. For LLDPE and relatively short slit dies ($L/2H = 2$ and 8, slit height $2H = 2.5$ mm) Sornberger et al. [68] found no effect on the development of the surface distortion. Moynihan et al. [69] showed for LLDPE that an increased length of the capillary die ($L/2R = 12.5$, 27, and 75; $2R = 0.69$ mm) shifts the onset to higher flow rates. Constantin [70] reported that the apparent critical shear rate for surface distortion onset is independent of R ($2R = 1.02$–2.5 mm) for LLDPE and increases with R for HDPE at 190°C. For LDPE, he finds that the apparent critical shear rate for the onset of volume distortion increases with wider dies. Wang et al. [71] found for LLDPE that the slope change associated with the onset of surface distortions is independent of the capillary die radius. It corresponds to the findings of Clegg [72] who also observed a diameter independence of the surface distortion onset for HDPE. For LDPE, Herranen and Savolainen [73] found for experiments with a single-screw extruder coupled to capillary dies that the severity of the volume distortion increases for smaller die lengths and vanishes for long dies. Several authors [74–80] reported that the severity of volume distortions decreased for PS, LDPE, and branched PDMS as the die length increased. Metzner et al. [81] indicated that in an infinite tube experiment, no extrudate distortions are found at extrusion rates well within the melt fracture regime. For LLDPE with capillary dies that change from an $L/2R = 15$ ($L = 30$ mm) to an $L/2R = 26$ ($L = 80$ mm), Herranen and Savolainen [73] found that the onset of surface distortion moved to lower shear rates.

Combeaud et al. [82] systematically studied volume distortions for a commercial linear PS in capillary and slit dies of various L and R, and H, respectively. They claimed that the apparent reduction in severity of the helical distortion with increasing L is related to a decreasing frequency of the helix with L. The frequency at $L/R = 0$ is about 3 Hz, whereas at $L/R = 32$, it is about 1.5 Hz at all apparent shear rates.

At constant L/R, some authors have reported the onset of LLDPE surface distortions to scale with linear exit velocity, which scales with Q/R^2. As shear rate scales with Q/R^3, the onset rate becomes proportional to $1/R$ and therefore decreases with increasing R. The presence of slip may alter this scaling picture. Howells and Benbow [8] were the first to report this rate/R scaling for the onset of surface distortions. The observations of Migler et al. [83] are also in line with these findings. Den Doelder [84] observed the same for a 1 MFR LLDPE. All are in agreement with the Cogswell theory [19] (See Chapter 7).

The effect of L/R change on τ_c is less clear in relation to the amplitude and frequency of surface distortions. According to Venet [10], at the same flow rates, the surface distortions are more intense (higher amplitude) with longer dies that have a fully developed steady-state flow near the die exit (Figures 6.5 through 6.7). All

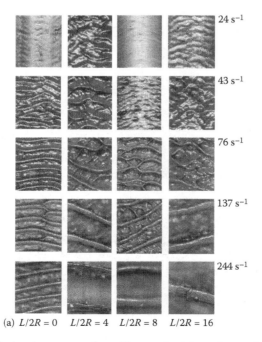

(a) $L/2R = 0$ $L/2R = 4$ $L/2R = 8$ $L/2R = 16$

FIGURE 6.5 Optical microscopy of capillary extrudates of a metallocene-catalyzed ULDPE (MFR = 1 g/10 min, $\rho = 870 \text{ kg/m}^3$) after cooling from 190°C. At higher L/R (0–16 with $2R = 1.39$ mm) and apparent shear rates, surface distortions have a higher amplitude and lower frequency (a).

(continued)

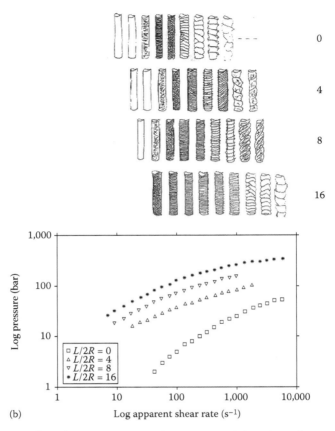

(b)   Log apparent shear rate (s⁻¹)

FIGURE 6.5 (continued)   The associated flow curves and the schematic extrudate distortion representation indicate that the melt fracture onset shear rate is independent of L/R (b). (From Venet, C., Propriétés d'écoulement et défauts de surface de résins polyéthylènes, PhD thesis, Ecole des Mines de Paris (CEMEF), Sophia Antipolis, France, 1996.)

three polymers (a metallocene-catalyzed ULDPE (MFR = 1 g/10 min, ρ = 870 kg/m³), a gamma-irradiated LLDPE (dose 5 kGy, MFR = 1 g/10 min, ρ = 917 kg/m³), and a metallocene-catalyzed HDPE (MFR = 1 g/10 min, ρ = 957 kg/m³)) have the same L/R and apparent shear rate dependence. The apparent shear rate for the onset of surface distortion is independent of the die length. For the same L/R ratio, at higher flow rates, the surface distortion amplitude increases while the frequency decreases. For longer dies, such an effect on the appearance of surface distortion occurs at lower apparent shear rates.

For linear PE and TFE-HFP copolymers, Tordella [67] observed that the surface roughness is independent of capillary diameter at constant L/R with $2R$ = 0.5–2.5 mm.

Vinogradov et al. [54] showed for PB and PI a flow curve with a wider transition zone at higher L/R. Uhland [85] reported for HDPE the same influence for $L/2R$ = 15

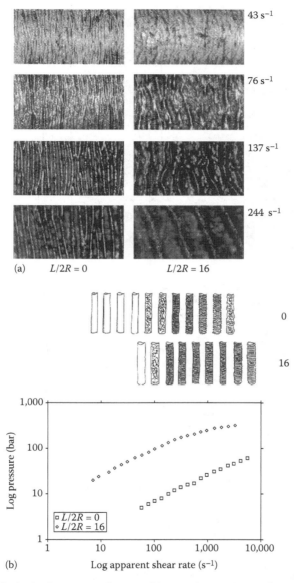

FIGURE 6.6 Optical microscopy of the capillary melt extrudates of a gamma-irradiated LLDPE (dose 5 kGy (kilo Gray), MFR = 1 g/10 min, ρ = 917 kg/m^3) after cooling from 190°C. At higher L/R (0 and 16 with $2R$ = 1.39 mm) and apparent shear rates, surface distortions have a higher amplitude and lower frequency (a). The associated flow curves and the schematic extrudate distortion representation indicate that the melt fracture onset shear rate is independent of L/R (b). (From Venet, C., Propriétés d'écoulement et défauts de surface de résins polyéthylènes, PhD thesis, Ecole des Mines de Paris (CEMEF), Sophia Antipolis, France, 1996.)

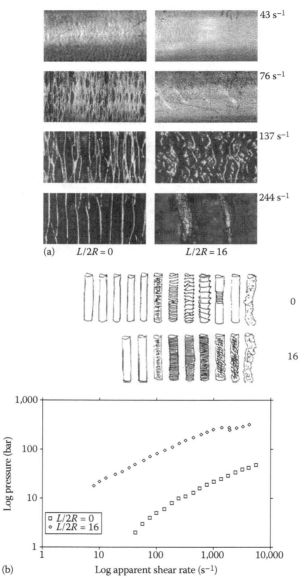

FIGURE 6.7 Optical microscopy of capillary melt extrudates of a metallocene-catalyzed HDPE (MFR = 1 g/10 min, ρ = 957 kg/m^3) after cooling from 190°C. At higher L/R (0 and 16 with $2R$ = 1.39 mm) and apparent shear rates, surface distortions have a higher amplitude and lower frequency (a). The associated flow curves and the schematic extrudate distortion representation indicate that the melt fracture onset shear rate is independent of L/R (b). (From Venet, C., Propriétés d'écoulement et défauts de surface de résins polyéthylènes, PhD thesis, Ecole des Mines de Paris (CEMEF), Sophia Antipolis, France, 1996.)

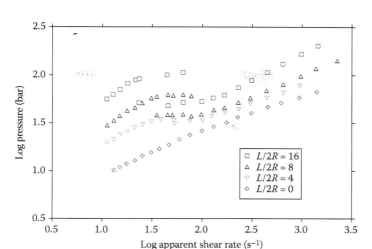

FIGURE 6.8 Flow curves for an HDPE homopolymer at 160°C for selected $L/2R$ ($2R = 1.3$ mm) between 0 and 16. (From Durand, V., Ecoulements et instabilité des polyéthylènes haute densité, PhD thesis, Ecole des Mines de Paris (CEMEF), Sophia Antipolis, France, 1993.)

and a decreasing R ($2R = 8.2$ to 2.09 mm) at 180°C. Drda and Wang [53,71] confirmed the results for a different set of HDPE. Durand [52] showed for HDPE homopolymers that the flow curve discontinuity disappeared when changing from an $L/2R$ of 16 to 0 with $2R = 1.39$ mm (Figure 6.8). The distinction between the continuous and discontinuous flow curves disappears although melt fracture persists (Figure 6.7). For the short dies, the pressure oscillation amplitude is reduced while the frequency increases (Figure 6.9).

For higher flow rates in the transition region between Branch I and II, the form of the pressure oscillation changes. One pressure oscillation period defining one hysteresis cycle shows a shorter time for the pressure to rise and a longer time for the pressure to fall (Figure 6.10). The pressure oscillation frequency also increases when the volume in the barrel decreases [16,52] (Table 6.2).

6.2.2 Die Entry and Exit Angle

For most commercially available polymers, typically a change in the die entry angle seems to have no effect on the onset and appearance of surface distortions [6,37,86–89]. For linear PE and TFE-HFP copolymers, Tordella [67] observed the surface roughness onset to be independent of the entry angle (varying from 180° to 20°). For LDPE, Tordella [67] observed that a smaller entry angle reduced the severity of the volume distortions. In agreement with Clegg [72] and Dennison [87], Den Otter [80] found no influence of entry-angle changes (from 180° to 60°) on the surface roughness onset or severity for HDPE. Sornberger et al. [68] reported no effect on surface distortion for the entry angle of slit dies (180° and 90°) for LLDPE.

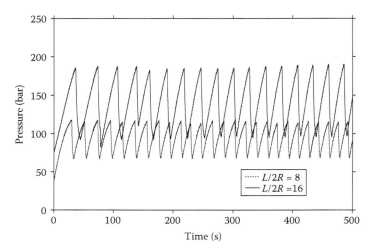

FIGURE 6.9 The pressure oscillations of an HDPE homopolymer at 160°C show higher amplitude but lower frequency when the $L/2R$ ($2R = 1.3$ mm) changes from 16 to 8. (From Durand, V., Ecoulements et instabilité des polyéthylènes haute densité, PhD thesis, Ecole des Mines de Paris (CEMEF), Sophia Antipolis, France, 1993.)

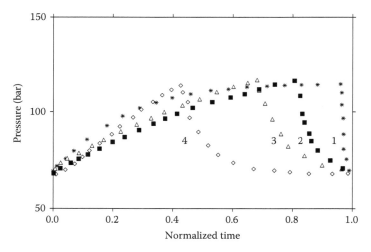

FIGURE 6.10 The form of a pressure oscillation for an HDPE homopolymer changes for higher flow rates (in order of 1 to 4) in the transition region between Branch I and II of the flow curve. (From Durand, V., Ecoulements et instabilité des polyéthylènes haute densité, PhD thesis, Ecole des Mines de Paris (CEMEF), Sophia Antipolis, France, 1993.)

Ramsteiner [79] indicated that the die entry shape has an influence on the volume distortions in PS and LDPE. For LDPE, Den Otter [80] found the frequency of volume distortions to be three times higher for conical entry (60°) versus flat entry (180°) dies. The conical entry diminished the distortion amplitude but did not reduce the melt fracture onset shear rate, in agreement with the findings of Bagley et al.

TABLE 6.2

Typical Influence of *L/R* on Surface Distortions for Most Polymers

Capillary Die		Melt Fracture			Flow Curve
L	*R*	Amplitude	Frequency	Onset shear rate	Breadth transition zone
↑	—	↑	↓	—	↑
—	↑			↑	↑

Note: Volume distortions disappear for long dies; —, unchanged.

[90]. Uhland [85] did not find any effect on the shape of the flow curve for HDPE when changing the entry angle from 180° to 90°, and 40° (*L/2R* = 15, 2*R* = 6.1 mm, at 180°C). Perez-Gonzalez et al. [3] indicated that for LLDPE, the amplitude of pressure oscillations decreased with reduced entry angle (nitrated steel + aluminum dies, *L/2R* = 15, 2*R* = 1 mm, 180°–30° in 30° steps). Combeaud et al. [82] studied PS volume distortions with slit-die designs with half-entry angles of 30°, 45°, and 90°, as well as a smooth convergent (trumpet shape). They find no effect on the flow curve, but a strong effect on the onset apparent shear rate of the distortions, especially going from smooth convergent via 30° to 45°. From 45° to 90°, the change is very limited. A factor-four increase is reported in critical shear rate for the smooth convergent entrance versus a flat entrance. Figure 6.11 shows the geometries and flow-induced birefringence patterns for steady flow at *L* = 20.75 mm, *T* = 180°C, and an apparent shear rate of 20 s^{-1}. At higher flow rates, instable flow-induced birefringence patterns are produced except for the trumpet-shaped die, although, in all these cases, a volume distorted extrudate is observed.

Earlier experiments on LDPE by Hürlimann and Knappe [91] indicated the natural formation of an entry vortex, presenting a kind of "self-assembled fluid" die entry geometry to optimize flow conditions. The mechanical "replacement" of the vortex

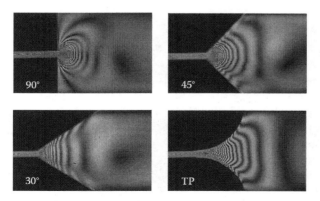

FIGURE 6.11 Flow-induced birefringence patterns in PS flowing through slit dies with different entry angle geometries (30°, 45°, 90° and trumpet shaped (TP)) in stable flow mode (*L* = 20.75 mm, $\dot{\gamma}$ = 20 s^{-1} at 180°C). (From Combeaud, C. et al., *J. Non-Newtonian Fluid Mech.*, 121(2–3), 175, 2004).

geometry may remove its natural development but does not prevent the development of a volume-distorted extrudate. The onset rate may change but not the critical extensional stress. Kamerkar and Edwards [92] experimented with straight-walled and semi-hyperbolically converging capillary dies for PP, LDPE, and HDPE. They reported on slip effects and hypothesized the formation of a highly oriented boundary layer but did not elaborate on the effects of die geometry on the extrudate shape.

Piau et al. [12] mentioned a possible influence of the die exit shape on the surface distortion onset point. For LLDPE at 200°C, Wang et al. [71] show an influence on the flow curve slope by changing the die exit angle from 180° to 60°. The 60° exit angle (for $L/2R = 15$, $2R = 1.04$ mm and 180° entry angle) gave in relation to the onset of surface distortions, a smaller flow curve slope change. Arda and Mackley [93] used slit dies with a flat exit and with rounded exits, with a curvature (radius) of 0.8 and 2.0 mm, at a die gap of 1.2 mm and die length of 8 mm. With increasing curvature, the polymer separation point from the die is less controlled (which may be a problem for practical extrusion when trying to control film dimension) and the birefringence stress pattern suggests lower exit stress levels. The critical die wall shear stress for the onset of surface distortions increased from 0.14 MPa for the sharp exit to 0.16 MPa at 0.8 mm radius and 0.17 MPa at 2.0 mm exit radius for a 1 MI LLDPE. Similar experimental results were found by Pol et al. [94]. They further support their findings by computational fluid dynamics calculations indicating overall lower elongation flow and the reduction of normal stresses when the die exit is tapered.

6.2.3 DIE CONSTRUCTION MATERIAL

Dies for processing polymers can be fabricated from different materials. The interaction between polymer and die wall is an important criterion to define the boundary condition for modeling melt fracture [95]. For LDPE, Benbow and Lamb [96] studied the influence of several capillary die construction materials on the onset of melt fracture. Small but significant changes in critical wall shear stress were found (Table 6.3).

For HDPE and TFE-HFP copolymers, Tordella [67] observed the onset of extrudate surface roughness to be independent of capillary construction material (polished and very rough stainless steel, glass, graphite, PTFE).

Ramamurthy [36] mentioned an influence of the die construction material both on the onset and the development of surface and volume distortion for LLDPE (Table 6.4). Especially with brass, the surface distortions could be made to disappear during extrusion-blown-film experiments [97].

Kalika and Denn [98] and Kurtz [99] described similar findings. However, Tordella [74,88] and many other researchers [38,87] do not seem to find any influence on the onset of melt fracture. Ramamurthy [100] explained that 30 min were necessary to establish the equilibrium of "chemical exchanges" at the polymer–die wall interface, during continuous extrusion, to observe an influence. In most capillary rheometers, such an experiment is not possible.

Ghanta et al. [101] repeated the Ramamurthy experiments with stainless steel and brass dies for an LLDPE. They observed with nitrogen-blanketed extrusion and

TABLE 6.3
Variation of Critical Stress for the Onset of Melt Fracture with Die Composition as Found by Benbow and Lamb for LDPE ($L/2R = 6.3$, $2R = 1$ mm, 150°C)

Die Composition	τ_c (MPa)
Brass	0.155
Nylon 66 + carbon black	0.155
Copper	0.150
Nylon + 30 fiberglass	0.145
Nickel silver	0.135
Mild steel	0.135
Phosphor bronze	0.120
Silver steel	0.920

Source: Benbow, J.J. and Lamb, P., New aspects of melt fracture, in SPE ANTEC, Los Angeles, CA, 3, 1963, 7–17.

TABLE 6.4
Influence of Die Material on Critical Wall Shear Stress for the Onset of Surface τ_c^*, and Volume Distortions τ_c as Found by Ramamurthy for a 1 g/10 min MFR LLDPE ($L/2R = 20$, $2R = 1$ mm, 220°C)

Die Material	τ_c^* (MPa)	τ_c (MPa)
Aluminum (6061-T6)	0.137	0.391
Beryllium copper	0.104	0.377
Carbon steel (SAE 4130)	0.144	0.435
Carbon steel (SAE 4340)	0.144	0.435
CDA 360[a]	0.172	0.413
CDA 464[b]	0.153	0.407
Bronze (Ampco 45)	0.146	0.434
Copper	0.132	0.415
Stainless steel (304)	0.152	0.441

Source: Ramamurthy A.V., J. Rheol., 30(2), 337, 1986.

[a] Copper Development Association designation: CDA 360 (Cu 61.5%, Zn 35.5%, Pb 3%).

[b] Copper Development Association designation: CDA 464 (Cu 60%, Zn 39.2%, Sn 0.8%).

carefully cleaned dies that brass dies eliminated the appearance of surface distortions and substantially enhanced the throughput. Furthermore, the flow curve discontinuity observed for stainless steel dies reduced significantly.

Halley and Mackay [102] observed for LLDPE a higher pressure for slit dies with the inserts of brass versus steel, aluminum, copper, and zinc. However, the appearance of the extrudate is not mentioned. Galt and Maxwell [103] reported experiments on time-averaged velocity profiles of 8 μm particles in slit dies using PTFE, glass, and steel walls. The observed velocity profile differences are small and not conclusive. Piau and coworkers [39,104] demonstrated for PDMS, PB, HDPE, and LLDPE that the use of PTFE dies delayed the onset and reduced the severity of the surface distortions.

Dao and Archer [62] used aluminum and stainless steel with PB and observed stick-slip transitions, which they related to flow-induced disentanglement of surface-adsorbed and bulk polymer molecules. Experiments with polished R-brass substrates led to much lower steady-state shear stresses without any stick-slip transition. Oxidation of the R-brass substrates at elevated temperatures dramatically increased steady-state shear stresses and the corresponding transient stresses showed a weak stick-slip process.

6.2.4 DIE SURFACE ROUGHNESS

Sandblasted surfaces, and those with tool marks transverse to the flow direction, appear to have little effect on the finish of the extrudate surface according to Tordella [74]. Benbow and Lamb [96] showed that no effects on the onset critical shear stress for volume distortion of LDPE can be observed for brass dies lapped with coarse, medium, and fine grades of carborundum. A very smooth surface created by pushing a ball bearing through an undersized hole gave an increase of 5% in the critical wall shear stress τ_c.

Wang and coworkers [71,105] compared the effects of threaded and smooth aluminum dies ($L/2R = 16$, $2R = 1.5$ mm, 60° entry angle) for HDPE and LLDPE. The stress for the onset of surface distortion remained the same. The stress for the onset of volume distortion was shifted to a higher value for the threaded die. Similar grooved dies have been used by Bergem [106], White et al. [107], Chauffoureaux et al. [108], and Knappe and Krumbock [109] to investigate possible slip phenomena in HDPE, rubber and PVC, respectively. They usually observed a higher shear stress for the grooved dies. For PP at 200°C, Kazatchkov et al. [17] showed that in comparison to smooth dies sandblasted slit dies give flow curves at higher wall shear stresses. For PDMS, El Kissi et al. [110] reported higher flow rates for the surface distortion onset when roughened steel die surfaces are used. Piau et al. [39] shifted for LLDPE at 190°C (steel die with 1 or 5 mm PTFE at the exit, $L/2R = 20$, $2R = 2$ mm), the appearance of surface distortions to higher flow rates by increasing the roughness of PTFE in the flow direction using sand paper. A perpendicular roughness showed the opposite effect. Arda and Mackley [93] studied the effect of surface roughness in a slit die for a 1 MFR LLDPE. The critical stress for the onset of surface distortions was found to increase from 0.16 to 0.19 MPa when the die insert surface roughness decreased from 1.5 to 0.05 μm.

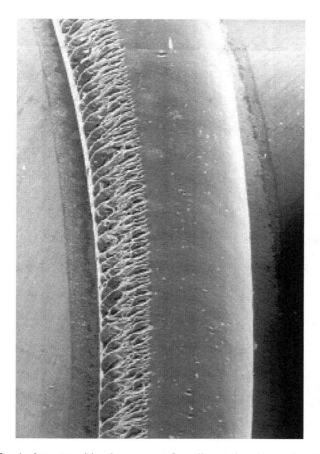

FIGURE 6.12 A sharp transition between surface-distorted and smooth extrudate for an ULDPE ($L/2R = 30$, $2R = 1\,\text{mm}$, 160°C, 180° entrance angle). At a higher flow rate a screw thread surface distortion covers the entire extrudate. Less severe cases are more common and typically referred to as die-lines.

Several authors [10,69,74] have observed that surface distortions can appear in a very narrow band along the circumference of the capillary extrudate (Figure 6.12). Typically, such localized melt fracture is associated with localized flow differences induced by either a die defect or uneven flow entering the die [10,74]. More often, the narrow bands become lines possibly showing the distinct surface distortion. Mostly, such lines do not show a surface distortion but are shape indentations induced by minute die defects or accumulated substances at the die exit. These lines are called die-lines as discussed in Chapter 4. Experiments by Piau et al. [39] with capillary dies composed of two halves, each with a different surface roughness, showed the same sharp extrudate distortion boundaries, as shown in Figure 6.12. Gulmus and Yilmazer [111] investigated the effect of surface roughness (R_a: 0.49–1.51 µm) for a PMMA particle suspension of average particle size of 121.20 µm in hydroxyl terminate polybutadiene. The wall slip velocity and the viscosity were found to be independent of the surface roughness and no effect of the surface construction material was identified. Glass and aluminum surfaces gave, respectively, high and low flow-slip velocities.

6.2.5 Die Surface Modifier

PTFE and elastomeric fluorocarbon die coatings are often used to totally suppress surface distortions [39,71,105,112]. Such coatings can also make pressure oscillations disappear in case polymers have a discontinuous flow curve [4,39,113]. The die surface coating is typically put in place by using sprays or by evaporating the solvent from a polymer solution. Alternatively, PTFE, elastomeric fluorocarbon, and other oligomers are added to the polymer. These materials migrate to the die wall during the processing operation. In such cases, a continuous processing is required to allow the additives to migrate to the die wall and adhere.

El Kissi et al. [110] showed for PDMS a significant delay in the onset of surface distortions when grafting a silica die with fluorinated trichlorosilane. Moynihan and Baird [114] and Piau et al. [39,104] demonstrated that coating a few millimeters of the die exit region with fluoroelastomers significantly delays melt fracture onset to higher flow rates for HDPE, LLDPE, and PB. Wang and Plucktaveesak [115] showed for an HDPE and capillary steel and aluminum dies with the bottom half coated with fluoroelastomer, that a "self-oscillation" in the extrudate could be made visible. In these constant-pressure experiments the extrudate showed alternating smooth and surface-distorted regions. For PP, Kazatchkov et al. [17] showed with a sliding-plate rheometer [116] with PTFE coated surfaces at 200°C, that the wall shear stress is significantly reduced compared to the "uncoated surface" experiments. Subsequent runs using the same coating showed a continued reduction in the wall shear stress. Establishing a steady-state situation seemed to take some time, similar to Ramamurthy's [36] findings for different die materials. Xing and coworkers [117,118] investigated the effects of fluoroelastomer-coated capillary dies ($L/2R = 20$, $2R = 1.75$ mm, 180° entry angle) for LLDPE, PS, chlorinate PE (CPE), PC, and an ionomer (sodium–ethylene-acrylic-acid copolymer). The melt viscosity was effectively lowered for LLDPE and CPE but less so for the other polymers. Kulikov and Hornung [119] proposed silicon rubber or fluorinated silicon rubber coatings to eliminate LLDPE surface distortions.

Migler et al. [83,120] showed how a coating created from the migration of masterbatch-containing hexafluoropropylene-vinylidene fluoride copolymer eliminated surface distortions at conventional rates. However, they also showed that the surface distortions emerged again at very high flow rates. Other surface-modifying additives that have been proposed as processing aids are boron nitride [121,122], nanoclay [123], and diatomite/PEG [124]. Yang and White [125] coated a capillary die with ethylene-butene copolymer (EB) and observed a significant flow-rate increase for polyamide 12 (PA12). Adding maleic anhydride–grafted ethylene-octene copolymer to the coating eliminated the effect. The effect indicated the importance of interfacial tension differences and the presence of interfacial slippage calculated to start at 0.045 MPa with an average velocity of 15 mm/s at 0.1 MPa for the EB–PA12 system.

6.2.6 TEMPERATURE

The Boltzmann time–temperature superposition principle is assumed to be valid for homogenous polymer melts [126]. The validity implies that the polymer relaxation times all have the same temperature dependence. The average polymer relaxation time λ_0 is proportional to the steady-state recoverable shear compliance J_e^0, which is inversely related to the applied stress in a constant stress experiment (see Chapter 3). As the melt fracture onset is observed at a critical shear stress, it is expected that the time–temperature superposition principle also applies for τ_c. As expected, it is observed that increasing the temperature delays the onset to higher flow rates and reduces the amplitude of surface distortions ($T\uparrow \propto \lambda_0\downarrow \propto J_e^0\downarrow \propto \tau_c\uparrow$) [4,6,10,72,87]. Ramamurthy [100], however, concluded that τ_c^* for surface (0.14 MPa) and τ_c for volume distortions (0.43 MPa) to be independent of temperature for LLDPE between 160°C and 260°C. The dependence of the critical true-wall shear rate for the onset of surface distortions on temperature and MFR was fit to the expression

$$\dot{\gamma}_s = 4.533 \times 10^5 (\text{MFR})^{1.182} \exp\left(-\frac{4360}{RT}\right), \tag{6.9}$$

where
  R is the universal gas constant
  T is the temperature in Kelvin

The Arrhenius temperature dependence is selected since the melt processing temperature for most polymers far exceeds their glass transition temperature [126].

In this spirit, Vlachopoulos and Alam modified Equations 6.3 through 6.6 to include a temperature dependency for τ_c. This resulted for PS, PP, and HDPE in

$$\frac{\tau_c}{T} = 1.717 \times 10^3 + \frac{2.67 \times 10^8}{M_w} \tag{6.10}$$

and for LDPE in

$$\frac{\tau_c}{T} = 1.317 \times 10^3 + \frac{1.005 \times 10^8}{M_w} \tag{6.11}$$

Wang and Drda [113] and Barone et al. [127] reasoned that the surface-distortion wavelength is proportional to the polymer relaxation time. The relaxation time is defined as the reciprocal of the shear rate where the loss (G'') and storage (G') moduli are equal [127]. For LLDPE, they presented some evidence that surface distortions depend on temperature in a similar fashion as the polymer relaxation time following a Williams–Landel–Ferry temperature dependence [126]. The surface distortion wavelength was observed to increase linearly with shear stress when normalized for a reference temperature of 200°C. Venet [10] had arrived at similar findings

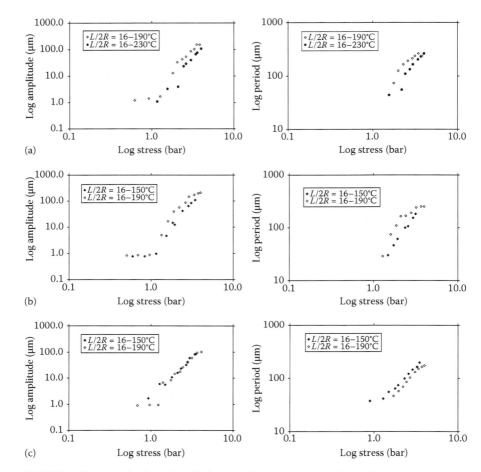

FIGURE 6.13 Amplitude and period of capillary extrudate surface distortions in relation to the wall shear stress normalized for temperature for LLDPE (a), 5 kGy gamma-irradiated LLDPE (b), metallocene HDPE (c) ($L/2R=16$, $2R=1.39$ mm, 180° entry angle, reference temperature 190°C). (From Venet, C., Propriétés d'écoulement et défauts de surface de résins polyéthylènes, PhD thesis, Ecole des Mines de Paris (CEMEF), Sophia Antipolis, France, 1996.)

for metallocene HDPE, LLDPE, and gamma-irradiated LLDPE. A master curve of surface distortion amplitude and period versus shear stress, normalized for temperature, gave a nearly complete superposition for each of the three resins (Figure 6.13).

Similar findings were reported by Beaufils et al. [41] for LLDPE. When the melt temperature rises from 145°C to 205°C, the surface roughness remained nearly the same except that it is shifted to higher shear rates.

Uhland [85] showed for HDPE an increasing critical shear stress and shear rate at higher melt temperature (170°C–200°C, $L/2R=15$, $2R=8.2$ mm, 180° entry angle). For the discontinuous flow curve, Branch I shifted to higher shear rates with increasing temperature in agreement with the time–temperature superposition principle.

Branch II, however, did not depend on temperature. Similar results were reported by Vinogradov et al. [52] for PB and PI. Temperature had no noticeable effect on the width of the transition region. For a 5 kGy dose, gamma-irradiated LLDPE, Venet [10] showed at 150°C a small flow curve discontinuity that reduced at 190°C to a change of the slope. In the absence of a flow curve discontinuity, Vinogradov et al. [128] reported for PP a 40% increase of surface distortion τ_c^* and a 25% increase of volume distortion τ_c when raising the temperature from 180°C to 240°C. Drda and Wang [86] showed with constant-pressure capillary rheometry for a low-MM commercial HDPE ($M_w = 130$ kg/mol), that the flow curve discontinuity totally disappeared when raising the temperature from 160°C to 200°C. Venet also observed, for a fixed capillary geometry, that the amplitude and period of the surface distortions shifted to higher shear rates when increasing the temperature. Consequently, the surface distortion severity (lower amplitude and period) was reduced at higher temperatures at a fixed flow rate (Figure 6.13).

In contrast, several authors [10,19,71,129] have reported surface distortions to disappear at low temperature close to the glass transition or melting point of polymers. For an LLDPE melt at 150°C, Venet [10] showed an onset of surface distortions at higher shear rates than at 190°C or 230°C (Figures 6.14 through 6.17). A flow curve discontinuity could only be measured at 190°C or 230°C.

Similar results were reported by Wang and Drda [29] for an LLDPE. For extruded high impact PS foams, Park et al. [130] demonstrated that for a 110°C die temperature, lowering the melt temperature from 150°C to 120°C removed the sharkskin surface distortions. In 1977, Cogswell [19,131] already indicated that local cooling of the die tip can reduce or delay the surface distortions to higher shear rates. A 20°C–50°C temperature difference seemed to be required (Table 6.5).

Cook et al. [129] reported similar results for butene- and hexene-based LLDPE copolymers. Independent of the melt temperature (185°C or 205°C), a smooth extrudate was obtained when the die temperature reached about 145°C. Ramsteiner [132] also performed experiments where the barrel and die temperature were controlled independently. For a 200°C PS melt, a die temperature increase from 170°C to 220°C

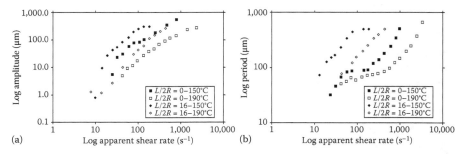

FIGURE 6.14 Amplitude (a) and period (b) as the function of apparent wall shear rate for an ULDPE. Higher temperature (190°C versus 150°C) and longer dies ($L/2R = 0$ versus 16, $2R = 1.39$ mm, 180° entry angle) shift the surface distortion severity to higher shear rates. (From Venet, C., Propriétés d'écoulement et défauts de surface de résins polyéthylènes, PhD thesis, Ecole des Mines de Paris (CEMEF), Sophia Antipolis, France, 1996.)

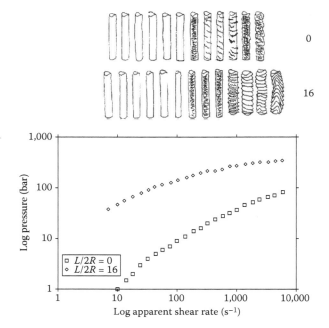

FIGURE 6.15 Flow curve of an LLDPE for $L/2R = 0$ or 16, with $2R = 1.39$ mm and at different temperatures. The extrudate appearance is schematically represented at each shear rate for 150°C. (From Venet, C., Propriétés d'écoulement et défauts de surface de résins polyéthylènes, PhD thesis, Ecole des Mines de Paris (CEMEF), Sophia Antipolis, France, 1996.)

shifts the onset of melt fracture to a higher flow rate and a lower pressure. For HDPE, the flow curve discontinuity appeared at a higher pressure and flow rate with increasing temperature (equal melt and die temperature). A 170°C melt temperature yielded a flow curve discontinuity at a lower pressure and flow rate when the die temperature was lowered from 200°C to 150°C. The most significant pressure and flow rate decrease happened for a die temperature of 150°C.

Barone et al. [127] extensively studied the effect of cooling the die exit wall. Two modified capillary rheometers were used, one equipped with a heat sink and one equipped with a water-cooled die. A range of temperature differences between melt and die exit were examined (aluminum die, $L/2R = 15$, $2R = 1.4$ mm, 60° entry angle) ($T_{melt} - T_{die}$: 220°C−205°C, 200°C−182°C, 180°C−166°C, 160°C−149°C, 150°C−140°C for the heat sink; 210°C−158°C, 200°C−159°C, 190°C−150°C, 180°C−141°C for the water-cooled die). The LLDPE surface distortion wavelength was observed to become longer compared to the uncooled die experiments.

Perez-Gonzalez et al. [133] also showed beneficial distortion-postponing effects for a mLLDPE at low T, which they related to flow-induced crystallization. Santamaria et al. [134] obtained similar results for a series of mPEs. They observed a region at low T where smooth extrudates are obtained, and related it to the absence of slip at the die exit. The temperature range is influenced by SCB level.

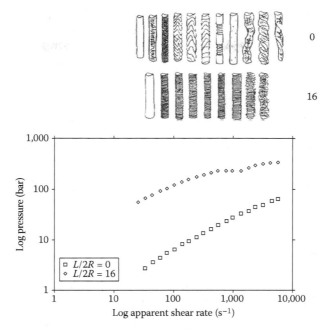

FIGURE 6.16 Flow curve of an LLDPE for $L/2R = 0$ or 16, with $2R = 1.39$ mm and at different temperatures. The extrudate appearance is schematically represented at each shear rate for 190°C. (From Venet, C., Propriétés d'écoulement et défauts de surface de résins polyéthylènes, PhD thesis, Ecole des Mines de Paris (CEMEF), Sophia Antipolis, France, 1996.)

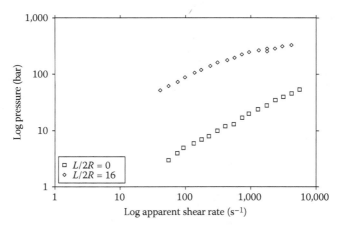

FIGURE 6.17 Flow curve of an LLDPE for $L/2R = 0$ or 16, with $2R = 1.39$ mm and at different temperatures. The extrudate appearance is schematically represented at each shear rate for 230°C. (From Venet, C., Propriétés d'écoulement et défauts de surface de résins polyéthylènes, PhD thesis, Ecole des Mines de Paris (CEMEF), Sophia Antipolis, France, 1996.)

TABLE 6.5

**Typical Die Tip and Melt Temperatures for Obtaining
Smooth Extrudates for Polymers Prone to Give
Surface Distortions as Reported by Cogswell**

Polymer	Melt Temperature (°C)	Die Tip Temperature (°C)
LDPE	170	130
HDPE	170	145
PVC	180	130
PMMA	200	150

Source: Cogswell, F.N., *J. Non-Newtonian Fluid Mech.*, 2(1), 37, 1977.
With permission.

Their work builds on similar findings by Kolnaar and Keller [135] and Pudjijanto and Denn [136].

Miller and Rothstein [137] used programmed die-exit cooling and heating to influence surface distortions via temperature gradients. In a subsequent paper [138], they used a rectangular slit die to impose temperature gradients across a flowing PE melt. It appeared that the amplitude and frequency of the surface distortion only depends on the wall temperature. Consequently, it is concluded that surface distortions are an extensional flow-induced interfacial (melt–die wall) effect near the die exit.

6.3   GENERAL OBSERVATION

Polymer architecture and polymer-processing conditions determine the observable melt fracture. Relatively small changes in polymer architecture and polymer-processing conditions may have a subtle impact on melt fracture that can be easily missed. Within the experimental limits of capillary rheometers, it is observed that polymer architecture changes may delay or advance the onset shear rate of melt fracture and change the extrudate distortion appearances significantly. The critical wall shear stress seems to be less sensitive to polymer architecture changes. Changes in the processing variables indicate that discontinuous and continuous flow curves are not uniquely related to polymer architecture. The experimental timescale in relation to the average molar relaxation time seems to be critical and suggests a scalability of the melt fracture phenomena. In that respect, surface distortions are particularly sensitive to any polymer architecture and polymer processing condition change. They are found to be initiated near or at the die exit region. Volume distortions are initiated at the die entrance region but seem less sensitive to these variables. For surface and volume distortion to occur, local extensional stresses either at the die entrance or the die exit are found to determine their onset. Accordingly, the polymer architectural response to such extensional stress fields will be detrimental for initiating polymer melt fracture. It implies that polymers of a high degree of polymerization, with a higher entanglement density, are more prone to melt fracture.

Changes in die construction material, the nature of the die surface, and the die temperature indicate that the boundary conditions play an important role in defining the appearance of melt fracture. They reflect the impact of each processing variable on the local extensional stress levels either at the die entry or at the die wall–polymer interface.

REFERENCES

1. Ball, P., Fluid dynamics: How coffee leaves its mark. *Nature*, **389(6653)**:788 (1997).
2. Yamane, H. and J. L. White, A comparative study of flow instabilities in extrusion, melt spinning, and tubular blown film extrusion of rheologically characterized high density, low density and linear low density polyethylene melts. *Nihon Reoloji Gakkaishi—J. Soc. Rheol. Jpn.*, **15**:131–140 (1987).
3. Perez-Gonzalez, J., L. Perez-Trejo, L. de Vargas, and O. Manero, Inlet instabilities in the capillary flow of polyethylene melts. *Rheol. Acta*, **36**:677–685 (1997).
4. Rudin, A., H. P. Schreiber, and D. Duchesne, Use of fluorocarbon elastomers as processing additives for polyolefins. *Polym. Plast. Technol. Eng.*, **29(3)**:199–234 (1990).
5. Petrie, C. J. S. and M. M. Denn, Instabilities in polymer processing. *AIChE J.*, **22(2)**:209–236 (1976).
6. Beaufils, P., Etude des défauts d'extrusion des polyéthylènes lineaires. Approche experimentale et modélisation des écoulements. PhD thesis, Ecole des Mines de Paris (CEMEF), Sophia Antipolis, France, 1989.
7. Weill, A., About the origin of sharkskin. *Rheol. Acta*, **19(5)**:623–632 (1980).
8. Howells, R. E. and J. J. Benbow, Flow defects in polymer melts. *Trans. Plast. Ind.*, **30**:240–253 (1962).
9. Karbashewski, E., A. Rudin, L. Kale, W. J. Tchir, and H. P. Schreiber, Effects of polymer structure on the onset of processing defects in LLDPE's. In *SPE ANTEC*, Montreal, Canada, 49, 1991, pp. 1378–1381.
10. Venet, C., Propriétés d'écoulement et défauts de surface de résins polyéthylènes. PhD thesis, Ecole des Mines de Paris (CEMEF), Sophia Antipolis, France, 1996.
11. Baik, J. J. and C. Tsoganakis, A study of extrudate distortion in controlled rheology polypropylene. *Polym. Eng. Sci.*, **38(2)**:274–281 (1998).
12. Piau, J. M., N. El Kissi, and B. Tremblay, Influence of upstream instabilities and wall slip on melt fracture and sharkskin phenomena during silicones extrusion through orifice dies. *J. Non-Newtonian Fluid Mech.*, **34(2)**:145–180 (1990).
13. Piau, J. M. and N. El Kissi, The different capillary flow regimes of entangled polydimethylsiloxane polymers: Macroscopic slip at the wall, hysteresis and cork flow. *J. Non-Newtonian Fluid Mech.*, **37**:55–94 (1990).
14. Spencer, R. S. and R. E. Dillon, The viscous flow of molten polystyrene-II. *J. Colloid Sci.*, 241–255 (1949).
15. Rosenbaum, E. E., S. G. Hatzikiriakos, and C. W. Stewart, The melt fracture behavior of Teflon resins in capillary extrusion. In *SPE ANTEC*, Boston, MA, 53, 1995, pp. 1111–1115.
16. Rosenbaum, E. E., S. G. Hatzikiriakos, and C. W. Stewart, Flow implications in the processing of tetrafluoroethylene/hexafluoropropylene copolymers. *Int. Polym. Process.*, **10(3)**:204–212 (1995).
17. Kazatchkov, I. B., S. G. Hatzikiriakos, and C. W. Stewart, Extrudate distortion in the capillary/slit extrusion of a molten polypropylene. *Polym. Eng. Sci.*, **35(23)**:1864–1871 (1995).
18. Hatzikiriakos, S. G. and J. M. Dealy, Role of slip and fracture in the oscillating flow of HDPE in a capillary. *J. Rheol.*, **36(5)**:845–884 (1992).

19. Cogswell, F. N., Stretching flow instabilities at the exits of extrusion dies. *J. Non-Newtonian Fluid Mech.*, **2(1)**:37–47 (1977).
20. Blyler, L. L. J. and A. C. J. Hart, Capillary flow instability of ethylene polymer melts. *Polym. Eng. Sci.*, **10(4)**:193–203 (1970).
21. Vinogradov, G. V., A. Y. Malkin, Y. G. Yanovskii, E. K. Borisenkova, B. V. Yarlykov, and G. V. Berezhnaya, Viscoelastic properties and flow of narrow distribution polybutadienes and polyisoprenes. *J. Polym. Sci., A2* **10**:1061–1084 (1972).
22. Vlachopoulos, J. and M. Alam, Critical stress and recoverable shear for polymer melt fracture. *Polym. Eng. Sci.*, **12(3)**:184–192 (1972).
23. Graessley, W. W., Viscosity of entangling polydisperse polymers. *J. Chem. Phys.*, **47(6)**:1942–1953 (1967).
24. Lin, Y. H., Explanation for slip-stick melt fracture in terms of molar dynamics in polymer melts. *J. Rheol.*, **29(6)**:605–637 (1985).
25. Lin, Y. H., Unified molar theories of linear and non-linear viscoelasticity of flexible linear polymers-explaining the 3.4 power law of the zero-shear viscosity and the slip-stick melt fracture phenomenon. *J. Non-Newtonian Fluid Mech.*, **23**:163–187 (1987).
26. Doi, M. and S. F. Edwards, Dynamics of rod-like macromolecules in concentrated solution. Part 1. *J. Chem. Soc. Faraday Trans. II*, **74**:1789–1802 (1978).
27. Doi, M. and S. F. Edwards, Dynamics of rod-like macromolecules in concentrated solution. Part 2. *J. Chem Soc. Faraday Trans. II*, **74**:1802–1818 (1978).
28. Doi, M. and S. F. Edwards, Dynamics of rod-like macromolecules in concentrated solution. Part 3. *J. Chem. Soc. Faraday Trans. II*, **74**:1818 (1978).
29. Wang, S. Q. and P. A. Drda, Stick-slip transition in capillary flow of polyethylene. 2. Molar weight dependence and low temperature anomaly. *Macromolecules*, **29**:4115–4119 (1996).
30. Wang, S. Q., P. A. Drda, and A. Baugher, Molar mechanisms for polymer extrusion instabilities: Interfacial origins. In *SPE ANTEC*, Toronto, Canada, 55, 1997, p. 1067.
31. Yang, X., H. Ishida, and S. Q. Wang, Wall slip and absence of interfacial flow instabilities in capillary flow of various polymer melts. *J. Rheol.*, **42(1)**:63–80 (1998).
32. Allal, A., A. Lavernhe, B. Vergnes, and G. Marin, Relationship between molecular structure and sharkskin defect for linear polymers. *J. Non-Newtonian Fluid Mech.*, **134(1–3)**:127–135 (2006).
33. Griffith, A. A., The phenomena of rupture and flow in solid. *Philos. Trans. R. Soc.*, **221**:163–198 (1921).
34. Vlachopoulos, J., Die swell and melt fracture. Effects of molar weight distribution. *Rheol. Acta*, **13(2)**:223–227 (1974).
35. Baik, J. J. and C. Tzoganakis, Melt fracture of controlled-rheology polypropylenes. In *SPE ANTEC*, Indianapolis, IN, 54(Vol. 1), 1996, pp. 1145–1149.
36. Ramamurthy, A. V., LLDPE rheology and blown film fabrication. *Adv. Polym. Technol.*, **6**:489–499 (1986).
37. Kurtz, S. J., Die geometry solutions to sharkskin melt fracture. In *Theoretical and Applied Rheology* (B. Mena, A. Garcia-Rejon, and C. Rangel Nafaile, Eds.). UNAM Press, Mexico City, Mexico, 1984, pp. 399–407.
38. Lim, F. J. and W. R. Schowalter, Wall-slip of narrow molar weight distribution polybutadienes. *J. Rheol.*, **33(8)**:1359–1382 (1989).
39. Piau, J.-M., N. El Kissi, F. Toussaint, and A. Mezghani, Distortions of polymer melt extrudates and their elimination using slippery surfaces. *Rheol. Acta*, **34**:40–57 (1995).
40. Fetters, L. J., D. J. Lohse, D. Richter, T. A. Witten, and A. Zirkel, Connection between polymer molar weight, density, chain dimensions, and melt viscoelastic properties. *Macromolecules*, **27**:4639–4647 (1994).
41. Beaufils, P., B. Vergnes, and J. F. Agassant, Characterization of the sharkskin defect and its development with the flow conditions. *Int. Polym. Process.*, **4(2)**:78–84 (1989).

42. Kurtz, S. J., Visualization of exit fracture in the sharkskin process. In *Polymer Processing Society, PPS10*, Akron, OH, 1994, pp. 8–9.

43. Ajji, A., S. Varennes, H. P. Schreiber, and D. Duchesne, Flow defects in linear low density polyethylene processing: Instrumental detection and molar weight dependence. *Polym. Eng. Sci.*, **33(23)**:1524–1531 (1993).

44. El Kissi, N. and J. M. Piau, Flow of entangled polydimethylsiloxanes through capillary dies: Characterisation and modelisation of wall slip phenomena. In *Third European Rheology Conference and Golden Jubilee of the British Society of Rheology* (D. R. Oliver, Ed.). Elsevier Applied Science, London, U.K., 1990, pp. 144–146.

45. Lai, S. Y. and G. W. Knight, Unique rheology and processing aspects of ULDPE. In *SPE ANTEC*, Detroit, MI, 50, 1992, pp. 2084–2087.

46. Dealy, J. M., *Rheometers for Molten Plastics*. Van Nostrand Reinhold, New York, 1982.

47. Goyal, S. K., I. B. Kazatchkov, N. Bohnet, and S. G. Hatzikiriakos, Influence of molar weight distribution on the rheological behavior of LLDPE resins. In *SPE ANTEC*, Toronto, Canada, 55, 1997, pp. 1076–1080.

48. Dickie, B. D. and R. J. Koopmans, Long-chain branching determination in irradiated linear low-density polyethylene. *J. Polym. Sci. C: Polym. Lett.*, **28(6)**:193–198 (1990).

49. Lai, S.-Y., J. R. Wilson, G. W. Knight, J. C. Stevens, and P.-W. S. Chum, Elastic substantially linear olefin polymers. U.S. Patent 6849704. The Dow Chemical Company, Patent, 2003.

50. Lai, S.-Y., J. R. Wilson, G. W. Knight, and P.-W. S. Chum, Elastic substantially linear olefin polymers. U.S. Patent 5272236. The Dow Chemical Company, 1991.

51. Lai, S.-Y., J. R. Wilson, G. W. Knight, and J. C. Stevens, Elastic substantially linear olefin polymers. U.S. Patent 5665800. The Dow Chemical Company, Patent, 1996.

52. Durand, V., Ecoulements et instabilité des polyéthylènes haute densité. PhD thesis, Ecole des Mines de Paris (CEMEF), Sophia Antipolis, France, 1993.

53. Wang, S. Q. and P. A. Drda, Superfluid-like stick-slip transition in capillary flow of linear polyethylene melts. 1. General features. *Macromolecules*, **29**:2627–2632 (1996).

54. Vinogradov, G. V., V. P. Protasov, and V. E. Dreval, The rheological behavior of flexible chain polymers in the region of high shear rates and stresses, the critical process of spurting, and supercritical conditions of their movement at T>Tg. *Rheol. Acta*, **23**:46–61 (1984).

55. Fernandez, M., J. F. Vega, A. Santamaria, A Munoz-Escalona, and P. Lafuente, The effect of chain architecture on "sharkskin" of metallocene polyethylenes. *Macromol. Rapid Commun.*, **21**:973–978 (2000).

56. Vega, J. F., M. Fernandez, A. Santamaria, A. Munoz-Escalona, and P. Lafuente, Rheological criteria to characterize metallocene catalyzed polyethylenes. *Macromol. Chem. Phys.*, **200**:2257–2268 (1999).

57. Kazatchkov, I. B., N. Bohnet, S. K. Goyal, and S. G. Hatzikiriakos, Influence of molecular structure on the rheological and processing behavior of polyethylene resins. *Polym. Eng. Sci.*, **39(4)**:804–815 (1999).

58. Wang, W. J., S. Kharchenko, K. Migler, and S. Zhu, Triple-detector GPC characterization and processing behavior of long-chain-branched polyethylene prepared by solution polymerization with constrained geometry catalyst. *Polymer*, **45**:6495–6505 (2004).

59. Perez, R., E. Rojo, M. Fernandez, V. Leal, P. Lafuente, and A. Santamaría, Basic and applied rheology of m-LLDPE/LDPE blends: Miscibility and processing features. *Polymer*, **46**: 8045–8053 (2005).

60. Delgadillo-Velazquez, O. and S. G. Hatzikiriakos, Processability of LLDPE/LDPE blends: Capillary extrusion studies. *Polym. Eng. Sci.*, **47(9)**:1317–1326 (2007).

61. Den Doelder, J., R. Koopmans, M. Dees, and M. Mangnus, Pressure oscillations and periodic extrudate distortions of long-chain branched polyolefins. *J. Rheol.*, **49(1)**:113–126 (2005).

62. Dao, T.T. and L.A. Archer, Stick-slip dynamics of entangled polymer liquids. *Langmuir*, **18**:2616–2624 (2002).

63. Ferri, D. and M. Canetti, Spurt and melt flow distortions of linear styrene-isoprene-styrene triblock copolymers. *J. Rheol.*, **50(5)**:611–624 (2006).

64. Fujiyama, M. and H. Inata, Melt fracture behavior of polypropylene-type resins with narrow molecular weight distribution. I. Temperature dependence. *J. Appl. Polym. Sci.*, **84**:2111–2119 (2002).

65. Fujiyama, M. and H. Inata, Melt fracture behavior of polypropylene-type resins with narrow molecular weight distribution. II. Suppression of sharkskin by addition of adhesive resins. *J. Appl. Polym. Sci.*, **84**:2120–2127 (2002).

66. Naguib, H. E. and C. B. Park, A study on the onset surface melt fracture of polypropylene materials with foaming additives. *J. Appl. Polym. Sci.*, **109**:3571–3577 (2008).

67. Tordella, J. P., Unstable flow of molten polymers: A second site of melt fracture. *J. Appl. Polym. Sci.*, **7**:215–229 (1963).

68. Sornberger, G., J. C. Quantin, R. Fajolle, B. Vergnes, and J. F. Agassant, Experimental study of the sharkskin defect in linear low density polyethylene. *J. Non-Newtonian Fluid Mech.*, **23**:123–135 (1987).

69. Moynihan, R. H., D. G. Baird, and R. Ramanathan, Additional observations on the surface melt fracture behavior of linear low-density polyethylene. *J. Non-Newtonian Fluid Mech.*, **36**:255–263 (1990).

70. Constantin, D., Linear-low-density polyethylene melt rheology: Extensibility and extrusion defects. *Polym. Eng. Sci.*, **24(4)**:268–274 (1984).

71. Wang, S. Q., P. A. Drda, and Y. W. Inn, Exploring molar origins of sharkskin, partial slip, and slope change in flow curves of linear low density polyethylene. *J. Rheol.*, **40(5)**:875–898 (1996).

72. Clegg, P. L., The flow of molten polymer and their effect on fabrication. *Br. Plast.*, **30**:535–537 (1957).

73. Herranen, M. and A. Savolainen, Correlation between melt fracture and ultrasonic velocity. *Rheol. Acta*, **23(4)**:461–464 (1984).

74. Tordella, J. P., Melt fracture—Extrudate roughness in plastic extrusion. *SPE J.*, **February**:36–40 (1956).

75. Bagley, E. B., S. H. Storey, and D. C. West, Post extrusion swelling of polyethylene. *J. Appl. Polym. Sci.*, **7**:1661–1672 (1963).

76. Kendall, V. G., The effect of fabrication technique on the properties of films and bottles made from polyethylene. *Trans. Plast. Inst.*, **June**:49–59 (1963).

77. Fujiki, T., M. Uemura, and Y. Kosaka, Flow properties of molten ethylene-vinyl acetate copolymer and melt fracture. *J. Appl. Polym. Sci.*, **12(2)**:267–279 (1968).

78. Ballenger, T. F., I. J. Chen, J. W. Crowder, G. E. Hagler, D. C. Bogue, and J. L. White, Polymer melt flow instabilities in extrusion: Investigation of the mechanisms and material and geometric variables. *Trans. Soc. Rheol.*, **15(2)**:195 (1971).

79. Ramsteiner, F., Effect of die geometry on flow resistance, extrudate dilation, and melt fracture of plastics melts. *Kunststoffe*, **62(11)**:766–772 (1972).

80. Den Otter, J. L., Melt fracture. *Rheol. Acta*, **10(2)**:200–207 (1971).

81. Metzner, A. B., E. L. Carley, and I. K. Park, Polymer melts: A study of steady-state flow, extrudate irregularities and normal stresses. *Mod. Plast.*, **37(July)**:133 (1960).

82. Combeaud, C., Y. Demay, and B. Vergnes, Experimental study of the volume defects in polystyrene extrusion. *J. Non-Newtonian Fluid Mech.*, **121(2–3)**:175–185 (2004).

83. Migler, K.B., Y. Son, F. Qiao, and K. Flynn, Extensional deformation, cohesive failure, and boundary conditions during sharkskin melt fracture. *J. Rheol.*, **46(2)**:383–400 (2002).

84. Den Doelder, C. F. J., Design and implementation of polymer melt fracture models. PhD thesis, TU Eindhoven, Eindhoven, the Netherlands, 1999. ISBN 90-386-0701-6.

85. Uhland, E., Das anomale Fliessverhalten von Polyathylene hoher Dichte. *Rheol. Acta*, **18**:1–24 (1979).

86. Drda, P. P. and S. Q. Wang, Stick-slip transition at polymer mel/solid interfaces. *Phys. Rev. Lett.*, **75(14)**:2698–2701 (1995).

87. Dennison, M. T., Flow instability in polymer melts: A review. *Plast. Inst. Trans. J.*, **35(120)**:803–808 (1967).

88. Tordella, J. P., Unstable flow of molten polymers. In *Rheology* (F. R. Erich, Ed.). Academic Press, New York, 1969, pp. 57–92.

89. Wang, X., B. Clement, P. J. Carreau, and P. G. Lafleur, HDPE and LLDPE extrudate roughness study by screening design. In *SPE ANTEC*, Detroit, MI, 50, 1992, pp. 1256–1258.

90. Bagley, E. B., I. M. Cabott, and D. C. West, Discontinuity in the flow curve of polyethylene. *J. Appl. Phys.*, **29**:109–110 (1958).

91. Hürlimann, H. P. and W. Knappe, Relation between the extensional stress of polymer melts in the die inlet and in melt fracture. *Rheol. Acta*, **11(3–4)**:292–301 (1972).

92. Kamerkar, P. A. and B. J. Edwards, Experimental study of slip flow in capillaries and semihyperbolically convergent dies. *Polym. Eng. Sci.*, **47(2)**:159–167 (2007).

93. Arda, D. R. and M. R. Mackley, The effect of die exit curvature, die surface roughness and a fluoropolymer additive on sharkskin extrusion instabilities in polyethylene processing. *J. Non-Newtonian Fluid Mech.*, **126(1)**:47–61 (2005).

94. Pol, H. V., Y. M. Joshi, P. S. Tapadia, A. K. Lele, and R. A. Mashelkar, A geometrical solution to the sharkskin instability. *Ind. Eng. Chem. Res.*, **46(10)**: 3048–3056 (2007).

95. Brochard, F. and P. G. de Gennes, Shear dependent slippage at polymer/solid interface. *Langmuir*, **8**:3033–3037 (1992).

96. Benbow, J. J. and P. Lamb, New aspects of melt fracture. In *SPE ANTEC*, Los Angeles, CA, 3, 1963, pp. 7–17.

97. Ramamurthy, A. V., Eliminating surface melt fracture when extruding ethylene polymers. U.S. Patent 4554120. Union Carbide Corp., Danbury, CT, 1985.

98. Kalika, D. S. and M. M. Denn, Wall slip and extrudate distortion in linear low-density polyethylene. *J. Rheol.*, **31(8)**:815–834 (1987).

99. Kurtz, S. J., The dynamics of sharkskin melt fracture: Effect of die geometry. In *Theoretical Applied Rheology, Proceedings of 11th International Congress on Rheology* (P. Moldenaers and R. Keunings, Eds.). Brussels, Belgium. Elsevier, Amsterdam, the Netherlands, 1992, pp. 377–379.

100. Ramamurthy, A. V., Wall slip in viscous fluids and influence of materials of construction. *J. Rheol.*, **30(2)**:337–357 (1986).

101. Ghanta, V. G., B. L. Riise, and M. M. Denn, Disappearance of extrusion instabilities in brass capillary dies. *J. Rheol.*, **43(2)**:435–442 (1999).

102. Halley, P. J. and M. E. Mackay, The effect of metals on the processing of LLDPE through a slit die. *J. Rheol.*, **38(1)**:41–51 (1993).

103. Galt, J. and B. Maxwell, Velocity profiles for polyethylene melts. *Mod. Plast.*, **42(4)**:115–189 (December 1964).

104. Piau, J. M., N. El Kissi, and A. Mezghani, Slip flow of polybutadiene through fluorinated dies. *J. Non-Newtonian Fluid Mech.*, **59**:11–30 (1995).

105. Wang, S. Q. and P. A. Drda, Stick-slip transition in capillary flow of linear polyethylene: 3. Surface conditions. *Rheol. Acta*, **36**:128–134 (1997).

106. Bergem, N., Visualization studies of polymer melt flow anomalies in extrusion. In *Seventh International Congress on Rheology* (C. Klason, Ed.). Goteborg, Sweden, 1976, pp. 50–54.

107. White, J. L., M. H. Han, N. Nakajima, and R. Brzoskowski, The influence of materials of construction on biconical rotor and capillary measurements of shear viscosity of rubber and its compounds and considerations of slippage. *J. Rheol.*, **35(1)**:167–189 (1991).

108. Chauffoureaux, J. C., C. Dehennau, and J. Van Rijckevorsel, Flow and thermal stability of rigid PVC. *J. Rheol.*, **23(1)**:1–24 (1979).

109. Knappe, W. and E. Krumbock, Evaluation of slip flow of PVC-compounds by capillary rheometry. In *Ninth International Congress on Rheology* (B. Mena, A. Garcia-Rejon, and C. Rangel Nafaile, Eds.). Acapulco, Mexico, 1984, p. 417.

110. El Kissi, N., L. Leger, J. M. Piau, and A. Mezghani, Effect of surface properties on polymer melt slip and extrusion defects. *J. Non-Newtonian Fluid Mech.*, **52(2)**:249–261 (1994).

111. Gulmus, S. A. and U. Yilmazer, Effect of the surface roughness and construction material on wall slip in the flow of concentrated suspensions. *J. Appl. Polym. Sci.*, **103(5)**:3341–3347 (2007).

112. Hatzikiriakos, S. G., P. Hong, W. Ho, and C. W. Stewart, The effect of Teflon coatings in polyethylene capillary extrusion. *J. Appl. Polym. Sci.*, **55(4)**:595–603 (1995).

113. Wang, S. Q. and P. Drda, Molar instabilities in capillary flow of polymer melts. Interfacial stick-slip transition, wall slip, and extrudate distortion. *Macromol. Chem. Phys.*, **198(3)**:673–701 (1997).

114. Moynihan, R. H. and D. H. Baird, The origin of sharkskin in the extrusion of LLDPE. In *Polymer Processing Society, PPS7*, Hamilton, Canada, 1991, pp. 242–243.

115. Wang, S. Q. and N. Plucktaveesak, Self-oscillations in capillary flow of entangled polymers. *J. Rheol.*, **43(2)**:453–460 (1999).

116. Giacomin, A. J., T. Samurkas, and J. M. Dealy, Novel sliding plate rheometer for molten plastics. *Polym. Eng. Sci.*, **29(8)**:499–504 (1989).

117. Xing, K. C., W. Wang, and H. P. Schreiber, Extrusion of non-olefinic polymers with fluoroelastomer processing aid. In *SPE ANTEC*, Toronto, Canada, 55, 1997, pp. 53–58.

118. Xing, K. C. and H. P. Schreiber, Fluoropolymers and their effect on processing linear low density polyethylene. *Polym. Eng. Sci.*, **36(3)**:387–393 (1996).

119. Kulikov, O. and K. Hornung, A simple way to suppress surface defects in the processing of polyethylene. *J. Non-Newtonian Fluid Mech.*, **124(1–3)**:103–114 (2004).

120. Migler, K. B., C. Lavallée, M. P. Dillon, S. S. Woods, and C. L. Gettinger, Visualizing the elimination of sharkskin through fluoropolymer additives: Coating and polymer–polymer slippage. *J. Rheol.*, **45(2)**:565–581 (2001).

121. Kazatchkov, I. B., F. Yip, and S. G. Hatzikiriakos, The effect of boron nitride on the rheology and processing of polyolefins. *Rheol. Acta*, **39**:583–594 (2000).

122. Lee, S. M., G. J. Nam, and J. W. Lee, The effect of boron nitride particles and hot-pressed boron nitride die on the capillary melt flow processing of polyethylene. *Adv. Polym. Technol.*, **22(4)**:343–354 (2003).

123. Hatzikiriakos, S. G., N. Rathod, and E. B. Muliawan, The effect of nanoclays on the processability of polyolefins. *Polym. Eng. Sci.*, **45(8)**:1098–1107 (2005).

124. Liu, X. and H. Li, Effect of diatomite/polyethylene glycol binary processing aid on the melt fracture and the rheology of polyethylenes. *Polym. Eng. Sci.*, **45(7)**:898–903 (2005).

125. Yang, J. and J. L. White, Interface slippage study between polyamide 12 and ethylene butene copolymer melt in capillary extrusion. *J. Rheol.*, **53(5)**:1121–1132 (2009).

126. Ferry, J. D., *Viscoelastic Properties of Polymers*. John Wiley & Sons, Inc., New York, 1980.

127. Barone, J. R., N. Plucktaveesak, and S. Q. Wang, Interfacial molar instability mechanism for sharkskin phenomenon in capillary extrusion of linear polyethylenes. *J. Rheol.*, **42(4)**:813–832 (1998).

128. Vinogradov, G. V., M. L. Friedman, B. V. Yarlykov, and A. Y. Malkin, Unsteady flow of polymer melts: Polypropylene. *Rheol. Acta*, **9(3)**:323–329 (1970).
129. Cook, D. G., R. Cooke, and A. Rudin, Use of chilled die lips to improve production rates in extrusion of polyethylene. *Int. Polym. Process.*, **4(2)**:73–77 (1989).
130. Park, C. B., A. H. Behravesh, and R. D. Venter, Low density microcellular foam processing in extrusion using CO_2. *Polym. Eng. Sci.*, **38(11)**:1812–1823 (1998).
131. Cogswell, F. N., *Sharkskin*. U.S. Patent 3920782. Imperial Chemical Industries Ltd., London, U.K., 1975.
132. Ramsteiner, F., Melt fracture behavior of plastics. *Rheol. Acta*, **16(6)**:650–651 (1977).
133. Perez-Gonzalez, J., L. de Vargas, V. Pavlinek, B. Hausnerova, and P. Saha, Temperature-dependent instabilities in the capillary flow of a metallocene linear low-density polyethylene melt. *J. Rheol.*, **44(3)**:441–451 (2000).
134. Santamaria, A., M. Fernandez, E. Sanz, P. Lafuente, and A. Munoz-Escalona, Postponing sharkskin of metallocene polyethylenes at low temperatures: The effect of molecular parameters. *Polymer*, **44**:2473–2480 (2003).
135. Kolnaar, J. W. H. and A. Keller, A singularity in the melt flow of polyethylene with wider implications for polymer melt flow rheology. *J. Non-Newtonian Fluid Mech.*, **67**:213–240 (1996).
136. Pudjijanto, S. and M. M. Denn, A stable "island" in the slip-stick region of linear low-density polyethylene. *J. Rheol.*, **38(6)**:1735–1744 (1994).
137. Miller, E. and J. P. Rothstein, Control of the sharkskin instability in the extrusion of polymer melts using induces temperature gradients. *Rheol. Acta*, **44**:160–173 (2004).
138. Miller, E., S. J. Lee, and J. P. Rothstein, The effect of temperature gradients on sharkskin surface instability in polymer extrusion through a slit die. *Rheol. Acta*, **45(6)**:943–950 (2006).

7 Understanding Melt Fracture

The overriding priority in science is prediction [1]. Understanding melt fracture is driven by the desire to predict. Prediction is possible when a logical relation is found between the polymer architecture and its flow behavior. The relation preferably takes the form of a mathematical equation, that is, a model that allows quantitative predictions about polymer melt fracture. In particular, research wants to quantify why and when melt fracture occurs. The premise of understanding why is to achieve control over when. The last 50+ years of research on melt fracture have seen an empirical and a theoretical approach to meet this challenge. The empirical approach resulted in a significant amount of literature describing melt fracture with different levels of accuracy and precision. Subsequently, many variables are identified that affect the appearance of melt fracture. The molecular variables are considered tools for designing new polymers that may either reduce melt fracture or shift its appearance to higher flow rates. Alternatively, the processing variables indicate possible routes for the development of improved flow channels, dies, and extruder operation. Simultaneously, and often inspired by experimental observation, theories have been forwarded that aim to explain certain aspects of melt fracture. The macroscopic phenomenology often suggests a continuum mechanical approach. The principles of the conservation of mass and momentum are used in combination with a constitutive equation that links the dynamic quantity stress with the kinematic quantity strain through a parameter or material function (e.g., viscosity) representing the characteristic response of the polymer. Alternative theoretical approaches and experimentation consider the interaction between polymer and flow channel at the interface on a microscopic scale [2–12]. Here, the incentive is the elucidation of the slip controversy, in particular, in relation to the observation of surface distortions. Chapters 8 and 9 will address these issues in more detail.

For understanding polymer melt fracture, the significant body of literature needs synthesis. The multifaceted but often partial knowledge should lead to a coherent insight on how melt fracture comes about. It should possibly provide the basis for relating the detailed polymer composition to a material function and associating this to the dynamics of stress and kinematics of strain. In the following, the flow curve shall refer to the experimentally obtained (apparent wall) shear stress–shear rate data, and the shear stress–shear rate curve to the curve as captured by a constitutive equation.

7.1 MELT FRACTURE MECHANISMS

7.1.1 REYNOLDS TURBULENCE

The very high viscosity (>10³ Pa s) of most commercially available polymer melts quickly makes it clear that melt fracture is something different than the Reynolds turbulence observed for liquids and gases [13,14]. Many original thoughts were formulated subsequently.

7.1.2 THERMAL CATASTROPHE

At one time, viscous heating and thermal effects were considered as a possible cause for melt fracture and, in particular, for the discontinuity in the flow curve observed for linear PE. The idea was that enough heat could be dissipated during viscous flow to significantly reduce the viscosity, thermally decompose, or chemically modify the polymer melt in some sort of form. Lupton and Regester [15] convincingly showed that viscous heating is limited to a few centigrade and cannot account for the observed melt fracture. Their calculations are based on obtaining the maximum temperature rise corresponding to a given pressure drop based on an adiabatic process. Thus, the notion of some sort of a thermal catastrophe can be refuted [14]. Similarly, the heat is insufficient for thermal decomposition or chemical modification and can be dismissed for most thermoplastics.

7.1.3 STRESS-INDUCED FRACTIONATION

Chen and Joseph [16] revisited an idea proposed by Busse in 1964 [17,18] suggesting a segregation due to high stresses near the capillary wall of low- and high-MM components. The low-MM component would accumulate at the die wall during flow and allow short wave hydrodynamic instabilities to develop. No evidence for such a "sponge effect" has been found. Laun [19] reported slip velocity measurements on PP with added low-MM wax that clearly demonstrate that no such polymer segregation takes place (see Figure 10.1). Low-MM wax only reduces the overall viscosity [20]. Furthermore, segregation would be a diffusion-controlled slow process. The suggested gradual development of melt fracture over time is not observed.

7.1.4 FRACTURE

The fracture hypothesis presented by Tordella [21] states that beyond a certain critical stress (or shear rate, or viscoelastic deformation), a melt breakdown occurs much in analogy to a tensile test failure of a solid polymer. This suggests that melt fracture is actually a cohesive failure of the polymer melt. Along the same lines Cogswell [22] developed the idea that surface distortions are caused by a rupturing of a highly stretched surface layer. Upon exiting the die, the surface layer accelerates from a minimal velocity (theoretically $v = 0$ at the die wall) to the average velocity of the emerging extrudate, while the bulk of the material in the core decelerates (Figure 7.1). This had been experimentally found by Gogos and Maxwell [23] in

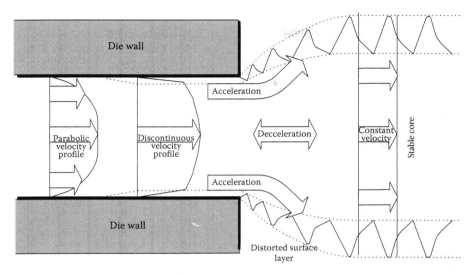

FIGURE 7.1 Macroscopic representation of surface distortions induced through a stretching and cohesive failure of a surface layer. The surface layer is accelerated while the bulk of the material in the core decelerates upon exiting the die.

relation to extrudate swell studies. Subsequent finite element studies indicated that there is a singularity at the point of separation between the melt and the die where the stretch ratio appears to be infinite [24,25].

As a consequence, at the die exit, the surface layer is stretched at a rate exceeding the critical break stress for the melt (e.g., ~1 MPa [22,26]) and causes rupture. However, the melt elasticity retards stress growth so that if the total deformation is small, even very rapid stretch rates may only produce low stresses and no surface distortions. Accordingly, Cogswell [22] concluded that more elastic (i.e., lower elastic modulus) polymers with broad MMD or branching permit more rapid acceleration without building up high stress levels and thus eliminate surface distortions at lower shear rates. Consequently, cooling of the die lips to achieve higher melt elasticity in the surface layer should also reduce the appearance of surface distortions. Cogswell and later investigators experimentally confirmed this [10,22,27].

In 1949, Spencer and Dillon [28] used a similar reasoning to explain the spiraling volume distortions of PS extrudates. They reasoned that the orientation in a melt stream has its maximum in a surface layer at the capillary wall and drops to zero at the center along with the shearing stress. As the shearing stress is raised, the surface attains the maximum elongation possible for the polymer. Further raising the shearing stress only forms a thicker surface layer of polymer at maximum elongation. At this stage, melt in the die consists of a surface layer at constant maximum elongation and a core of constant average elongation, all of the core being oriented less than the surface layer. When the melt emerges out of the die it cannot cool fast enough to prevent the recovery of orientation under experimental conditions. The outside skin would like to contract more than the less-oriented core and thus compresses it.

As conjecture, the system is considered as unstable and the core buckles and thus spirals. This idea no longer finds any support because these types of distortions are damped out by increasing $L/2R$ ratio. Flow visualization experiments (on PS, LDPE, PP) indicated that the origin of the spiraling motions lies at the die entry [29,30].

In the case of linear PE, Ramsteiner [31] suggested stress-induced crystallization, as observed by Mackley et al. [32], as a possible cause for melt fracture. Lower melt temperatures favor earlier crystallization for the same induced stress field at the die entry, and thus explain a flow curve discontinuity at lower shear rates. However, at temperatures well above the crystallization temperature, melt fracture still occurs.

Kurtz [33–35] combined the fracturing of the surface layer with a slip mechanism to explain the formation of surface distortions (Figure 7.2). The ensuing stick-slip mechanism complemented the explanation provided by El Kissi and Piau [36] and Piau et al. [37]. They concluded that surface distortions are the consequences of a cyclical fracture of the surface layer and a stress relaxation at the die exit. Extrudate observations close to the die indicate the formation of a superficial fracture that sticks to the edge of the die, forming a melt ridge. While the extrudate core flows continuously, the melt ridge grows and then detaches (slips [35]) from the die edge and a new melt ridge starts forming. The extrudate surface in between two ridges is rough but may regain a certain smoothness due to the relaxation of the melt.

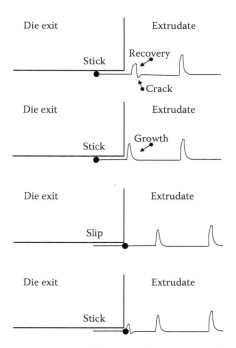

FIGURE 7.2 A visual representation of the formation of surface distortions resulting from a stick-slip mechanism. (Adapted from Kurtz, S.J., Visualization of exit fracture in the sharkskin process, in *Polymer Processing Society, PPS10*, Akron, OH, 1994, pp. 8–9.)

Vinogradov et al. [38] measured the flow curves of nearly uniform PI and PB using a constant-pressure capillary rheometer. The flow curve discontinuity was associated with a hysteresis loop in relation to the ensuing pressure oscillations and the repetitive and periodic change of the extrudate distortion patterns. A first approach to model the pressure and flow rate transients was presented that included the polymer compressibility as a relevant variable in relation to earlier considerations by Lupton and Regester [15]. Vinogradov et al. [38] theorized that this melt fracture observation was due to the transition to another physical state rather than to the viscous properties of the polymer. In subsequent papers, the Vinogradov school [39,40] proposed two transition states for the polymer in a hysteresis loop: from a fluid state to a forced high-elastic state (behaving more or less as a cross-linked polymer), to a forced leathery state. They concluded that the polymer deformability is limited by the melt strength, while the fracture process is determined by the recoverable strain. Furthermore, the presence of strong slip should be attributed to a phase transition of the polymer near the wall. The transient nature of the transitions is considered responsible for the equivalence between temperature and shear rate effects on melt fracture. Also, a correlation is established for uniaxial extension and low-amplitude oscillatory shear deformation to predict fracture phenomena.

All these ideas relate the origin of the various types of melt fracture to the elastic nature of the polymer. The notion that the stored elastic energy during the flow of a viscoelastic polymer melt is responsible for melt fracture requires the identification of a quantitative measure (Weissenberg or Deborah number, or recoverable strain). As discussed before (see Chapter 5) the correlation of these macroscopic measures to melt fracture is either not general enough, too crude to distinguish between the various types of surface and volume distortions, or difficult to measure.

7.1.5 CAVITATION

In relation to surface distortions, cavitation was considered as a possible explanation on how the polymer fractures by Tremblay [41]. Cavitation (a gas bubble–forming process) is expected to occur in a manner similar to that thought to be involved during the process of crazing [42–44]. The bubbles are supposed to form during extrusion at the die exit. They eventually coalesce to form the characteristic surface distortions. Flow simulations for inelastic fluids show that high stresses at the die exit produce negative hydrostatic pressures in a small zone extending a few microns from the die lip. The only experimental indication for such a mechanism seems to be provided by Venet and Vergnes [27,45]. For certain polymers, they detect, as the very first signs of surface distortions, a pattern of holes (pits) that grows larger with increasing flow rate and eventually turns into an overall low-amplitude high-frequency fluctuation (see Figures 1.3 and 1.4).

7.1.6 INTERFACIAL SLIP

Benbow and Lamb [46] were the first to argue that slip along the die wall must be associated with the appearance of extrudate distortions. The experimental evidence however does not always seem to support the presence of slip [47–50] for all types of melt fracture in an equally convincing manner. In particular, in the case of polymers that give experimentally discontinuous flow curves. The onset of melt fracture, usually referred to as "spurt," could also be related to the maximum stress in a discontinuous constitutive equation without the need to invoke a nonzero velocity at the die wall [44]. Although the slip and constitutive mechanism are different in principle, in practice they are difficult to distinguish because both predict the same macroscopically observable phenomenon [44]. The study by Ramamurthy [51] and Kalika and Denn [52] on LLDPE melt fracture associated a change of slope with the observable onset of surface distortions on the extrudates. The calculation of a measurable slip velocity using the Mooney approach started a decade of scientific controversy. Many experimental data and arguments for and against slip are forwarded. Eventually, the constitutive hypothesis becomes difficult to maintain. However, the suggestion that slip causes the appearance of surface distortions is inconsistent with the fact that a thin layer of fluoro-polymer at the die wall reduces the severity of melt fracture even though it enhances slip [10,53–57]. Furthermore, experimental evidence exists on the use of fluoropolymers that either reduce or promote slip, while they continue to enhance the extrudate smoothness [58,59]. These experiments indicate that the slip–surface distortions relationship is not well understood, and suggests that melt fracture may not be associated with slip only [60].

The emphasis on slip as a mechanism associated with melt fracture inspired research into the interfacial interactions between the polymer and the die wall. Brochard and de Gennes [2] provided a theoretical framework for such interfacial interactions. de Gennes originally theorized [2,61] that polymer melts flowing on a smooth, non-adsorbing solid surface should always slip at any shear rate. The observation that macromolecules anchored to the die wall strongly reduce the slippage [62] leads to the definition of an extrapolation length b:

$$b = \frac{v_s}{\dot{\gamma}_w} \tag{7.1}$$

The extrapolation length b is defined as the ratio of two absolute values, the slip velocity v_s and the shear rate at the wall $\dot{\gamma}_w$. The ratio measures the relative importance of slip. Accordingly, b is essentially a measure for the level of entanglement between bulk macromolecules and a layer of strongly adsorbed macromolecules ($b=0$ implies no slip) (Figure 7.3). The slip velocity at an interface is determined by the shear stress:

$$v_s = \frac{\eta \dot{\gamma}_w}{\beta} = \frac{\tau_w}{\beta} \tag{7.2}$$

where
β is the interfacial friction coefficient
τ_w is measured at the wall

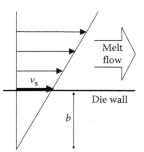

FIGURE 7.3 Extrapolation length b as defined by Brochard and de Gennes [2] is a material characteristic parameter and a measure for slip.

Accordingly, the extrapolation length b can be considered as a fundamental material property depending on the bulk viscosity η and the friction coefficient β at a certain stress. The theory does not predict the role of MM or mass distribution. The use of this approach brings about issues of measuring the theoretical variables. Moreover, picturing a molecular mechanism for slip suggests several possibilities. Polymer adsorption–desorption between polymer and die wall and/or entanglement–disentanglement between adsorbed and bulk macromolecules can be envisioned.

7.1.6.1 Microscopic Mechanisms—Cohesive Failure

In 1976, Bergem [63] already suggested that macromolecular disentanglements may be responsible for melt fracture in reference to the observation of a discontinuous flow curve. Drda and Wang [64] quantify this for HDPE. At the critical stress τ_c associated with the flow curve discontinuity onset in a constant pressure experiment, the surface layer experiences a structural transition along the die wall. The transition may be thought of as a flow-induced disentanglement between adsorbed and bulk macromolecules at the polymer melt–die wall interface. This leads to an interfacial layer of entanglement-free macromolecules and results in an extremely low surface-layer shear viscosity in comparison to the bulk shear viscosity away from the surface layer. The apparent wall shear rate difference between slip (disentanglement) and stick (entanglement) at the critical stress τ_c is related to the slip velocity v_s at the wall by the Mooney relation:

$$\dot{\gamma}_{\text{slip}} - \dot{\gamma}_{\text{stick}} = \frac{4v_s}{R} \tag{7.3}$$

Experimentally, τ_c is observed to increase with temperature indicating that a time–temperature superposition principle applies to the mechanism. This leads Drda and Wang [7,8,64–66] to the conclusion that the interfacial transition arises most likely from macromolecular disentanglement and not from macromolecular desorption. The critical stress would decrease with increasing temperature if the breakdown of interfacial interaction is due to stress-induced macromolecular desorption. The onset of "spurt" is thus caused by a disentanglement mechanism that is macroscopically measurable in terms of a slip velocity.

A similar argument is used by Barone et al. [10] to define the interfacial origin of surface distortions. A local conformational transition occurs at the die wall near the exit where the adsorbed macromolecules entrap a layer of bulk macromolecules. This layer oscillates between entanglement and disentanglement states. The associated reversible coil-stretch transition leads to cycles of local stress relaxation and growth and causes periodic perturbations of the extrudate. Accordingly, surface distortions are caused by a local stick-slip at the die exit. When a macromolecular disentanglement occurs at $\tau_{\text{exit}} > \tau_c$, the local stress reduces to a certain stress and gives rise to a periodic oscillation in shear rate. Due to the local origin, no overall fluctuations in pressure drop or volumetric flow over the die can be observed. During the transition, disentangled macromolecules slip over a layer of adsorbed ones! Support for this local mechanism is found in terms of the significantly higher stresses at the die exit

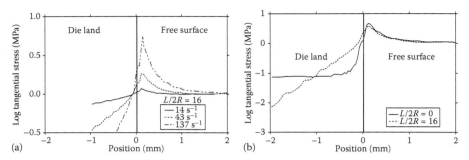

FIGURE 7.4 Finite element calculations of the tangential stress along a flow line close to the die wall indicate a significant higher stress at the die exit than expected from stress calculations based on pressure drop measurements. The influence of the shear rate (14, 43, and 137 s^{-1}) (a), and the $L/2R$ at 137 s^{-1} ($L/2R=0$ and 16, $2R=1.39$, 90° entry angle) (b) are calculated for an LLDPE using a Phan-Tien–Tanner constitutive equation with five relaxation times. (From Venet, C. and Vergnes, B., *J. Non-Newtonian Fluid Mech.*, 93(1), 117, 2000. With permission from Elsevier.)

than in the die land, as obtained from the overall pressure drop (Figure 7.4). Many authors reported on this die exit feature [12,22,24,27,67,68] while some indicated [27,68,69] that the high exit stress contains a shear and a significant elongational component due to the boundary discontinuity. The microscopic disentanglement mechanism does not appear to be mutually exclusive versus the original macroscopic mechanism proposed by Cogswell and followed by others [22,33,67,70–72]. Essentially, microscopic disentanglement can be thought to manifest itself as a macroscopic cohesive failure of the surface layer because the local stresses can exceed the melt strength of the polymer in an oscillating fashion. Barone et al. confirmed this by repeating capillary die experiments similar to those performed by Bergem [63] upon a suggestion by Cogswell [73] who referred to experiments performed by Benbow and Lamb [46].

Barone et al. [10] observed for LLDPE that the surface distortion wavelength θ depends on applied stress. The period p relates to the relaxation time defined by the reciprocal of the oscillating frequency at which storage (G') and loss (G'') modulus have the same value, and is related to the wavelength via

$$\theta = pv = R\left(p\dot{\gamma}\right)\frac{\rho}{4\rho'}\left(\frac{R}{R'}\right)^2 \tag{7.4}$$

where
    v is the extrudate velocity
    ρ and ρ' are the density of melt and solid phase
    R and R' are the radius of capillary die and cooled extrudate

The period decreases with stress and obeys the time–temperature superposition principle. Accordingly, the nature of the surface distortion is controlled by the stress relaxation and growth dynamics, as well as the critical stresses for the entangled polymer to undergo a reversible coil-stretch transition. Therefore, the lowering of surface distortions (in terms of longer wavelengths or total disappearance) by lowering the die lip temperature [22] is understandable through the longer period, which is due to the slower local melt dynamics. Also, the effect of reduced surface distortions from the use of additives promoting either slip or adhesion may be understood [59,75]. When slip is promoted, the stress levels near the die exit are lowered and extrudate surface distortions are reduced. When adhesion is enhanced, the number of adsorbed macromolecules increases and the high exit stress is distributed over many macromolecules, reducing disentanglement [56,66]. Similarly, an explanation is found for the causes of massive slip [56,66] and no sharkskin [9,72] induced by only a local fluoro-polymer coating 1–2 mm from the die exit. At the coating, the bulk cannot establish an entanglement. Only in the die region before the coating, entanglements occur resulting in a coil-stretch transition. The oscillations relax completely within the time it takes the melt to travel to the die exit and yield a smooth extrudate. Barone et al. [10] also rationalize the proposal by Kurtz [33] for introducing a critical extrudate velocity in addition to a critical stress for the onset of surface distortions. Surface roughness as characterized by wavelength and amplitude is found to increase linearly with the die radius R [7,9]. Both determine how noticeable surface distortions appear. For a certain spatial resolution, surface distortions are more observable for dies with larger R and less observable for smaller R. As the onset of surface distortions is controlled by the overall applied stress and the extrudate velocity is proportional to the capillary radius, a critical velocity can be inferred [10].

7.1.6.2 Microscopic Mechanisms—Adhesive Failure

As an alternative to a cohesive failure mechanism, an adhesive failure mechanism finds some experimental support. The influence of the material of the construction of the die on surface distortion onset as observed by Benbow and Lamb [46], Ramamurthy [51], Person and Denn [76], and Ghanta et al. [77] indicates that its microscopic origin may be related to polymer detachment (desorption) from the die wall. Further support is found when Vinogradov et al. [78] measured an electrostatic charge buildup upon extruding PB and PI in a melt fracture regime through steel, glass, and PTFE dies. The polarity depended uniquely on the chemical nature of the polymer–die combination.

Macromolecular detachment from the wall requires stresses that overcome the work of adhesion W_a. Hatzikiriakos et al. [79] calculated a relation between the critical stress for the onset of slip and the work of adhesion. Anastasiadis and Hatzikiriakos [60] further expanded and confirmed these results using a sessile drop method [80,81]. This method determines the shape of the profile of a sessile drop of one fluid into the matrix of the other (or in air). The measured surface tension γ_L and contact angle φ can be used to determine the work of adhesion:

$$W_a = \gamma_L (1 + \cos \varphi) \tag{7.5}$$

For several fluoro-polymer coatings on a stainless steel substrate, the work of adhesion is found to be related to the critical stress (in MPa) as

$$\tau_c = 5.118 \times W_a - 2.674 \times 10^{-2} \tag{7.6}$$

Equation 7.6 indicates that a finite value of adhesive strength is required for no slip (set $\tau_c = 0$ in Equation 7.6). Above this value, slip occurs at a certain finite shear stress. Accordingly, for interfaces having the work of adhesion less than this finite value, the no-slip-boundary condition may apply.

Lau and Schowalter [4] were the first to propose a kinetic rate model for capturing the adsorption–desorption mechanism based on Eyring's theory. Using a similar approach, Hatzikiriakos and Dealy [59] developed a model for slip assuming that the polymer at equilibrium is adsorbed at the interface. Under the influence of shear, they are extended and when the stored elastic energy can overcome a fraction f_a of the work of adhesion, the polymer jumps a certain length d comparable to the radius of the gyration of the polymer. This approach also allowed expressing the critical stress for the onset of slip with the work of adhesion as

$$\tau_c = \frac{2 f_a W_a}{d}. \tag{7.7}$$

Hill et al. [5] gave an alternative interpretation of equation where d is considered as an elastic pseudo-layer thickness. The estimates of the layer thickness are in the order of 1 μm or less [60]. Stewart [82] modified the Lau and Schowalter [4] model to incorporate proportionality between the bonding free energy and the work of adhesion. The experimental slip velocity as the function of shear stress is correctly represented for the same LLDPE resins as independently measured by Ramamurthy [51] and Kalika and Denn [52] in relation to the appearance of surface distortions. The work of adhesion is determined via an expression proposed by Wu [83]:

$$W_a = 2\sqrt{\gamma_p \gamma_b} \tag{7.8}$$

where γ_p and γ_b are the surface tensions of polymer and die wall.

The slip equations derived by Hill [6], Lau and Schowalter [4], Stewart [82], and Hatzikiriakos and coworkers [12,79] are monotonic and do not compare well with experimental findings by Migler et al. [3] and Henson and Mackay [84]. Migler et al. [3] measures for PDMS, a sigmoidal slip velocity as the function of shear rate indicating two slip regions. (No direct reference is made to the type of extrudate distortion that may occur but it is implicitly assumed to be surface distortion.) The transition from weak to strong slip falls within a narrow shear rate range (Figure 5.26). These results show a functional similarity with the measurements by Beaufils and coworkers [67,85] and Venet and coworkers [27,45] of surface distortion in terms of amplitude and period as the function of apparent shear rate and apparent shear stress (Figures 5.24, 6.13, and 6.14).

The remaining discrepancies between experiment and model predictions inspired further research. Hill [6] presented a more physically plausible model for slip that should enable to estimate the model parameters independently and allow for a quantitative slip description. At low slip, the model takes on the same form as proposed by de Gennes [61,86] but displaying an additional dependence of the work of adhesion. The total loss of adhesion occurs at a critical stress that depends on the difference between the work of adhesion and the work of cohesion. Dhori et al. [87] proposed that the die exit boundary condition oscillates between slip and no slip due to wetting (adsorption) and de-wetting (desorption) in steel dies.

7.2 THE CONSTITUTIVE APPROACH

A critique for the above melt fracture mechanisms is their qualitative nature. For all practical purposes, the gained physical insights need to be captured into mathematical equations. The model should encompass all the relevant variables to allow the quantitative prediction of the experimental observations. Ultimately, the model contains the physics to understand why melt fracture occurs for any polymer and predicts when it occurs accurately. Only then scientific knowledge exists as a useful tool to determine how to avoid melt fracture.

The method of choice for arriving at a quantitative melt fracture model is often continuum mechanics because it provides the necessary mathematical rigor. The use of the conservation principles of mass and momentum in combination with a constitutive equation implies that melt fracture is totally defined by the viscoelastic nature of the melt. Important issues with this approach are to consider the relationship between polymer composition and material functions as used in constitutive equations that govern the polymer–wall interaction, and the relation between the constitutive equation, the flow curve, and the associated extrudate distortions. The latter is of particular importance as the flow curve gives *no* indication of the nature of melt fracture.

7.2.1 PHENOMENOLOGY

The observation of Bagley et al. [88] of a discontinuous flow curve for linear PE inspired Huseby [89] to use a viscoelastic model developed by Pao [90,91] for describing the shear stress as a double valued function of shear rate. The Pao constitutive equation is formulated in terms of the polymer relaxation spectrum and the velocity gradient (shear rate) in a capillary. For linear PEs it is concluded that melt fracture must be a natural consequence of the viscoelastic behavior arising from the structure of the polymer melt. The approach also suggested that no slip-boundary condition should be assumed a priori to describe melt fracture phenomena. In contrast, Pearson and Petrie [92] introduced a slip-boundary condition in their linear stability analysis of incompressible Newtonian Poiseuille flow. The linear stability analysis of a set of equations defined by the conservation equations and a (non-)linear constitutive equation supports the association of melt fracture with the presence of a negative slope in the shear stress–shear rate curve [92,93]. However, experimental confirmation is not available because of the impossibility to measure the negative slope. Stability analysis is a procedure for determining whether a solution of the conservation and stress equations corresponding to steady operation (that is a solution that is constant or periodic in time) can be maintained in the face of disturbances entering the system. The analytical methods used in a stability analysis are treated in specialized texts [94,95]. There are three classes of analysis: those that establish conditions under which a process is absolutely unstable to any disturbance, no matter how small; those that determine the effect of small but finite disturbances near conditions corresponding to absolute instability; and those that establish conditions under which a process is absolutely stable regardless of the magnitude of the upset. Linear stability to infinitesimal disturbances is studied by obtaining the set of linear partial differential equations, which describe the transient behavior of the process near the steady state [94].

7.2.2 RELAXATION OSCILLATIONS

Weill [96–98] followed in the footsteps of Huseby [89], and theorized that the discontinuous nature of the flow curve can, on its own, predict the experimental observations of HDPE melt fracture. As a consequence of that hypothesis, he concluded that surface distortions and spurt must have the same physical origin. He accepted the existence of a discontinuous constitutive equation as such and focused on modeling the transient flow in capillary experiments. To that purpose, the idea of a system that successively stores and relaxes potential energy is used. The resulting cyclic phenomenon is referred to as a relaxation oscillation, first described by Van der Pol [99] in 1926.

However, Weill only considered one first-order differential equation (based on the conservation of mass including compressibility) instead of a second-order differential equation (or alternatively, two coupled first-order differential equations) necessary to fully describe the relaxation oscillation as shown by Molenaar and Koopmans [100]. This approach showed that for the discontinuous flow curve, a discontinuity in the velocity profile appears when the critical stress τ_c is reached and a sudden jump in shear rate results (Figure 7.5). Weill concluded that experimentally no distinction can be made between the velocity profile, as shown in Figure 7.6 (bottom), and the one resulting from a pure slip of the polymer at the wall. He further theorized that the appearance of surface distortions is explained as originating from a high-frequency oscillatory flow created at the die entry. The oscillation is envisioned to propagate through the die and finally to impose its frequency to the stretching flow that ruptures the skin layer at the die exit as suggested by Cogswell [22]. No experimental evidence exists, however, to support that particular mechanism for the initiation of surface distortions at the die entry. Moreover, the possibility to average a global flow process in a 1D modeling approach is insufficient to capture a local phenomenon.

Molenaar and Koopmans [100] expanded the Weill model [96–98] to a system of two coupled first-order differential equations. Such a 1D relaxation oscillation model allowed integrating a non-monotonic constitutive equation. The quantitative predictions of the transient pressure and flow rates are made possible as demonstrated by Durand et al. [101] for HDPE. However, the experimental discontinuous flow curve is still required as input for the model. For practical reasons, this is unacceptable. Unfortunately, the introduction of a Johnson–Segalman–Oldroyd or K-BKZ

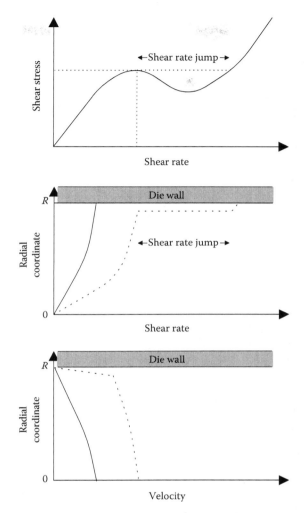

FIGURE 7.5 A discontinuous flow curve (top) indicating a transition region that leads to a shear rate (middle) and velocity (bottom) profile discontinuity as proposed by Weill. The shear rate range of the transition region is reflected in the shear rate profile discontinuity. (From Weill, A., *J. Non-Newtonian Fluid Mech.*, 7(4), 303, 1980. With permission.)

non-monotonic constitutive equation into the relaxation oscillation model does not lead to quantitative flow curve predictions. Den Doelder et al. [102,103] showed that a stress-dependent slip equation is needed to attain a quantitative agreement with the experiment (see Chapter 9).

The relaxation oscillation model can be expanded to capture a local pressure oscillation that can be associated with the onset of surface distortions [104]. Such an approach considers a core-annular coextrusion with different rheological characteristics. The existence of a non-monotonic shear stress–shear rate suggests this

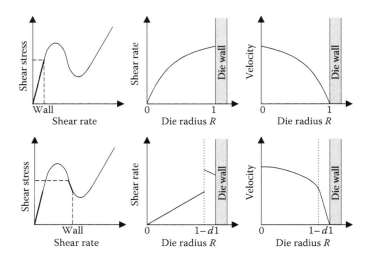

FIGURE 7.6 Stable and unstable solutions to the relaxation oscillation model using a non-monotonic constitutive equation: the shear stress–shear rate curve (left), the shear rate profile (middle), and the velocity profile (right).

possibility (Figure 7.6). However, the mathematical framework also indicates that an infinite number of steady-state solutions are possible in a certain shear rate range (in line with findings by McLeish [105] and Georgiou and Vlassopoulos [106]). A unique steady-state solution for each imposed shear stress (or shear rate) can only be obtained when additional selection criteria are introduced. Potentially, the principle of energy minimization may provide this required criterion [107]. The unstable solution in Figure 7.6 suggests the existence of an oscillating thin surface layer. It has been shown to be a valid solution to the system of equations [104]. Accordingly, it is suggestive to think that a macroscopic mathematical formalism exists that captures the dynamics of a coil-stretch molecular mechanism as forwarded by Wang and Drda [66]. A high shear rate surface layer can be considered as the macroscopic equivalent to the slip.

The appearance of the zones of different shear rates is also known as shear banding [108]. Ovaici et al. [109] successfully modeled melt fracture of chocolate using a two-layer approach in combination with the relaxation oscillation model. The combination of the nonlinear plastic flow behavior and the compressibility of chocolate in the reservoir controls melt fracture. The surface rupture of chocolate was quantitatively described in terms of core-annular coextrusion of two Bingham-plastic chocolate components having different yield stress. The analogy with polymer melts is suggestive [102]. Inn et al. [110] used this approach to model surface distortions for polymer melts.

7.2.3 NUMERICAL SIMULATIONS

The above constitutive modeling approach focuses on the flow curve. The added conjecture is that viscoelastic materials that yield a discontinuous shear stress–shear rate curve give rise to melt fracture phenomena. This does not address the experimental evidence of melt fracture for materials that yield continuous flow curves. Furthermore, the nature of melt fracture can only be derived at the various processing conditions in association with experimental observation.

To obtain the shape of the extrudate, Georgiou and Crochet [111] and Georgiou [112] numerically solved a time-dependent compressible Newtonian Poiseuille flow with slip along the die wall. The compressibility acts as the storage of elastic energy. It sustains the oscillation of the pressure drop and the mass flow rate and generates waves on the extrudate surface [106]. The compressibility and nonlinear slip represent the underlying mechanism for spurt in various 1D phenomenological models. Georgiou and Vlassopoulos [106], Kumar and Graham [113], and Brasseur et al. [114] showed that compressibility can be replaced by elasticity using an Oldroyd-B model in combination with a non-monotonic slip equation to yield self-sustained oscillations. The amplitude and period of the oscillations increased as the function of the elasticity. These approaches indicate that extrudate perturbations can be calculated although the results still remain to be experimentally verified.

Kolkka et al. [115] considered simple shear and plane Poiseuille flows using a Johnson–Segalman model with added Newtonian viscosity. They showed that piecewise smooth solutions with a local shear rate corresponding to the descending portion of the steady shear stress–shear rate curve are linearly unstable. Their numerical calculations for pressure-controlled capillary flow showed that the time-dependent solutions jump to one of the two stable positive-slope branches. The negative-slope branch is unattainable. Malkus et al. [116–118] obtained numerical time-dependent results for the start-up of Poiseuille flow at fixed flow rates. For certain values of the material parameters, the calculated pressure drop appeared to be oscillatory but the existence of a true oscillatory regime is not certain. Den Doelder et al. [102] confirmed earlier criticism by Denn [119] that the calculated oscillations do not correspond to the ones experimentally observed. Espanol et al. [120] solved a 2D plane shear flow of a Johnson–Segalman model with added Newtonian viscosity. They found that the flow separated into two layers with the lower shear rate adjacent to the fixed wall. Georgiou and Vlassopoulos [106] further explored the Johnson–Segalman model with added Newtonian viscosity by studying the stability of the piece-wise steady-state solutions and investigated the dynamics of the time-dependent flow problem. They found that in addition to the standard linear solution for the velocity, there exists an uncountable number of piece-wise linear steady-state solutions.

7.2.4 MOLECULAR CONSIDERATIONS

The non-monotonic shear stress–shear rate behavior can physically be thought of as a separation of flow into two dynamic regimes: one with a history of high deformation rate (near the wall) and one with a low deformation rate (in the core) [105,121]. This implies a short and a long characteristic time for the viscoelastic polymer. The physical interpretation is, respectively, to associate the Rouse relaxation time (see Chapter 8), which controls macromolecular conformation with short times (high shear rate), and the reptation relaxation time, which controls the motion of macromolecules with longer times (low shear rate) [122]. Based on this concept, the

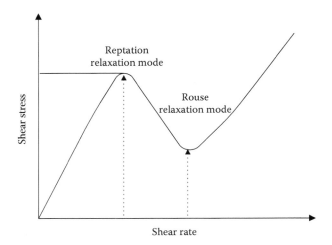

FIGURE 7.7 The hypothesized relationship between shear stress and shear rate for polymer melts based on the modified Doi–Edwards model taking into account reptation and Rouse relaxation modes.

association of a non-monotonic constitutive equation with two relaxation times to the spurt effect has been established [122,123] (Figure 7.7).

McLeish [105] and McLeish and Ball [121] investigated the hysteresis phenomenon in discontinuous shear stress–shear rate curves. They used a modified Doi–Edwards model that included the shorter relaxation times to account for the Rouse equilibration processes. Above a critical stress, the Poiseuille flow separates into two phases. The surface layer may be defined by shear rate conditions of Branch II while the core is determined by the lower shear rates at Branch I depending on the imposed overall flow rate (Figure 7.8).

The high shear rate surface layer is associated with an apparent slip. Many possible steady-state solutions are found to exist that correspond to a range of possible positions of the interface between the two phases. A linear stability analysis indicated that under typical flow conditions, the amount of material with a high-deformation-rate history is maximized. The stability of the perturbations is governed by the normal stress effects and may be related to certain types of melt fracture [105].

In a similar fashion, Lin [124,125] modified the Doi–Edwards model to include the Rouse relaxation mode to arrive at a general linear viscoelastic theory for uniform linear polymers. The maximum of the shear stress–shear rate curve (Figure 7.7) is an unstable point and the negative slope is associated with "spurt." The flow curves of uniform PI and PS and a commercial linear PE are measured and compared with theory. Despite some quantitative differences between theory and experiment at high shear rate, Lin [124,125] claims that the theory correctly shows the effect of the two relaxation modes. Numerical calculations show that the minimum in the shear stress–shear rate curve vanishes by broadening the MMD. As indicated in Equation 7.9, the reptation (λ_{rep}) and Rouse (λ_{Rouse}) relaxation modes will separate further apart when the MM M increases:

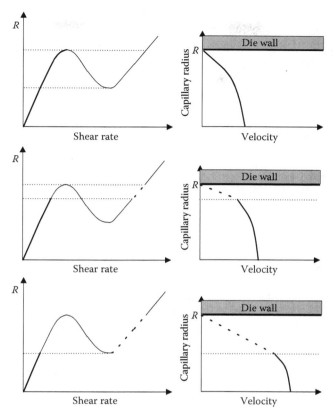

FIGURE 7.8 The propagation of an interface from a high-shear region into a low-shear region, showing the shear rates and resulting velocity profile as function of capillary radius R. (From McLeish, T.C.B. and Ball, R.C., *J. Polym. Sci., Part B: Polym. Phys.*, 24, 1735, 1986. With permission from John Wiley & Sons.)

$$\frac{\lambda_{rep}}{\lambda_{Rouse}} = 3\frac{M}{M_e}\left[1 - \sqrt{\frac{M_e}{M}}\right]^2 \tag{7.9}$$

where
M_e is the entanglement MM
M is the MM of the (uniform or nearly uniform) polymer

The flow curve of a nonuniform polymer is considered equal to the weighted sum of the individual uniform polymer contribution. The maximum in the discontinuous shear stress–shear rate curve will start to appear for $M > 10M_e$.

Molecular considerations for constitutive equations have been applied explicitly to the solid boundary–polymer interface. Relaxation mechanisms pertaining to bulk constitutive behavior can be applied near walls. A key difference is that the relaxation of adsorbed macromolecules cannot occur via reptation. So-called arm retraction mechanisms, as proposed for branched polymers, can be used to model near-wall relaxation of even linear polymer architectures. Constitutive equations for such systems have been derived by Joshi and Lele [126], Tchesnokov et al. [127], and Stepanyan et al. [128]. A detailed treatment of both bulk and interfacial macroscopic constitutive equations based on polymer dynamics considerations can be found in Chapter 8.

7.3 GENERAL UNDERSTANDING

From the previous pages, it is possible to conclude that melt fracture can be understood at two levels. On the macroscopic level, the elastic character of the polymer melt and the concept of slip at the polymer–flow channel interface are commonly accepted. Slip however is *not* its cause and is only associated with melt fracture as a macroscopic measurable observation. On the microscopic level, an understanding of slip is developed via a dynamic process of either entanglement–disentanglement (stick-slip) between macromolecules attached to the flow channel wall and the bulk macromolecules, or via a macromolecular attachment–detachment from the flow channel wall. Experimental evidence tends to favor the former stick-slip mechanism. The microscopic mechanisms are conjectures and difficult to prove. However, they can assist in building quantitative melt fracture models. It should also be kept in mind that real surfaces are not perfectly smooth as implied in the theoretical model developments. The surface topology, that is, roughness in relation to the molecular dimensions should be a factor of consideration.

The two main types of extrudate distortions can be understood in terms of the same basic mechanism, that is, excessive (elongational) stress build-up. The extrudate distortions originate either on a local scale or a global scale. The former relates to surface distortions that originate at the die exit and deform the surface of the extrudate in various ways. The latter relates to volume distortions that originate in the reservoir die entry region and propagate through the die deforming the entire extrudate. An intermediate form of extrudate distortion is "spurt." The extrudate appears as being sequentially smooth, surface distorted and volume distorted in a periodic fashion. The minute compressibility of polymer melts is a sufficient condition for its appearance. Absolute, quantitative predictions for the onset and type of melt fracture based on independent measurements of polymer composition remain elusive. Experimental verification is still required. The mathematical modeling focuses on macroscopic constitutive approaches. The models indicate that the concepts of slip and thin surface layer development are equivalent. The mechanistic distinction may only be relevant on a microscopic level although its precise understanding may yield different pragmatic solutions for avoiding polymer melt fracture.

With this knowledge, a general understanding of polymer melt fracture can be derived. A distinction is made between the macroscopic and microscopic level. The flow channel is defined by a reservoir and a capillary die. A constant-flow rate experiment is imagined (Figure 7.9).

Initially, at low flow rates, the extrudates are smooth, and a laminar Poiseuille flow situation exists (Figure 7.10). The stress along the wall and throughout the flow

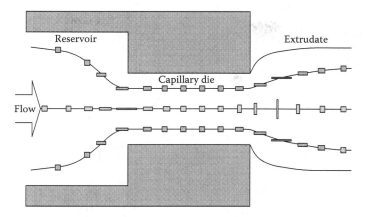

FIGURE 7.9 A flow channel representation that consists of a reservoir and a capillary die as used for developing a general understanding of melt fracture during a constant-flow-rate experiment. The rectangles represent polymer domains that deform during flow. Along the center line, the stretch, or, equivalently, the elongational stress is most pronounced at the capillary entry while a contraction occurs at the die exit. In the outer layers near the die wall, the stretch is most pronounced at the capillary exit.

channel is lower than the critical stress for melt fracture. On a microscopic level, the macromolecules adhere to the flow channel wall and entangle with the bulk polymer. The length (two to three time M_e [129]) and flexibility (relaxation modulus G_0) of the macromolecule is critical to the dynamics of entanglement formation (polymer relaxation time λ).

At a certain flow rate, surface distortions appear (Figure 7.10). The local stress at the die exit in a very thin surface layer exceeds the critical stress and slip stands to occur. In practice, the complex 3D stress field has shear and elongational components although the latter is considered dominant near and at the die exit. In addition, practical stress values are based on pressure measurements and associated calculations having the implicit assumption of laminar flow, as discussed in Chapter 5. The observed critical stress may thus not represent the actual local stress. The slip process increases flow rate locally and reduces the stress to a value lower than the critical stress. The level of slip will depend on the detailed die surface topology and the physical interactions of polymer and flow channel at the interface. Subsequently, the stress grows again above the critical stress and the process repeats itself. The stress oscillation in the thin surface layer is re-enforced by an acceleration force with a strong elongational character upon exiting the die. This either leads to a low-amplitude high-frequency surface distortion or a fracture of the surface layer. Either situation occurs depending on the nature of the polymer, associable with the relaxation modulus G_0. If the total stress exceeds the melt strength or maximum extensional strain of cohesive failure, a surface fracture occurs. The fracture event relieves the local stress and the process of local stress buildup starts again. The critical stress for melt fracture on a microscopic level is determined by the stress required to disentangle the bulk macromolecules from the ones attached to the flow channel

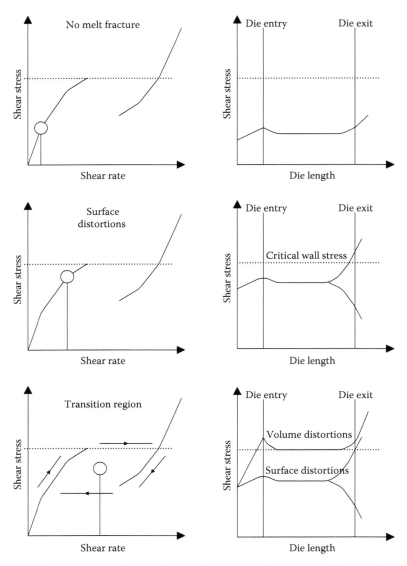

FIGURE 7.10 Summary of the understanding of melt fracture. At different shear rates (left) the tangential stress along a streamline close to the wall may reach values larger than the critical wall stress τ_c and yield melt fracture. For stress values below τ_c the extrudate is smooth (top). If τ_c is surpassed at the die exit, surface distortions are observed on the extrudate (middle). For shear rates in the transition region (bottom), spurt is observed and the stress profile changes periodically to yield surface as well as volume distortions of the extrudate. Volume distortions appear if the overall stress at the die entry exceeds τ_c being highest along the central flow lines.

wall. The fact of such a brush-like system induces a stretching of the interfacial bulk macromolecules. When these stretched macromolecules disentangle, they slip over the attached ones, recoil, and re-entangle. The polymer relaxation time defines the dynamics of this process and determines the surface distortion period [10,130]. Depending on the polymer nature and the processing conditions, the local stress buildup is insufficient for fracture to occur and only generates a periodic undulation at the surface.

For highly entangled polymers (e.g., HDPE), at higher flow rate, the overall stress at the die wall becomes sufficiently high to give rise to massive slip (Figure 7.10). The entire volume in the capillary slips at the die wall. The flow rate exceeds the imposed flow rate and causes an overall drop in stress. The flow rate reduces and the overall stress gradually builds up again while the melt compresses in the reservoir. The flow rate is smaller than the imposed flow rate. When the stress buildup reaches the critical stress, the process starts over again. The cyclic flow rate corrections induced to match the imposed flow rate causes extrudate swell differences and a sequence of surface and volume distortions (Figure 5.7). The extrudate appears smooth during the initiation of slip and is followed by a volume distortion during the decompression and stress reduction phase facilitated through slip. The stress drop over the die corrects the flow rate difference and slows down the flow rate yielding a reduced extrudate swell. The extrudate is smooth or shows surface distortions depending on the local stress at the die exit. Finally, the extrudate swell is lowest and slowly increases. When surface distortions are present, the severity increases too when the stress starts to build up. The flow curve is discontinuous and characterized by two "stable" steady-state branches separated by a distance depending on the flow rate transition. This process reflects the coupling between stresses and flow rates in combination with melt compressibility, which is required for relaxation oscillations to occur [100].

The compressibility is an essential part of the mechanism as it stores energy in the reservoir. The sequence in the size of the extrudate distortions (i.e., the pressure oscillation frequency) depends on the volume of polymer melt in the reservoir and the length of the flow channel. Reducing the length of the flow channel to an orifice makes the flow curve discontinuity disappear [27]. The extrudate only shows steady-state surface or volume distortions at the highest flow rates [27,45]. In the case of constant pressure experiments, the flow rate difference cannot rearrange itself through compression and no hysteresis cycle exists. On a microscopic level, the stress becomes high enough to disentangle all the interfacial bulk macromolecules from the ones attached to the capillary die wall. A stick-slip mechanism starts in the entire capillary die causing stress oscillations in the reservoir. The oscillation originates in the stretch, disentanglement, and recoil dynamics of the polymer. For lowly entangled polymers (e.g., PS, PDMS) [7]), a strongly reduced number of entanglements exist. The limited coil-stretch transitions will reduce the interfacial stress oscillations and the associated extrudate surface distortions. The flow curve becomes continuous and only volume distortions are possible. Polymers that induce an intermediate entanglement level may still show some surface distortion. The degree of entanglement can be expressed in terms of the extrapolation length b [8]. Its value must be sufficiently high compared to the dimensions of the die for an effective coil-stretch mechanism to occur. In the limit of the zero-shear rate viscosity, it can be shown based on the reptation theory [8,61,131] that $b(0)$ can be derived as

$$b(0) = n\left(\frac{N}{N_e}\right)^2 a \qquad (7.10)$$

Equation 7.10 indicates that b scales strongly with the number of monomers n, the Kuhn length a [132] and the number of entanglement points per macromolecule N/N_e, with N the number of monomers per macromolecule and N_e the number of monomers between adjacent entanglements. The Kuhn length is a statistical concept referring to a short section of a macromolecule containing several atoms so that the ends of that section can be considered as free joints in an imagined "pearl necklace." Alternatively, b being a function of the shear stress, τ can be expressed as the ratio of the bulk viscosity η and the effective viscosity at the interfacial layer η_i [8,133]:

$$b(\tau) = \frac{\eta(\tau)}{\eta_i} a \qquad (7.11)$$

The interfacial viscosity can be as high as the bulk viscosity and as low as the monomeric viscosity. For shear thinning polymer melts, the bulk viscosity reduces as the flow rate increases. The extrapolation length b will become small. Accordingly, three factors control the interfacial slip behavior [133]: (a) the inherent level of macromolecular entanglement in the bulk, (b) the level of applied (shear) stress or (shear) rate, and (c) the nature of the melt/wall interfacial interaction. The latter indicates that in the case of strong polymer adsorption to the flow channel, wall slip will be determined by the presence of interfacial macromolecular disentanglement and result into melt fracture observation. In the case of no adsorption or stress-induced detachment of adsorbed macromolecules, slip can be measured but no melt fracture determined. In both cases, b is non-negligible. Yang et al. [133] demonstrated for capillary experiments with PS, PP, EVA, and LDPE using dies without a fluoropolymer coating that the extrapolation length b is small. No surface distortions can be observed. For experiments with fluoro-polymer coated dies, slip can be measured but no surface distortions are observed. It should be noted that these polymers are known as not prone to surface distortion formation.

The stick-slip (coil-stretch) mechanism implies that a permanent slip occurs when no entanglements can be formed between the bulk macromolecules and those adsorbed to the flow channel wall. Similarly, die material composition changes can reduce the adhesion, subsequently limit entanglements, and reduce melt fracture. Alternatively, the polymer compositions may be altered by reducing the macromolecular length (or M_w) or increasing M_e to reduce melt fracture.

Polymers (PS, PP, LDPE, EVA) that do not seem to show stick-slip, locally at the die exit or globally throughout the die, still give rise to volume distorted extrudates. The origin of volume distortions is typically related to the presence of, and the instability of edge vortices in the dead spots of the reservoir (Figure 5.12). However, their absence does not avoid the presence of volume distorted extrudates at sufficiently high flow rates. The types of polymers that show formation of vortices seem to be "more elastic" polymers (lower G_0) of higher M_e.

As an hypothesis, the instability development of the vortices may be associated with the same stick-slip mechanism as described before. Imagine a coextrusion experiment with layers the size of the macromolecule. The stick-slip mechanism may now take place between entangled macromolecule "layers" in the bulk, instead of between the macromolecules attached to the wall and those in the bulk. Each laminar flow layer experiences a different shear rate when flowing through a flow channel (Figure 7.9). At the transition from the reservoir to the capillary, the flow channel contraction induces a strong extensional stress mainly on the central layer and to a lesser extent on the layers at the edges of the die entry (Figure 7.9 and Chapter 5). As the macromolecules between layers are entangled, the varying degrees of extensional stresses may induce different levels of stretching. When the stress at the edge layers attains the critical stress for disentanglement, the stick-slip mechanism occurs (Figure 7.11) and volume distortions are induced as there is no solid wall to prevent the layers from buckling and distorting.

In particular, in the layers along the edge vortex that make up the newly formed entry angle (Figure 7.12), the macromolecules stretch, disentangle, slip, recoil, and re-entangle. This induces a vortex oscillation that only becomes visible when volume-distorted extrudates are large enough to propagate throughout the length of the die. When the overall flow rate increases, more layers will give stick-slip and result in a total disruption of the homogeneous flow field in the reservoir [21,29,134–136]. Consequently, the originally stable vortices become unstable and will be pulled temporarily into the die (Figure 5.12). In the reservoir, a globally oscillating flow (with a full 3D character) will result that is sufficiently intense to propagate throughout the entire die and give volume distorted extrudates. Combeaud et al. [137] demonstrated for PS in line with earlier findings by Hürlimann et al. for LDPE [138] that the elongational stress at the die entrance is the critical element for inducing volume distortions even in the absence of a vortex.

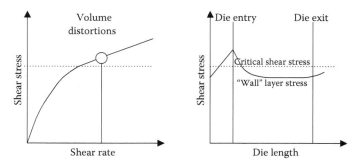

FIGURE 7.11 In laminar flow, volume distortions can be understood in terms of a stick-slip mechanism between vortex and adjacent layers at the die entry. When a critical stress is exceeded the volume distortion is initiated.

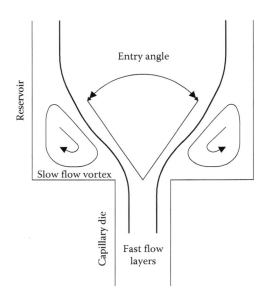

FIGURE 7.12   The origin of volume distortions could be associated with a stick-slip mechanism that induces vortex instability at the die entry. When the high extensional stresses in the polymer layers along the edge vortices surpass a critical stress for stretch-coil disentanglements to occur, volume distortion can be initiated. The polymer layers between the slow (vortex) and fast (bulk) flow start to buckle and destabilize the vortex inducing a swirling and oscillating motion reflected in a volume-distorted extrudate.

At very high flow rates, the vortices may develop a higher order of organization to achieve a periodic flow-regulating pattern as observed for Boger fluids [139]. The resulting irregular die inflow results in an observable volume distortion. For long dies, the degree of volume distortion is known to be reduced.

In the case of discontinuous flow curves at flow rates on Branch II but at overall bulk stresses below τ_c, sometimes the extrudate may have an overall smooth appearance instead of the expected volume distortions. In such a case, a steady-state situation is reached after an initial stress overshoot (Figure 5.6). The polymer melt has reached a state of permanent slip at the die wall. Following the above reasoning, the central core layers, however, will have exceeded τ_c at the die entry but the overall stress level is not sufficient to give a stick-slip mechanism at the die wall. The situation is highly labile and any disturbance of the flow field at the die entry will yield volume-distorted extrudates. The irregular flow initiated at the die entry will surpass any stick-slip mechanism and give rise to a permanent slip in the die.

In certain cases, volume-distorted extrudates show surface distortions as well. This implies that the local stresses at the die exit have reached τ_c and a stick-slip mechanism takes place locally. This should become even more evident for longer die as the die entry–initiated flow instability dampens out through the pressure drop over the die.

Figures 7.13 and 7.14 summarize the main types of melt fracture and their relation to the essential features of the extrusion system. A distinction is made

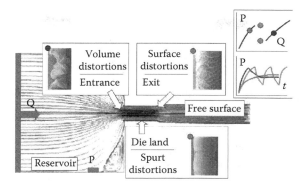

FIGURE 7.13 Polymers that predominantly give rise to surface distorted extrudates typically have a discontinuous flow curve. The presence of an entry vortex is not a sufficient condition for volume distortions to occur.

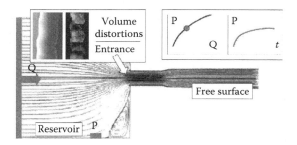

FIGURE 7.14 Polymers that predominantly give rise to volume distorted extrudates typically have a continuous flow curve. The presence of an entry vortex is not a sufficient condition for volume distortions to occur.

between systems that lead to a discontinuous and those that lead to a continuous flow curve.

7.4 GENERAL OBSERVATION

The above-formulated hypothesis provides a coherent understanding of polymer melt fracture. It is, at least qualitatively, supported by experimental evidence. Still, the hypothesis is conceptual in nature and is difficult to develop into a robust and coherent mathematical frame that captures all features of polymer melt fracture. Essential is the presence of extensional flow fields induced by work exerted onto a polymer melt inducing at one point in time an instability that, depending on its place of occurrence, will manifest itself in various forms as a distorted extrudate. The key challenge is coupling the local behavior with its global manifestation. The ultimate research objective is to predict the onset of melt fracture and to devise tools, methods, and polymers to circumvent it. Some of these aspects are addressed in Chapters 9 and 10.

REFERENCES

1. Ruhla, C., *The Physics of Chance*. Oxford University Press, Oxford, U.K., 1993.
2. Brochard, F. and P. G. de Gennes, Shear dependent slippage at polymer/solid interface. *Langmuir*, **8**:3033–3037 (1992).
3. Migler, K. B., H. Hervet, and L. Leger, Slip transition of a polymer melt under shear stress. *Phys. Rev. Lett.*, **70(3)**:287–290 (1993).
4. Lau, H. C. and W. R. Schowalter, A model for adhesive failure of viscoelastic fluids during flow. *J. Rheol.*, **30(1)**:193–206 (1986).
5. Hill, D. A., T. Hasegawa, and M. M. Denn, On the apparent relation between adhesive failure and melt fracture. *J. Rheol.*, **34(6)**:891–918 (1990).
6. Hill, D. A., Wall slip in polymer melts: A pseudo-chemical model. *J. Rheol.*, **42(3)**:581–601 (1998).
7. Wang, S. Q. and P. A. Drda, Superfluid-like stick-slip transition in capillary flow of linear polyethylene melts. 1. General features. *Macromolecules*, **29**:2627–2632 (1996).
8. Wang, S. Q. and P. A. Drda, Stick-slip transition in capillary flow of polyethylene. 2. Molecular weight dependence and low temperature anomaly. *Macromolecules*, **29**:4115–4119 (1996).
9. Wang, S. Q., P. A. Drda, and Y. W. Inn, Exploring molecular origins of sharkskin, partial slip, and slope change in flow curves of linear low density polyethylene. *J. Rheol.*, **40(5)**:875–898 (1996).
10. Barone, J. R., N. Plucktaveesak, and S. Q. Wang, Interfacial molecular instability mechanism for sharkskin phenomenon in capillary extrusion of linear polyethylenes. *J. Rheol.*, **42(4)**:813–832 (1998).
11. Hatzikiriakos, S. and J. M. Dealy, The effect of interface conditions on wall slip and melt fracture of high-density polyethylene. In *SPE ANTEC*, Montreal, Canada, 49, 1991, pp. 2311–2314.
12. Hatzikiriakos, S. G. and J. M. Dealy, Wall slip of molten high density polyethylenes. II. Capillary rheometer studies. *J. Rheol.*, **36(4)**:703–741 (1992).
13. Bird, R. B., W. E. Stewart, and E. N. Lightfoot, *Transport Phenomena*. John Wiley & Sons Inc., New York, 1960.
14. Pearson, J. R. A., Mechanisms for melt flow instability. *Plast. Polym.* **37(August)**:285–291 (1969).
15. Lupton, J. M. and J. W. Regester, Melt flow of polyethylene at high rates. *Polym. Eng. Sci.*, **5(4)**:235–245 (1965).
16. Chen, K. P. and D. D. Joseph, Elastic short wave instability in extrusion flows of viscoelastic liquids. *J. Non-Newtonian Fluid Mech.*, **42(1–2)**:189–211 (1992).
17. Busse, W. F., Two decades of polymer physics. A survey and forecast. *Phys. Today*, **17(9)**:32–41 (1964).
18. Busse, W. F., Mechanical structures in polymer melts. II. Roles of entanglements in viscosity and elastic turbulence. *J. Polym. Sci., A2*, **5**:1261–1281 (1967).
19. Laun, H. M., Squeezing flow rheometry to determine viscosity, wall slip and yield stresses of polymer melts. In *Polymer Processing Society, PPS12*, Sorrento, Italy, 1996, pp. 31–33.
20. Blyler, L. L. J. and A. C. J. Hart, Capillary flow instability of ethylene polymer melts. *Polym. Eng. Sci.*, **10(4)**:193–203 (1970).
21. Tordella, J. P., Melt fracture—Extrudate roughness in plastic extrusion. *SPE J.*, **February**:36–40 (1956).
22. Cogswell, F. N., Stretching flow instabilities at the exits of extrusion dies. *J. Non-Newtonian Fluid Mech.*, **2(1)**:37–47 (1977).
23. Gogos, C. G. and B. Maxwell, Velocity profiles of the exit region of molten polyethylene extrudates. *Polym. Eng. Sci.*, **6(4)**:353–358 (1966).

24. Lipscomb, G. G., R. Keunings, and M. M. Denn, Implications of boundary singularities in complex geometries. *J. Non-Newtonian Fluid Mech.*, **24(1)**:85–119 (1987).
25. Tanner, R. I., *Engineering Rheology*. Clarendon Press, Oxford, NY, 1985.
26. Wagner, M. H., V. Schulze, and A. Göttfert, Rheotens mastercurves and drawability of polymer melts. *Polym. Eng. Sci.*, **36**:925–935 (1996).
27. Venet, C., Propriétés d'écoulement et défauts de surface de résins polyéthylènes. PhD thesis, Ecole des Mines de Paris (CEMEF), Sophia Antipolis, France, 1996.
28. Spencer, R. S. and R. E. Dillon, The viscous flow of molten polystyrene. II. *J. Colloid Sci.*, **4(3)**:241–255 (1949).
29. Ballenger, T. F. and J. L. White, The development of the velocity field in polymer melts in a reservoir approaching a capillary die. *J. Appl. Polym. Sci.*, **15**:1949 (1971).
30. White, J. L., Critique on flow patterns in polymer fluids at the entrance of a die and instabilities leading to extrudate distortion. *Appl. Polym. Symp.*, **20**:155–174 (1973).
31. Ramsteiner, F., Melt fracture behaviour of plastics. *Rheol. Acta*, **16(6)**:650–651 (1977).
32. Mackley, M. R., F. C. Frank, and A. Keller, Flow induced crystallization of polyethylene melts. *J. Mater. Sci.*, **10(9)**:1501–1509 (1975).
33. Kurtz, S. J., Die geometry solutions to sharkskin melt fracture. In *Theoretical and Applied Rheology* (B. Mena, A. Garcia-Rejon, and C. Rangel Nafaile, Eds.). UNAM Press, Mexico City, Mexico, 1984, pp. 399–407.
34. Kurtz, S. J., The dynamics of sharkskin melt fracture: Effect of die geometry. In *Theoretical Applied Rheology: Proceedings of the 11th International Congress on Rheology* (P. K. Moldenaers and R. Keunings, Eds.), Brussels, Belgium. Elsevier, Amsterdam, the Netherlands, 1992, pp. 377–379.
35. Kurtz, S. J., Visualization of exit fracture in the sharkskin process. In *Polymer Processing Society*, PPS10, Akron, OH, 1994, pp. 8–9.
36. El Kissi, N. and J. M. Piau, Flow of entangled polydimethylsiloxanes through capillary dies: Characterisation and modelisation of wall slip phenomena. In *Third European Rheology Conference and Golden Jubilee of the British Society of Rheology* (D. R. Oliver, Ed.). Elsevier Applied Science, London, U.K., 1990, pp. 144–146.
37. Piau, J. M., N. El Kissi, and A. Mezghani, Slip flow of polybutadiene through fluorinated dies. *J. Non-Newtonian Fluid Mech.*, **59**:11–30 (1995).
38. Vinogradov, G. V., A. Y. Malkin, Y. G. Yanovskii, E. K. Borisenkova, B. V. Yarlykov, and G. V. Berezhnaya, Viscoelastic properties and flow of narrow distribution polybutadienes and polyisoprenes. *J. Polym. Sci.*, A2 **10**:1061–1084 (1972).
39. Borisenkova, E. K., V. E. Dreval, G. V. Vinogradov, M. K. Kurbanaliev, V. V. Moiseyev, and V. G. Shalganova, Transition of polymers from the fluid to the forced high-elastic and leathery states at temperatures above the glass transition temperature. *Polymer*, **23(1)**:91–99 (1982).
40. Vinogradov, G. V., V. P. Protasov, and V. E. Dreval, The rheological behaviour of flexible polymers in the region of high shear rates and stresses, the critical process of spurting, and supercritical conditions of their movement at T>Tg. *Rheol. Acta*, **23**:46–61 (1984).
41. Tremblay, B., Sharkskin defects of polymer melts: The role of cohesion and adhesion. *J. Rheol.*, **35(6)**:985–998 (1991).
42. Gent, A. N., Crazing and fracture of glassy plastics. *J. Macromol. Sci., Phys. B*, **8(3)**:597–603 (1973).
43. Argon, A. S., Physical basis of distortional and dilatational flow in glassy plastics. *J. Macromol. Sci., Phys. B*, **8(3)**:573–596 (1973).
44. Larson, R. G., Instabilities in viscoelastic flows. *Rheol. Acta*, **31**:213–263 (1992).
45. Venet, C. and B. Vergnes, Experimental characterization of sharkskin in polyethylenes. *J. Rheol.*, **41(4)**:873–892 (1997).
46. Benbow, J. J. and P. Lamb, New aspects of melt fracture. In *SPE ANTEC*, Los Angeles, CA, 3, 1963, pp. 7–17.

47. Den Otter, J. L., J. L. S. Wales, and J. Schijf, The velocity profiles of molten polymers during laminar flow. *Rheol. Acta*, **6(3)**:205–209 (1967).
48. Den Otter, J. L., Mechanisms of melt fracture. *Plast. Polym.*, **38(135)**:155–168 (1970).
49. Den Otter, J. L., Melt fracture. *Rheol. Acta*, **10(2)**:200–207 (1971).
50. Den Otter, J. L., Rheological measurements on two uncrosslinked, unfilled synthetic rubbers. *Rheol. Acta*, **14**:329–336 (1975).
51. Ramamurthy, A. V., Wall slip in viscous fluids and influence of materials of construction. *J. Rheol.*, **30(2)**:337–357 (1986).
52. Kalika, D. S. and M. M. Denn, Wall slip and extrudate distortion in linear low-density polyethylene. *J. Rheol.*, **31(8)**:815–834 (1987).
53. Rudin, A., A. T. Worm, and J. E. Blacklock, Fluorocarbon elastomer aids polyolefin extrusion. *Plast. Eng.*, **42(3)**:63–66 (1986).
54. Rudin, A., J. E. Blocklock, S. Nam, and T. Work, Improvements in polyolefin processing with fluorocarbon elastomer processing aid. In *SPE ANTEC*, Boston, MA, 32, 1986, pp. 1154–1157.
55. Athey, R. J., R. C. Thamm, R. D. Souffie, and G. R. Chapman, The processing behaviour of polyolefins containing fluoroelastomer additive. In *SPE ANTEC*, Boston, MA, 32, 1986, pp. 1149–1152.
56. Wang, S. Q., P. A. Drda, and A. Baugher, Molecular mechanisms for polymer extrusion instabilities: Interfacial origins. In *SPE ANTEC*, Toronto, Canada, 55, 1997, p. 1067.
57. Cook, D. G., R. Cooke, and A. Rudin, Use of chilled die lips to improve production rates in extrusion of polyethylene. *Int. Polym. Process.*, **4(2)**:73–77 (1989).
58. Hatzikiriakos, S. G., Wall slip of linear polyethylenes and its role in melt fracture. PhD thesis, McGill University, Montreal, Canada, 1991, p. 248.
59. Hatzikiriakos, S. G. and J. M. Dealy, Effects of interfacial conditions on wall slip and sharkskin melt fracture of HDPE. *Int. Polym. Process.*, **8(1)**:36–43 (1993).
60. Anastasiadis, S. H. and S. G. Hatzikiriakos, The work of adhesion of polymer/wall interfaces and its association with the onset of wall slip. *J. Rheol.*, **42(4)**:795–812 (1998).
61. de Gennes, P. G., Ecoulements viscometriques de polymeres enchevetres. *C. R. Acad. Sci. Ser. B*, **288**:219–220 (1979).
62. Brochard-Wyart, F., P. G. de Gennes, and P. Pincus, Suppression of sliding at the interface between incompatible polymer melts. *C. R. Acad. Sci., Ser. II*, **314(9)**:873–878 (1992).
63. Bergem, N., Visualization studies of polymer melt flow anomalies in extrusion. In *Seventh International Congress on Rheology* (C. Klason, Ed.). Göteborg, Sweden, 1976, pp. 50–54.
64. Drda, P. P. and S. Q. Wang, Stick-slip transition at polymer melt/solid interfaces. *Phys. Rev. Lett.*, **75(14)**:2698–2701 (1995).
65. Wang, S. Q. and P. A. Drda, Stick-slip transition in capillary flow of linear polyethylene: 3. Surface conditions. *Rheol. Acta*, **36**:128–134 (1997).
66. Wang, S. Q. and P. Drda, Molecular instabilities in capillary flow of polymer melts. Interfacial stick-slip transition, wall slip, and extrudate distortion. *Macromol. Chem. Phys.*, **198(3)**:673–701 (1997).
67. Beaufils, P., Etude des défauts d'extrusion des polyéthylènes linéaires. Approche expérimentalé et modélisation des scoulements. PhD thesis, Ecole des Mines de Paris (CEMEF), Sophia Antipolis, France, 1989.
68. Rutgers, R. P. G., An experimental and numerical study of extrusion surface instabilities for polyethylene melts. PhD thesis, Department of Chemical Engineering, University of Cambridge, Cambridge, U.K., 1998.
69. Richardson, S., A stick-slip problem related to the motion of a free jet at low Reynolds numbers. *Proc. Camb. Philos. Soc.*, **67**:477–489 (1970).
70. Sornberger, G., J. C. Quantin, R. Fajolle, B. Vergnes, and J. F. Agassant, Experimental study of the sharkskin defect in linear low density polyethylene. *J. Non-Newtonian Fluid Mech.*, **23**:123–135 (1987).

71. Piau, J. M., N. El Kissi, and B. Tremblay, Influence of upstream instabilities and wall slip on melt fracture and sharkskin phenomena during silicones extrusion through orifice dies. *J. Non-Newtonian Fluid Mech.*, **34(2)**:145–180 (1990).

72. Moynihan, R. H., D. G. Baird, and R. Ramanathan, Additional observations on the surface melt fracture behavior of linear low-density polyethylene. *J. Non-Newtonian Fluid Mech.*, **36**:255–263 (1990).

73. Watson, J. H., The mystery of the mechanism of sharkskin. *J. Rheol.*, **43(1)**:245–252 (1999).

74. Venet, C. and B. Vergnes, Stress distribution around capillary die exit: An interpretation of the onset of sharkskin defect. *J. Non-Newtonian Fluid Mech.*, **93(1)**:117–132 (2000).

75. Hatzikiriakos, S. G. and J. M. Dealy, Wall slip of molten high density polyethylene. I. Sliding plate rheometer studies. *J. Rheol.*, **35(4)**:497–523 (1991).

76. Person, T. J. and M. M. Denn, The effects of die materials and pressure dependent slip on the extrusion of linear low density polyethylene. *J. Rheol.*, **41**:249–265 (1997).

77. Ghanta, V. G., B. L. Riise, and M. M. Denn, Disappearance of extrusion instabilities in brass capillary dies. *J. Rheol.*, **43(2)**:435–442 (1999).

78. Vinogradov, G. V., V. E. Dreval, and V. P. Protasov, The static electrifications of linear flexible-chain polymers extruded through ducts above the glass transition temperature. *Proc. R. Soc. Lond., A* **409(1837)**:249–270 (1987).

79. Hatzikiriakos, S. G., C. W. Stewart, and J. M. Dealy, Effect of surface coatings on wall slip of LLDPE. *Int. Polym. Process.*, **8(1)**:30–35 (1993).

80. Anastasiadis, S. G., J.-K. Chen, J. T. Koberstein, A. F. Siegel, J. E. Sohn, and J. A. Emerson, The determination of interfacial tension by video image processing of pendent fluid drops. *J. Colloid Interface Sci.*, **119**:55–66 (1987).

81. Anastasiadis, S. H., Interfacial tension of immiscible polymer blends. PhD thesis, Princeton University, Princeton, NJ, 1988.

82. Stewart, C. W., Wall slip in the extrusion of linear polyolefins. *J. Rheol.*, **37(3)**:499–513 (1993).

83. Wu, S., *Polymer Interface and Adhesion.* Marcel Dekker, New York, 1982.

84. Henson, J. D. and M. E. Mackay, Effect of gap on viscosity of monodisperse polystyrene melts: Slip effects. *J. Rheol.*, **33**:38–47 (1995).

85. Beaufils, P., B. Vergnes, and J. F. Agassant, Characterization of the sharkskin defect and its development with the flow conditions. *Int. Polym. Process.*, **4(2)**:78–84 (1989).

86. de Gennes, P. G., Wetting: Statics and dynamics. *Rev. Mod. Phys.*, **57(3, Pt. 1)**:827–863 (1985).

87. Dhori, P. K., R. S. Jeyaseelan, A. Jeffrey Giacomin, and J. C. Slattery, Common line motion III: Implications in polymer extrusion. *J. Non-Newtonian Fluid Mech.*, **71(3)**:231–243 (1997).

88. Bagley, E. B., I. M. Cabott, and D. C. West, Discontinuity in the flow curve of polyethylene. *J. Appl. Phys.*, **29**:109 (1958).

89. Huseby, T. W., Hypothesis on a certain flow instability in polymer melts. *Trans. Soc. Rheol.*, **10(1)**:181–190 (1966).

90. Pao, Y.-H., Hydrodynamic theory for the flow of a viscoelastic fluid. *J. Appl. Phys.*, **28(5)**:591–598 (1957).

91. Pao, Y.-H., Theories for the flow of dilute solutions of polymers and of nondiluted liquid polymers. *J. Polym. Sci.*, **61**:413–448 (1962).

92. Pearson, J. R. A. and C. Petrie, In *Fourth International Congress on Rheology*, Providence, RI (E. H. Lee, Ed.). Interscience, New York, 1965, pp. 265–282.

93. Petrie, C. J. S. and M. M. Denn, Instabilities in polymer processing. *AIChE J.*, **22(2)**:209–236 (1976).

94. Pearson, J. R. A., *Mechanics of Polymer Processing.* Elsevier Applied Science, London, U.K., 1985.

95. Chandrasekhar, S., *Hydrodynamic and Hydromagnetic Stability*. Oxford University Press, Oxford, U.K., 1961.

96. Weill, A., Oscillations de relaxation du polyéthylène de haute densité et défauts d'extrusion. PhD thesis, Université Louis Pasteur, Strasbourg, France, 1978.

97. Weill, A., About the origin of sharkskin. *Rheol. Acta*, **19(5)**:623–632 (1980).

98. Weill, A., Capillary flow of linear polyethylene melt: Sudden increase of flow rate. *J. Non-Newtonian Fluid Mech.*, **7(4)**:303–314 (1980).

99. Van der Pol, B., On relaxation oscillation. *Philos. Mag.*, **2(11)**:978–994 (1926).

100. Molenaar, J. and R. Koopmans, Modeling polymer melt flow instabilities. *J. Rheol.*, **38(1)**:99–109 (1994).

101. Durand, V., B. Vergnes, J. F. Agassant, E. Benoit, and R. J. Koopmans, Experimental study and modeling of oscillating flow of high density polyethylenes. *J. Rheol.*, **40(3)**:383–394 (1996).

102. Den Doelder, C. F. J., R. J. Koopmans, J. Molenaar, and A. A. F. Van de Ven, Comparing the wall slip and the constitutive approach for modelling spurt instabilities in polymer melt flows. *J. Non-Newtonian Fluid Mech.*, **75(1)**:25–41 (1998).

103. Den Doelder, C. F. J., R. J. Koopmans, and J. Molenaar, Quantitative modelling of HDPE spurt experiments using wall slip and generalised Newtonian flow. *J. Non-Newtonian Fluid Mech.*, **79(2–3)**:503–514 (1998).

104. Molenaar, J., R. J. Koopmans, and C. F. J. Den Doelder, Onset of the sharkskin phenomenon in polymer extrusion. *Phys. Rev. E*, **58(4)**:4683–4691 (1998).

105. McLeish, T. C. B., Stability of the interface between two dynamic phases in capillary flow of linear polymer melts. *J. Polym. Sci., Part B: Polym. Phys.* **25(11)**:2253–2264 (1987).

106. Georgiou, G. C. and D. Vlassopoulos, On the stability of the simple shear flow of a Johnson-Segelman fluid. *J. Non-Newtonian Fluid Mech.*, **75**:77–97 (1998).

107. Koopmans, R. J. and J. Molenaar, The sharkskin effect in polymer extrusion. *Polym. Eng. Sci.*, **38(1)**:101–107 (1998).

108. Mair, R. W. and P. T. Callaghan, Shear flow of wormlike micelles in pipe and cylindrical Couette geometries as studied by nuclear magnetic resonance microscopy. *J. Rheol.*, **41(4)**:901–924 (1997).

109. Ovaici, H., M. R. Mackley, G. H. McKinley, and S. J. Crook, The experimental observation and modelling of an "Ovaici" necklace and stick-spurt instability arising during the cold extrusion of chocolate. *J. Rheol.*, **42(1)**:125–157 (1998).

110. Inn, Y. W., R. J. Fischer, and M. T. Shaw, Development of sharkskin melt fracture at the die exit in polybutadiene extrusion. In *70th Annual Meeting of Society of Rheology*, Monterey, CA, 1998, p. 32.

111. Georgiou, G. C. and M. J. Crochet, Compressible viscous flow in slits with slip at the wall. *J. Rheol.*, **38**:639–654 (1994).

112. Georgiou, G. C., Extrusion of a compressible Newtonian fluid with periodic inflow and slip at the wall. *Rheol. Acta*, **35**:531–544 (1996).

113. Kumar, K. A. and M. D. Graham, Effect of pressure-dependent slip on flow curve multiplicity. *Rheol. Acta*, **37**:245–255 (1998).

114. Brasseur, E., M. M. Fyrillas, G. C. Georgiou, and M. J. Crochet, The time-dependent extrudate-swell problem of an Oldroyd-B with slip along the wall. *J. Rheol.*, **42(3)**:549–566 (1998).

115. Kolkka, R. W., D. S. Malkus, M. G. Hansen, G. R. Ierley, and R. A. Worthing, Spurt phenomena of Johnson-Segalman fluid and related models. *J. Non-Newtonian Fluid Mech.*, **29**:303–335 (1988).

116. Malkus, D. S., J. A. Nohel, and B. J. Plohr, Dynamics of shear flow of a non-Newtonian fluid. *J. Comput. Phys.*, **87**:464–487 (1990).

117. Malkus, D. S., J. A. Nohel, and B. Plohr, Analysis of new phenomena in shear flow of non-Newtonian fluids. *SIAM J. Appl. Math.*, **51(4)**:899–929 (1991).

118. Malkus, D. S., J. A. Nohel, and B. J. Plohr, Oscillations in piston driven shear flow of a non-Newtonian fluid. In *IUTAM Symposium on Numerical Simulation of Non-Isothermal Flow of Viscoelastic Liquids* (J. F. Dijksman and G. C. D. Kuiken, Eds.). Kerkrade, the Netherlands. Kluwer, Dordrecht, the Netherlands, 1994, pp. 57–74.

119. Denn, M. M., Issues in viscoelastic fluid mechanics. *Annu. Rev. Fluid Mech.*, **22**:13–34 (1990).

120. Espanol, P., X. F. Yuan, and R. C. Ball, Shear banding flow in Johnson-Segelman fluid. *J. Non-Newtonian Fluid Mech.*, **65**(1):93–109 (1996).

121. McLeish, T. C. B. and R. C. Ball, A molecular approach to the spurt effect in polymer melt flow. *J. Polym. Sci., Part B: Polym. Phys.*, **24**:1735–1745 (1986).

122. Deiber, J. A. and W. R. Schowalter, On the comparison of simple non-monotonic constitutive equations with data showing slip of well-characterized polybutadienes. *J. Non-Newtonian Fluid Mech.*, **40**:141–150 (1991).

123. Vlassopoulos, D. and S. G. Hatzikiriakos, A generalized Giesikus constitutive model with retardation time and its association to the spurt effect. *J. Non-Newtonian Fluid Mech.*, **57**:119–136 (1995).

124. Lin, Y. H., Explanation for slip-stick melt fracture in terms of molecular dynamics in polymer melts. *J. Rheol.*, **29**(6):605–637 (1985).

125. Lin, Y. H., Unified molecular theories of linear and non-linear viscoelasticity of flexible linear polymers-explaining the 3.4 power law of the zero-shear viscosity and the slip-stick melt fracture phenomenon. *J. Non Newtonian Fluid Mech.*, **23**:163–187 (1987).

126. Joshi, Y. M. and A. K. Lele, Dynamics of end-tethered chains at high surface coverage. *J. Rheol.*, **46**(2):427–453 (2002).

127. Tchesnokov, M. A., J. Molenaar, and J. J. M. Slot, Dynamics of molecules adsorbed on a die wall during polymer extrusion. *J. Non-Newtonian Fluid Mech.*, **126**:71–82 (2005).

128. Stepanyan, R., J. J. M. Slot, J. Molenaar, and M. Tchesnokov, A simple constitutive model for a polymer flow near a polymer-grafted wall. *J. Rheol.*, **49**(5):1129–1151 (2005).

129. Colby, R. H., L. J. Fetters, and W. W. Graessley, The melt viscosity-molecular weight relationship for linear polymers. *Macromolecules*, **20**(9):2226–2237 (1987).

130. Deeprasertkul, C., C. Rosenblatt, and S. Q. Wang, Molecular character of sharkskin phenomenon in metallocene linear low density polyethylenes. *Macromol. Chem. Phys.*, **199**:2113–2118 (1998).

131. Doi, M. and S. F. Edwards, *The Theory of Polymer Dynamics*. Clarendon Press, Oxford, U.K., 1986.

132. Kuhn, W., Uber die Gestalt fadenformiger Molekule in Losungen. *Kolloid-Z*, **68**:2–15 (1934).

133. Yang, X., H. Ishida, and S. Q. Wang, Wall slip and absence of interfacial flow instabilities in capillary flow of various polymer melts. *J. Rheol.*, **42**(1):63–80 (1998).

134. Oyanagi, Y., Irregular flow behavior of high density polyethylene. *Appl. Polym. Symp.*, **20**:123–136 (1973).

135. Ballenger, T. F., I. J. Chen, J. W. Crowder, G. E. Hagler, D. C. Bogue, and J. L. White, Polymer melt flow instabilities in extrusion: Investigation of the mechanisms and material and geometric variables. *Trans. Soc. Rheol.*, **15**(2):195–215 (1971).

136. Tordella, J. P., Unstable flow of molten polymers. In *Rheology* (F. R. Erich, Ed.). Academic Press, New York, 1969, pp. 57–92.

137. Combeaud, C., B. Vergnes, A. Merten, D. Hertel, and H. Münstedt, Volume defects during extrusion of polystyrene investigated by flow induced birefringence and laser-Doppler velocimetry. *J. Non-Newtonian Fluid Mech.*, **145**:69–77 (2007).

138. Hürlimann, H. P. and W. Knappe, Relation between the extensional stress of polymer melts in the die inlet and in melt fracture. *Rheol. Acta*, **11**(3–4):292–301 (1972).

139. Boger, D. V. and K. Walters, *Rheological Phenomena in Focus*. Elsevier, Amsterdam, the Netherlands, 1993.

8 Advanced Polymer Rheology

In Chapter 3, relatively simple constitutive equations have been introduced. The validity of such a mathematical construct, that is, a model, in rheology depends on the number of experimentally observed flow phenomena that can be described adequately. A rheological model can be highly successful when applied to one specific type of experiment, but may only poorly describe other experimental observations. Accordingly, the value of a constitutive equation depends on the needs of the user and the extent of modeling accuracy that is desired. For most polymer melts, sophisticated nonlinear constitutive equations are required to obtain sufficient predictive accuracy.

Typically, the versatility of a constitutive equation is "judged" by its ability to describe the shear stress τ_{12} in relatively simple rheological experiments. Sometimes, the first and second normal stress differences

$$N_1 = \tau_{11} - \tau_{22}, \quad N_2 = \tau_{22} - \tau_{33} \tag{8.1}$$

are used too, although they are more difficult to obtain accurately. If measured in laminar flow, these quantities only depend on shear rate. In Equation 3.90, the viscosity η expresses the relationship between shear stress and shear rate:

$$\tau_{12} = \eta \dot{\gamma}, \tag{8.2}$$

with the shear rate as defined in Equation 3.30. Similarly, the first and second normal stress coefficients are introduced. They express the dependence of the normal stress differences on shear rate:

$$N_1 = \psi_1 \dot{\gamma}^2, \quad N_2 = \psi_2 \dot{\gamma}^2. \tag{8.3}$$

The viscosity together with these coefficients are referred to as the *viscometric functions*.

It is typical for polymers to show shear thinning or shear thickening (also referred to as *strain softening* and *shear hardening*), that is, a strong dependence of the viscometric functions on shear rate. In practice, the rate of shear thinning is the most important polymer melt characteristic. In the Newtonian model (Equation 3.89), these functions are constants, which is in contrast to the experimental evidence. This shortcoming is repaired heuristically and leads to the generalized Newtonian model dealt with in Equation 3.94. However, in such an approach, the functional form of the shear rate dependence of the viscosity does not follow from any theoretical hypothesis, but from fitting experimental data. From a theoretical point of view, the challenge is to develop models that inherently predict characteristic polymer properties like shear thinning.

In this chapter, the approach started in Chapter 3 is continued. In the literature, a great variety of constitutive equations has been published. Surveys are given in, for example, Refs. [1–5]. All the proposed constitutive equations have their advocates. Some of them are attractive from an analytical point of view; others are easily tractable in a numerical approach. No equation has yet proven to describe all polymer melt flow situations accurately so that it could be recommended for modeling any flow situation. At this stage, the advantages and disadvantages of the models have to be balanced in relation to their relevance to the application under consideration. Here, only a few characteristic ones are discussed in view of their use in conjunction with the topic of polymer melt fracture.

A distinction can be made between continuous or macroscopic models in which the polymer melt is described as a continuum, and discrete or microscopic models in which the macromolecules are considered, individually or in a coarse-grained way. First, however, it is important to present some empirical findings regarding polymer architecture and rheology.

8.1 MOLAR MASS, ZERO-SHEAR VISCOSITY, AND RECOVERABLE COMPLIANCE

The relationships between macroscopically derived material functions and polymer architecture are dominated by empirical findings [6,7]. MM and MMD are the more easily accessible polymer characteristics. The material functions zero-shear viscosity η_0 and the steady-state compliance J_e^0 show a dependence on MM (Figures 8.1 and 8.2).

An abrupt change in their functional form occurs at certain critical MM values. The experimentally found M_c and M_c' are not identical, although their physical interpretation follows the same general reasoning (Table 8.1). Low-MM polymer melts ($M < M_c$) are characterized by a few or no macromolecular entanglements. Their resistance to flow is controlled by the monomeric frictional coefficient times the macromolecular length—being proportional to MM M for linear homopolymers. High-MM polymers ($M > M_c$) are characterized by many entanglements. These cause a constraint on the possible macromolecular diffusion mechanism yielding a stronger MM dependence of the zero-shear viscosity. The zero-shear viscosity reflects an internal macromolecular friction and is a measure for the dissipated energy during flow. A constant J_e^0 value for $M > M_c$ implies that the number of macromolecular entanglements per unit melt volume is constant and independent of MM. The steady-state compliance determines the amount of elastically stored energy that can be recovered from the polymer melt after a deformation takes place. The critical MM values M_c and M_c' can be associated with the minimum macromolecular length needed to generate entanglements. Their numerical difference follows from the different experimental methodologies used to obtain them, suggesting different macromolecular dynamics to establish effective entanglements. The type of deformation

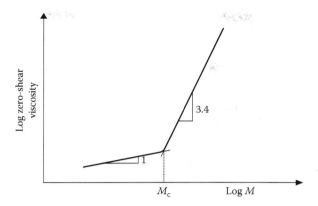

FIGURE 8.1   Schematic representation on double logarithmic axes of the zero-shear viscosity dependence on the MM. At a critical MM M_c a dramatic increase in slope from 1 to 3.4 occurs. (After Graessley, W.W., Viscoelasticity and flow in polymer melts, in *Physical Properties of Polymers*, Mark, J.E. (Ed.), American Chemical Society, Washington, DC, 1984.)

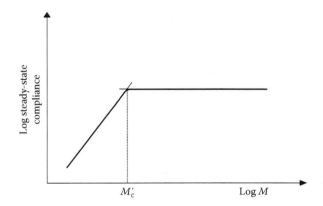

FIGURE 8.2   Schematic representation on double logarithmic axes of the steady-state compliance dependence on the MM. At a critical MM M_c' the steady-state compliance becomes independent of the MM.

history can induce different entanglements. Therefore, entanglements can be understood as temporal topological interactions, that is, constraints, between the macromolecules. The equivalency arises from the experimental observation that at high deformation rates, the rheological response of the polymer melt is similar to that of a fully chemically fixed, that is, cross-linked, macromolecular network (Figure 2.2). For these materials, the macromolecules are chemically bonded with on average a MM M_e between cross-links. Cross-linked networks yield an elastic modulus G_0 that is nearly independent of the deformation rate.

The entanglement MM M_e is related to M_c by a factor of about 2 ($M_c \sim 2M_e$) [9–11]. It can be calculated from the experimentally determined plateau (Figure 8.3)

TABLE 8.1
Critical MM Values (kg/mol) for Linear Polymers

Polymer Name	M_e	M_c	M_c'
Polyethylene	1.25	3.8	14.4
1,4-Polybutadiene	1.7	5.0	11.9
cis-Polyisoprene	6.3	10.0	28.0
Polyvinylacetate	6.9	24.5	86.0
Polydimethylsiloxane	8.1	24.4	56.0
Polystyrene	19.0	36.0	130.0

Source: After Graessley, W.W., Viscoelasticity and flow in polymer melts, in *Physical Properties of Polymers*, Mark, J.E. (Ed.), American Chemical Society, Washington, DC, 1984.

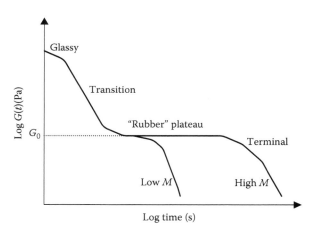

FIGURE 8.3 Schematic diagram of the relaxation modulus as function of time for a high and low MM (with $M > M_c$) polymer melt on logarithmic axes.

denoted as the elastic-, plateau-, or relaxation modulus G_0, making use of the classical "rubber-elasticity" theory [6]. This leads to Equation 8.4,

$$M_e = \frac{\rho RT}{G_0},\tag{8.4}$$

where
  ρ is the (polymer melt) density
  R is the universal gas constant
  T is the absolute temperature

In analogy with Equation 3.124, the average relaxation time λ_0 for $G(t)$ can be considered as a measure of the time required for final equilibration following a step-strain experiment. λ_0 is approximately proportional to the product $\eta_0 J_e^0$. Following the experimental findings represented in Figures 8.1 and 8.2 this implies for $M < M_c$

$$\begin{cases} \eta_0 \propto M \\ J_e^0 \propto M \end{cases} \Rightarrow \lambda_0 \propto M^2. \tag{8.5}$$

For $M > M_c$ it is found that

$$\begin{cases} \eta_0 \propto M^{3.4} \\ J_e^0 \approx \text{constant} \end{cases} \Rightarrow \lambda_0 \propto M^{3.4}. \tag{8.6}$$

Accordingly, the flow in each of the two MM regions is governed by different flow mechanisms. In the low MM region, the frictional forces at a very local scale control the flow, whereas in the high MM region, entanglement (topological constraint) plays the dominant role.

The effect of MMD on these two key rheological properties has also been studied empirically. For η_0, the effect of MMD is considered negligible when the weight-average MM M_w is selected to replace M in the entangled regime [6]. For J_e^0, a strong effect of MMD is demonstrated [6,12–14] with a clear role for the higher MM moments of the distribution:

$$\begin{cases} \eta_0 \propto M_w^{3.4} \\ J_e^0 \propto \left(M_z / M_w \right)^{3-3.7} \\ J_e^0 \propto \left(M_{z+1} M_z \right) / \left(M_w M_n \right) \end{cases}. \tag{8.7}$$

8.2 CONTINUOUS MODELS AND FRAME INVARIANCE

The derivation of continuous models is restricted by the rules of continuum mechanics. As for constitutive relations, not all mathematically possible relationships between local stress and strain are admissible from a physical point of view. In particular, such a relationship, when applied to a fluid element, must be independent of any (possibly time-dependent) rotation of the element. This is because the microscopic origin of the stress only depends on the relative orientations of the molecules in the element, and not on the orientation of the element as a whole. This independence of rotations is referred to as *frame invariance*. In Section 3.4.1, this point was discussed by noting that the strain tensor **A** in Equation 3.26 is not frame invariant, but the Finger tensor **B** in Equation 3.27 is. In addition, the rate of deformation tensor **D**, defined in Equation 3.42, is not frame invariant.

In order to construct frame-invariant constitutive equations, it is convenient to use coordinate systems that are not rigid, but embedded in and deforming with

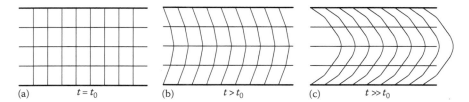

FIGURE 8.4 Level lines of contravariant coordinates that deform as time increases in the case of laminar flow in a slit die.

the flow. Such coordinates are called *convected coordinate frames*. The study of convected frames is the subject of differential geometry, a classical field of interest in mathematics. The application of the insights of differential geometry to continuum mechanics was stimulated by the work of Lodge [15,16] and Oldroyd [17]. The details of the subject are fairly complex and can be found in the specialized literature [2]. In what follows, the concept of contravariant coordinates, which are quite easy to imagine, will be used. Choose at the origin of time an orthogonal coordinate frame in the fluid. The level lines of this frame are straight and parallel or orthogonal. The position of each fluid particle is characterized as the crossing point of two orthogonal level lines. If the flow evolves, the particles follow their material lines. In the case of contravariant coordinates, it is assumed that the coordinates transform such that the level lines remain attached to the particles. While moving and deforming, they remain neither straight nor parallel or orthogonal (Figure 8.4).

In such a frame, the fluid particles always have the same coordinates and the Finger tensor **B** equals the unit matrix. Accordingly, the time derivative of **B** with respect to contravariant coordinates vanishes, which is denoted as

$$\overset{\triangledown}{\mathbf{B}} = \mathbf{0}. \tag{8.8}$$

The form of this so-called *upper-convected derivative* with respect to a fixed Cartesian coordinate frame is not easy to derive [2]. Its action on an arbitrary tensor **T** is given by

$$\overset{\triangledown}{\mathbf{T}} = \frac{D\mathbf{T}}{Dt} - \mathbf{C}^{\mathrm{T}} \cdot \mathbf{T} - \mathbf{T}^{\mathrm{T}} \cdot \mathbf{C}. \tag{8.9}$$

The material derivative D/Dt, defined in Equation 3.65, takes into account that the coordinate frame translates with the flow. The terms including the displacement tensor **C**, defined in Equation 3.40, account for the local deformations of the flow, that is, rotation and stretching.

Using definition (3.27) and property (3.54), it can be checked that Equation 8.8 is satisfied, so **B** is frame invariant. In the same way, one can check that the rate of deformation tensor **D**, defined in Equation 3.42, does not satisfy Equation 8.8, so this tensor is not frame invariant.

8.2.1 Upper-Convected Maxwell Model

Since the tensor $\mathbf{D}$ is not frame invariant, the Maxwell model, given in Equation 3.106 (differential form) and Equation 3.107 (integral form), is also not frame invariant. A straightforward modification for repair is to rewrite Equation 3.107 replacing $\mathbf{D}$ by the frame invariant strain tensor $\mathbf{B}$:

$$\tau(t) = \frac{\eta_0}{\lambda_0^2} \int_{-\infty}^{t} e^{-(t-t')/\lambda_0} \mathbf{B}(t', t)\, dt'. \tag{8.10}$$

The extra factor λ_0 in the denominator is introduced for later convenience. This expression is a special case of the Lodge equation for τ

$$\tau(t) = \int_{-\infty}^{t} M(t - t')\mathbf{B}(t', t)\, dt', \tag{8.11}$$

which is the frame invariant analogue of Equation 3.117. In Equation 3.117, it is shown that in laminar flow, the extension from Equation 3.117 to Equation 8.11 does not make any difference for the shear stress. However, these models still have a very different nature. Model (3.117) is linear because the stress and the displacement rates, measured by γ, are linearly related. The model implied by Equation 8.11 is called *quasi-linear*. The stress and $\mathbf{B}$ are linearly related, but $\mathbf{B}$ itself may be a quadratic function of the displacements measured by the tensor $\mathbf{A}$, as follows from definition (3.27). The integral formulation (Equation 8.10) has a differential counterpart given by

$$\tau + \lambda_0 \overset{\triangledown}{\tau} = \frac{\eta_0}{\lambda_0} \mathbf{I} = G_0 \mathbf{I}. \tag{8.12}$$

Since the upper-convected derivative enters here, the present model is referred to as the *upper-convected Maxwell* (UCM) model. When verifying the equivalence of the integral form in Equation 8.10 and the differential form in Equation 8.12 by substituting the former into the latter, it should be noted that the time t appears at three places in the integral: in the upper limit of integration, in the exponential, and in $\mathbf{B}$. The $\partial/\partial t$ term in the upper-convected derivative acts on these three places, whereas the other terms act on $\mathbf{B}$ only. In the verification, properties (3.54) and (8.8) can be used.

It seems as if the differential UCM equation (Equation 8.12) differs from the Maxwell equation (Equation 3.106) not only in the type of derivative, but also on the right-hand side. This is not true, as can be shown as follows. If the upper-convected derivative (Equation 8.9) is applied to the unit tensor $\mathbf{I}$, the following relation is obtained:

$$\overset{\triangledown}{\mathbf{I}} = -\mathbf{C}^{\mathrm{T}} - \mathbf{C} = -2\mathbf{D}. \tag{8.13}$$

Here, use is made of Equation 3.42. From this result it is easily verified that the stress tensor

$$\boldsymbol{\tau}' = \boldsymbol{\tau} - G_0\mathbf{I} \tag{8.14}$$

satisfies the equation

$$\boldsymbol{\tau}' + \lambda_0 \overset{\triangledown}{\boldsymbol{\tau}}' = 2\eta_0\mathbf{D}, \tag{8.15}$$

which has the same structure as Equation 3.106. For incompressible fluids, the stress tensors $\boldsymbol{\tau}'$ and $\boldsymbol{\tau}$ are identical, because they only differ a constant isotropic tensor. As is common, the prime is omitted in the following.

Example 8.1:   The UCM Model in Laminar Flow

For laminar flow, the strain tensor $\mathbf{B}$ is given by Equation 3.35. The substitution of its matrix elements into Equation 8.10 yields for the shear stress

$$\tau_{12}(t) = \tau_{21}(t) = \frac{\eta_0}{\lambda_0^2} \int_{-\infty}^{t} e^{-(t-t')/\lambda_0}\gamma(t', t)\, dt', \tag{8.16}$$

and for the normal stress differences

$$\tau_{11}(t) - \tau_{22}(t) = \frac{\eta_0}{\lambda_0^2} \int_{-\infty}^{t} e^{-(t-t')/\lambda_0}\gamma^2(t', t)\, dt', \tag{8.17}$$

$$\tau_{22}(t) - \tau_{33}(t) = 0.$$

In steady laminar flow, the shear rate $\dot{\gamma}$ is constant and the strain is given by

$$\gamma(t',t) = \dot{\gamma} \cdot (t - t'). \tag{8.18}$$

Substitution of this in (8.16) yields for the shear stress

$$\tau_{12} = \tau_{21} = \eta_0\dot{\gamma}, \tag{8.19}$$

while substitution into Equation 8.17 leads to

$$\psi_1 = \eta_0\lambda_0, \quad \psi_2 = 0, \tag{8.20}$$

where use is made of the notations in (8.1) and (8.3).

From Example 8.1, it is seen that the UCM model does not show shear thinning in any of the viscometric functions, which is a serious shortcoming. It is interesting to realize that the UCM model was derived from molecular arguments many years before the macroscopic argument of frame invariance was put forward [18]. The UCM model is widely used in viscoelastic flow calculations, mostly because of its simplicity. A nice feature of this model is that its governing equation (8.15) is identical to the equation of motion of a so-called *dumbbell model*. In the dumbbell model, a macromolecule is represented (i.e., coarse grained) by two beads connected by an elastic spring. It is a special case of the Rouse theory dealt with in Section 8.3.1.

Similar to the Maxwell model in Equation 3.125, the UCM model can be extended by including more than one exponential:

$$\tau(t) = \sum_{i=0}^{\infty} \frac{\eta_i}{\lambda_i^2} \int_{-\infty}^{t} e^{(t-t')/\lambda_i} \mathbf{B}(t', t) \, dt'. \tag{8.21}$$

This generalized UCM model, also referred to as the Lodge network model, includes a spectrum of relaxation times λ_i, the relative importance of which is measured by the parameters η_i. Both the Rouse theory, discussed in Section 8.3.1, and the reptation model, discussed in Section 8.3.2, give rise to constitutive equations of type (8.21). The latter theories have a molecular basis and yield explicit expressions for the parameters λ_i and η_i in terms of the parameters of the mechanical model describing the molecular interactions [2,19–23].

8.2.2 JOHNSON–SEGALMAN–(OLDROYD) MODELS

8.2.2.1 Johnson–Segalman Model

Johnson and Segalman [24–26] proposed a model that can be considered as an extension of the UCM model. They modified the upper-convected derivative in Equation 8.9 and arrived at

$$\overset{\circ}{\mathbf{T}} = \frac{D\mathbf{T}}{Dt} - \mathbf{C}^{\mathrm{T}} \cdot \mathbf{T} - \mathbf{T}^{\mathrm{T}} \cdot \mathbf{C} + \xi(\mathbf{D} \cdot \mathbf{T} + \mathbf{T}^{\mathrm{T}} \cdot \mathbf{D}), \tag{8.22a}$$

for an arbitrary tensor **T**. In view of Equation 3.41, this derivative can also be written as

$$\overset{\circ}{\mathbf{T}} = \frac{D\mathbf{T}}{Dt} - \mathbf{V}^{\mathrm{T}} \cdot \mathbf{T} - \mathbf{T}^{\mathrm{T}} \cdot \mathbf{V} - a(\mathbf{D} \cdot \mathbf{T} + \mathbf{T}^{\mathrm{T}} \cdot \mathbf{D}), \tag{8.22b}$$

with $a = 1 - \xi$ and $-1 \leq a \leq 1$. This is the so-called *Gordon–Schowalter* (GS) convected derivative [27]. Compared to the upper-convected derivative in Equation 8.9, the presence of an extra term with coefficient ξ is recognized. Gordon and

Schowalter derived a constitutive equation from a kinetic theory for dilute polymer solutions in which the zero-shear viscosity is given by $\eta_s + (1-\xi)\eta_p$. In this expression, the contributions of the solvent viscosity η_s and of the polymer viscosity η_p are recognized. The latter contribution is reduced by the factor $1-\xi$. This expresses that the macromolecule in a dilute solution may "slip" with respect to the solvent. Each macromolecule therefore transmits only a fraction of its tension to the surrounding solvent. The strength of the slip is measured by the parameter ξ. Although the GS convected derivative has been derived for polymer solutions, it is applied to polymer melts in the present context. The Johnson–Segalman (JS) model reads as

$$\tau + \lambda_0 \overset{\circ}{\tau} = \mu'\mathbf{I}. \tag{8.23}$$

In terms of the stress tensor $\tau' = \tau - \mu'\mathbf{I}$, this reads as

$$\tau' + \lambda_0 \overset{\circ}{\tau}' = 2\mu\mathbf{D}. \tag{8.24}$$

Since the GS derivative stems from a molecular theory, explicit expressions for μ' (and μ) are available:

$$\mu' = a^2 \nu kT, \quad \mu = a\mu'\lambda_0, \tag{8.25}$$

where
  ν is the number of entangled macromolecules per unit volume
  k is the Boltzmann constant
  T is the absolute temperature

The JS model can be put into integral form:

$$\tau(t) = \sum_{i=0}^{\infty} \frac{\eta_0}{\lambda_0^2} \int_{-\infty}^{t} e^{-(t-t')/\lambda_0} \mathbf{E}(t', t)\, dt', \tag{8.26}$$

provided that the newly introduced tensor $\mathbf{E}$, referred to as the *elastic strain measure*, satisfies the equations

$$\overset{\circ}{\mathbf{E}} = 0, \quad \mathbf{E}(t, t) = \mathbf{I}. \tag{8.27}$$

To show that the differential form in Equation 8.23 and the integral form in Equation 8.26 are identical, the same derivation applies as used to identify Equations 8.10 and 8.12 of the UCM model.

Example 8.2: JS Model for Laminar Flow

The JS model in Equation 8.24 can be worked out for laminar flow through a slit die as follows. The expressions for the tensors **C** and **D** of this type of flow are given by Equations 3.45 and 3.46, respectively. Because the system is essentially 2D, the (symmetric) stress tensor will have the form

$$\boldsymbol{\tau'} = \begin{bmatrix} \tau'_{11} & \tau'_{12} & 0 \\ \tau'_{21} & \tau'_{22} & 0 \\ 0 & 0 & \tau'_{33} \end{bmatrix} \tag{8.28}$$

with the nonvanishing elements depending on t and x_2 only.

After substitution of Equations 3.45 and 3.46 in Equation 8.24 and working out the matrix multiplications, one obtains the following set of differential equations for the diagonal elements:

$$\begin{cases} \lambda_0 \dfrac{\partial \tau'_{11}}{\partial t} = -\tau'_{11} + (a+1)\lambda_0 \dot{\gamma}\tau'_{12} \\[2mm] \lambda_0 \dfrac{\partial \tau'_{22}}{\partial t} = -\tau'_{22} + (a-1)\lambda_0 \dot{\gamma}\tau'_{12} \\[2mm] \lambda_0 \dfrac{\partial \tau'_{33}}{\partial t} = -\tau'_{33} \end{cases} \tag{8.29}$$

For the non-diagonal element τ_{12}, one finds the equation

$$\lambda_0 \frac{\partial \tau'_{12}}{\partial t} = -\tau'_{12} + \frac{1}{2}\lambda_0 \dot{\gamma}((a-1)\tau'_{11} + (a+1)\tau'_{22}) + \mu\dot{\gamma} \tag{8.30}$$

In Section 9.2, these equations will be used to model the flow curve in relation to melt fracture phenomena predictions. Here, only the calculation of the steady state of the system is considered. In this case, all differentiations with respect to time vanish, and the system reduces to a set of algebraic equations. After some algebra, one obtains for the shear rate–dependent viscosity, defined in Equation 8.2,

$$\eta(\dot{\gamma}) = \frac{\mu}{1 + (1 - a^2)\lambda_0^2 \dot{\gamma}^2}. \tag{8.31}$$

For the other viscometric functions one finds in a similar way

$$\begin{cases} \psi_1(\dot{\gamma}) = \dfrac{2\lambda_0 \mu}{1 + (1 - a^2)\lambda_0^2 \dot{\gamma}^2}, \\[3mm] \psi_2(\dot{\gamma}) = \dfrac{(a-1)\lambda_0 \mu}{1 + (1 - a^2)\lambda_0^2 \dot{\gamma}^2} \end{cases} \tag{8.32}$$

Their ratio is thus given by

$$\frac{\Psi_2}{\Psi_1} = -\frac{(1-a)}{2}. \tag{8.33}$$

The negative sign of this ratio agrees with the experimental findings. In experiments, its value varies from -0.05 to -0.30; this suggests a to lie between 0.4 and 0.9, and thus ξ to have a value between 0.1 and 0.6.

The JS model is a member of a family referred to as the *Oldroyd-8-constant model*. This is a generalization of the UCM model presented in Equations 8.10 and 8.12. The extension is such that first, frame invariance is retained, and second, the resulting equation is at most linear in τ and at most quadratic in $\mathbf{D}$ and τ combined. These conditions are at the level of continuum mechanics and no molecular arguments are introduced. Such a general approach leads to an equation with eight parameters. These parameters have to be calculated on a molecular basis or obtained from experimental data through fitting. Dealing with the full model is quite tedious. In the practice, only special cases of this family are applied.

From Example 8.2, it is seen that the JS model describes shear thinning. Although this is clearly an argument in favor of the model, it predicts unrealistic behavior under the application of step strain. The resulting viscometric functions show an unrealistic oscillatory dependence on the shear rate. This can be repaired by replacing the GS convected derivative by the so-called *Seth convected derivative* [3]. The latter approach is beyond the scope of this book.

8.2.2.2 Johnson–Segalman–Oldroyd Model

The JS model in Section 8.2.2.1 can be modified [28–32] by introducing an extra viscous term in the total stress tensor. This implies that instead of Equation 3.25, the total stress $\mathbf{S}$ is written as the sum of a pressure part, a Newtonian viscous part, and a JS–Oldroyd (JSO) part:

$$\mathbf{S} = -p\mathbf{I} + 2\eta_0\mathbf{D} + \boldsymbol{\sigma}. \tag{8.34}$$

It is assumed that $\boldsymbol{\sigma}$ (and not $\boldsymbol{\tau}'$, as in the JS model) satisfies an equation similar to Equation 8.24:

$$\boldsymbol{\sigma} + \lambda_0 \overset{\circ}{\boldsymbol{\sigma}} = 2\mu\mathbf{D} \tag{8.35}$$

with μ being a shear modulus. Following Aarts [33], this model is referred to as the JSO model.

Equation 8.35 constitutes a set of partial differential equations because derivatives with respect to both time and space are present via the material derivatives. In laminar flow through a slit die, the stress depends on time t and the vertical coordinate x_2. The material derivative D/Dt reduces to the partial time derivative only, and one obtains ordinary differential equations. Flow calculations first require solving the Equations 8.29 for the stress components, with σ replacing τ'. Second, the stresses have to be substituted into the equation of motion (Equation 3.83). For the JSO model, the factor τ in Equation 3.83 must be replaced by $2\eta_0 \mathbf{D} + \sigma$ defined in (8.34). The solution of the complete set of equations then yields the velocity field. This procedure is worked out in Section 9.2.

8.2.3 KAYE–BERNSTEIN–KEARSLEY–ZAPAS MODEL

The JSO model is a generalization of the UCM model in Equation 8.12. The UCM integral form (Equation 8.11) may now be generalized. In the UCM model, the Finger tensor $\mathbf{B}$ is present in a linear fashion. It is possible to go beyond this linearity by including higher powers of $\mathbf{B}$. An alternative method to generalize Equation 8.11 is to make the memory function M dependent on $\mathbf{B}$. Since M is a scalar function, it cannot depend on $\mathbf{B}$ itself, but only on its invariants I_B, II_B, and III_B introduced in Equations 3.10 through 3.12:

$$I_B = Tr(\mathbf{B}), \tag{8.36}$$

$$II_B = \frac{1}{2}(I_B^2 - Tr(\mathbf{B}^2)) = Tr(\mathbf{B}^{-1}), \tag{8.37}$$

$$III_B = Det(\mathbf{B}) = 1. \tag{8.38}$$

The last equality in Equation 8.38 holds for incompressible fluid flow only. The last equality in Equation 8.37 can be proven using Equation 3.16 and the last equality in Equation 8.38. Since III_B is taken to be constant, it can be omitted in the following.

Following the above ideas of generalization, Kaye [34,35] and Bernstein et al. [36] proposed a heuristic extension of the expression for the stress tensor of an ideal elastic solid. In such a solid, the interaction between stress and strain is instantaneous. To incorporate viscous effects, an integral over the past is introduced leading to the so-called *K-BKZ model* [2–5]:

$$\tau(t) = \int_{-\infty}^{t} \left[\varphi_1(t - t', I_B, II_B)\mathbf{B} + \varphi_2(t - t', I_B, II_B)\mathbf{B}^{-1} \right] dt'. \tag{8.39}$$

Equation 8.39 entails a very broad class of constitutive equations if the memory functions φ_1 and φ_2 can be chosen arbitrarily. Therefore, the inventors proposed a subclass of Equation 8.39 by demanding that φ_1 and φ_2 are not independent, but stem from a common potential function V:

$$\varphi_1 = \frac{\partial V(t-t', I_\mathrm{B}, II_\mathrm{B})}{\partial I_\mathrm{B}},$$

$$\varphi_2 = \frac{\partial V(t-t', I_\mathrm{B}, II_\mathrm{B})}{\partial II_\mathrm{B}}.$$

$$(8.40)$$

The introduction of this restriction has the additional advantage that it excludes some physically unreasonable predictions following from the full model for certain fast strain experiments [37].

The restricted class of K-BKZ models is still quite broad and nearly always further restrictions are introduced. Often, the potential function V is considered to be factorable into a time-dependent and a strain-dependent part:

$$V(t-t', I_\mathrm{B}, II_\mathrm{B}) = M(t-t') \cdot W(I_\mathrm{B}, II_\mathrm{B}). \tag{8.41}$$

If the restriction (Equation 8.41) holds, the K-BKZ model is called *separable* or *factorable*. The memory function M in Equation 8.41 is associated with the memory function in Equation 3.117, which implies that

$$M(t) = \frac{dG(t)}{dt}, \tag{8.42}$$

with the relaxation function G given by Equation 3.125. The parameters G_i and λ_i in the latter expression are to be fitted from experimental data. No standard procedure is available to fit the potential function W from experiment. It has to satisfy some limiting conditions in order to assure that the K-BKZ model reduces to the linear models presented in Chapter 3 in case of small displacements. If the strain is very small, the approximation $\mathbf{B} \approx \mathbf{B}^{-1} \approx \mathbf{I}$ holds, and thus $I_\mathrm{B} \approx II_\mathrm{B} \approx 3$. This may be easily verified by taking the very small strain γ in Equations 3.35 and 3.36. It leads to the condition

$$\lim_{I_\mathrm{B}, II_\mathrm{B} \to 3} \left(\frac{\partial W}{\partial I_\mathrm{B}} + \frac{\partial W}{\partial II_\mathrm{B}} \right) = 1. \tag{8.43}$$

Example 8.3: Viscometric Functions in the K-BKZ Model for Laminar Flow

From Example 3.2, it is clear that in laminar flow, both $\mathbf{B}$ and $\mathbf{B}^{-1}$, and thus also I_B and II_B only depend on the strain γ. The K-BKZ model yields for this case

$$\tau_{12}(t) = \int_{-\infty}^{t} M(t-t') \cdot \gamma(t', t) \cdot \left(\frac{\partial W}{\partial I_\mathrm{B}} - \frac{\partial W}{\partial II_\mathrm{B}} \right) dt', \tag{8.44}$$

where the '12' elements of Equations 3.35 and 3.36 are used. In steady laminar flow, the shear rate $\dot\gamma$ is constant in time and the strain is given by

$$\gamma(t',t) = \dot\gamma \cdot (t - t'). \tag{8.45}$$

Substitution of Equation 8.45 into Equation 8.44 leads to the K-BKZ result

$$\tau_{12}(t) = \dot\gamma \int_{-\infty}^{t} M(t - t') \cdot (t',t) \cdot \left(\frac{\partial W}{\partial I_B} - \frac{\partial W}{\partial II_B} \right) dt'. \tag{8.46}$$

The application of the transformation $s = t - t'$ yields an expression for the viscosity defined in Equation 8.2:

$$\eta(\dot\gamma) = \int_{0}^{\infty} M(s) \cdot s \cdot \left(\frac{\partial W}{\partial I_B} - \frac{\partial W}{\partial II_B} \right) ds, \tag{8.47}$$

In a similar way, the first and second normal stress coefficients introduced in Equation 8.3 are obtained:

$$\psi_1 = \int_{0}^{\infty} M(s) \cdot s^2 \cdot \left(\frac{\partial W}{\partial I_B} - \frac{\partial W}{\partial II_B} \right) ds,$$

$$\psi_2 = -\int_{0}^{\infty} M(s) \cdot s^2 \cdot \frac{\partial W}{\partial II_B} ds. \tag{8.48}$$

The latter equation shows that ψ_2, and thus N_2, vanish if the potential function W does not depend on II_B.

Example 8.4: Specific Choices for the K-BKZ Potential Function Applied to Laminar Flow

Since in laminar flow W is only a function of the strain, Wagner [38] expressed the terms with W in Equation 8.47 directly in terms of the strain:

$$\frac{\partial W}{\partial I_B} - \frac{\partial W}{\partial II_B} = e^{-\alpha\gamma}. \tag{8.49}$$

This exponential can be interpreted as a damping function, which induces a shear thinning correction on the relaxation function. The parameter α must be fitted from experimental data.

An alternative suggestion comes from Papanastasiou et al. [39]. In their approach, W is assumed to be independent of II_B. As shown in Example 8.3, this has the consequence that $N_2 = 0$. They propose

$$\frac{\partial W}{\partial I_B} = \frac{c}{c - 3 + I_B},$$ (8.50)

with c a dimensionless material constant to be fitted from the experimental data. It is easily checked that this choice satisfies condition (8.43). In laminar flow, one has

$$I_B = 3 + \gamma^2,$$ (8.51)

so that this model reduces to

$$\frac{\partial W}{\partial I_B} = \frac{c}{c + \gamma^2}.$$ (8.52)

For the stress components the following expressions are obtained

$$\tau_{11}(t) = \int_{-\infty}^{t} M(t - t') \cdot \frac{c}{c + \gamma^2(t', t)} \cdot (1 + \gamma^2(t', t)) \, dt',$$

$$\tau_{22}(t) = \tau_{33}(t) = \int_{-\infty}^{t} M(t - t') \cdot \frac{c}{c + \gamma^2(t', t)} \, dt',$$ (8.53)

$$\tau_{12}(t) = \tau_{21}(t) = \int_{-\infty}^{t} M(t - t') \cdot \frac{c}{c + \gamma^2(t', t)} \cdot \gamma(t', t) \, dt'.$$

These stresses depend on the spatial coordinate x_2, too. When substituted in Equation 3.84, the K-BKZ equation of motion in a slit die is obtained. It is an integro-differential equation. The steady state solution can be found in analytical form, whereas the calculation of the transient behavior requires a numerical approach [33,40].

8.3   MICROSCOPIC MODELS

Microscopic modeling faces the challenge of representing the polymer architecture, that is, the individual macromolecules, in a manner that captures their essential characteristics without the need of resorting to detailed atomistic descriptions of configurations and conformations. Configurations are the permanent geometrical arrangements of chemical bonds into space. Conformations are distinct geometrical shapes attainable by bond rotations. For a fixed configuration, multiple conformations are possible. In addition, the macromolecular representations need to hold sufficient information to enable the mathematical formalism to describe the dynamics of individual macromolecules and their contributions to the total stress. Polymers are very large flexible molecules for which a huge number of conformations are possible [41]. In the melt, this leads to a conformation distribution that is equivalent to that of a random walk (see Chapter 2). At equilibrium, the molecules remain unoriented even though the Brownian forces continuously change the conformation. It leads to the conclusion that the time-averaged mean-square distance $\langle S^2 \rangle_0$ separating one end of the molecule from the other obeys the random walk formula:

$$\left\langle S^2 \right\rangle_0 = n b_n^2 \tag{8.54}$$

where
    n is the number of random walk steps
    b_n is the length of such a step

If each covalent bond in the macromolecule connecting two neighboring atoms would be a rigid link but completely rotationally flexible at the joints, then b_n would be the length of that link. This representation of a macromolecule is called a *freely jointed chain*. In real polymers, the rotational freedom is restricted. However, as the orientation of a bond is unrelated to the one many bonds away, the conformational properties of a macromolecule are well represented by a random walk provided Equation 8.54 is satisfied. Still bond angle restrictions make $\langle S^2 \rangle_0$ larger than for a freely jointed chain,

$$\left\langle S^2 \right\rangle_0 = C_\infty n l^2 \tag{8.55}$$

with l the bond length and C_∞ the characteristic ratio given by

$$C_\infty \equiv \frac{b_n^2}{l^2} \tag{8.56}$$

At equilibrium, a polymer, that is, many macromolecules of different MM, has a distribution of conformations that is represented by a set of random walks or equivalently by the conformations of freely jointed chains. The probability function of such a distribution is approximately Gaussian or normal and the conformations are referred to as induced by Gaussian chains.

In polymer flow, work is exerted on the melt and induces a deformation that changes the conformations of the macromolecules. This implies that the stress depends on the conformation distribution. As the orientation and degree of the stretch of a macromolecule increases so will stress. The total stress in a fluid element depends on the contribution of all macromolecules in that element each having its own conformation. Microscopic theories for stress must therefore account for the distribution of conformations [3].

Since the number of macromolecules is huge, use has to be made of an averaging procedure and a much simpler representation of the macromolecules (coarse-graining). Some calculations are rather technical and it suffices to point out the basic ideas and formulae with reference to standard textbooks for details [4,20].

A standard concept is to describe a macromolecule as a series of beads connected by rigid rods or flexible springs. The simplest representation is the beads-and-springs chain. The springs mimic the elastic forces when the macromolecule is stretched while the beads produce hydrodynamic drag. The simplest of such representations is the elastic dumbbell, which contains a single spring connecting two beads [9]. In such a description, or for the generalized version of multiple beads and springs, an expression is needed for the stress tensor in terms of the internal (in-between beads) and external forces. Such an expression exists and forms the starting point of the molecular theory dealt with here.

Consider a box of volume V filled with beads. The position of bead i is denoted by $\mathbf{R}_i$ and the force of bead i exerted on bead j by $\mathbf{F}_{i,j}$. The total force on bead j is then given by

$$\mathbf{F}_j = -\frac{1}{V}\sum_{i \neq j}\mathbf{F}_{i,j}. \tag{8.57}$$

The viscous stress tensor is given by the Kirkwood stress formula [20]

$$\tau = -\frac{1}{V}\sum_j\langle\mathbf{F}_j\mathbf{R}_j\rangle. \tag{8.58}$$

Here, the notation $\langle\ldots\rangle$ means to average over all possible conformations of the beads. Between the brackets the dyadic product of the two vectors $\mathbf{F}_j$ and $\mathbf{R}_j$ is present as introduced in Equation 3.5.

8.3.1 ROUSE MODEL

In the Rouse model, a long, unbranched macromolecule is represented as a series of N beads, connected by "linear" springs that have no spatial restrictions on the angles between neighboring springs. The Rouse model is particularly useful to describe the dynamics of macromolecules in a solvent. If the concentration of macromolecules is low, they will hardly affect each other and only interact with the continuum of solvent molecules. The effect of this environment on the beads is twofold. Friction slows down movements, but also Brownian forces keep the beads moving in a stochastic way. When focusing on one particular bead j with time-dependent bead-position $\mathbf{R}_j(t)$, the equation of motion of this bead reads as

$$m\ddot{\mathbf{R}}_j = -\varsigma(\dot{\mathbf{R}}_j - \mathbf{v}_j) + \mathbf{F}_{j,j+1} + \mathbf{F}_{j,j-1} + \mathbf{f}_j, \tag{8.59}$$

where

    m is the bead mass
    ς is the friction coefficient
    $\mathbf{v}_j$ is the velocity of the solvent near bead j
    $\mathbf{f}_j$ represents the Brownian force exerted on bead j

The spring force $\mathbf{F}_{j,j+1}$ is linearly proportional to the distance between beads j and $j+1$ with spring constant k. For the end beads of the chain, not two, but only one neighboring force applies. The next step typically taken is to replace the discrete beads by an elastic string in which the mass is smeared out over the length of the chain. This requires a limiting procedure in which the number of beads is gradually increased and the mass per bead is decreased such that the total mass is conserved. The result is that the position $\mathbf{R}_j(t)$ of bead j transforms into a function $\mathbf{R}(x, t)$ with the discrete index j replaced by a continuous variable x. This variable x may be understood by imagining that the chain is given its equilibrium length L and that it is positioned along the x-axis and fills the interval $[0, L]$. The specific chain element which in that situation has coordinate x is indicated by x at all times, even if the chain attains an arbitrary position in shape, thus given by the function $\mathbf{R}(x, t)$.

After applying the transition to a continuous string, Equation 8.59 transforms into

$$\rho\frac{\partial^2 \mathbf{R}}{\partial t^2} = -\varsigma\left(\frac{\partial \mathbf{R}}{\partial t} - \mathbf{v}\right) + k\frac{\partial^2 \mathbf{R}}{\partial x^2} + \mathbf{g}, \qquad (8.60)$$

where

    $\rho(x, t)$ is the density of the chain mass
    k is the elastic constant
    $\mathbf{v}(x, t)$ is the solvent velocity around the xth element of the chain
    $\mathbf{g}(x, t)$ is the Brownian force at chain element x and time t

In polymer solutions, the friction is dominating the inertia term on the left-hand side, making the behavior in time of a chain in good approximation governed by

$$\varsigma\frac{\partial \mathbf{R}}{\partial t} = \varsigma\mathbf{v} + k\frac{\partial^2 \mathbf{R}}{\partial x^2} + \mathbf{g}. \qquad (8.61)$$

However, for stress calculations in a melt, this approximation may be too crude. Since Equation 8.61 contains a second order spatial derivative, two boundary conditions need to be specified. Standard conditions are

$$\frac{\partial \mathbf{R}}{\partial x}(0, t) = \frac{\partial \mathbf{R}}{\partial x}(L, t) = \mathbf{0}, \qquad (8.62)$$

which imply that the chain has no internal stress at the endpoints.

A somewhat unnatural property of the Rouse model is that the elastic internal force tends for the chain to contract unto a point. This is counteracted however by the Brownian forces if the chain is embedded in a solvent. That is why this peculiarity in the model is accepted. The feature could be remedied by introducing artificial forces at the end points, such that the chain gets a prescribed equilibrium length.

Since in a dilute melt the chains are assumed not to interact, the contribution to the stress from that chain can be calculated by applying Equation 8.58 with V being a typical volume that contains only one chain. In the continuous version

$$\tau = -\frac{1}{V}\frac{N}{L}\left\langle \int_0^L F(x,t)R(x,t)dx \right\rangle, \tag{8.63}$$

with the force $F(x, t)$ given by the right-hand side of Equation 8.60.

To calculate the shear stress, the velocity field of laminar flow in the x-direction is taken as in Equation 3.44:

$$v(x,t) = (\dot{\gamma}_0 R_2(x,t),0,0), \tag{8.64}$$

where R_2 is the second component of the position vector $R(x, t)$. The evaluation of (8.62) could readily make use of the following expansion in which variables are separated:

$$R(x,t) = \sum_{n=1}^{n=\infty} r_n(t)\phi_n(x). \tag{8.65}$$

Substituting this expansion in Equation 8.61 and using the boundary conditions (8.62) provides explicit expressions for the coefficient vector functions $r_n(t)$ and the so-called normal modes $\varphi_n(x)$. The outcome is that the stress can be expressed in terms of $r_n(t)$:

$$\tau = -\frac{kN}{VL}\sum_{n=1}^{n=\infty}\langle r_n(t)\, r_n(t)\rangle\left(\frac{n\pi}{L}\right)^2. \tag{8.66}$$

Interest lies with $\tau_{1,2}$ of the stress tensor. The (1,2) component of $\langle r_n(t)\, r_n(t)\rangle$ is given by

$$\langle r_n(t)r_n(t)\rangle_{1,2} \equiv \langle r_{1,n}(t)r_{2,n}(t)\rangle = 2\frac{k_B T}{\varsigma}\dot{\gamma}_0\int_{-\infty}^{t} e^{-\frac{2k\mu_n}{\varsigma}(t-t')}(t-t')\,dt', \tag{8.67}$$

with μ_n given by $\mu_n = -(n\pi/L)^2$. If Equations 8.66 and 8.67 are compared with Equations 3.115 and 3.116, respectively, it can be concluded that the relaxation function of the Rouse model is given by

$$G(t - t') = \frac{Nk_BT}{VL} \sum_{n=1}^{n=\infty} \int_{-\infty}^{t} e^{-\frac{2k_j\mu_n}{\varsigma}(t-t')} (t - t') \, dt'. \tag{8.68}$$

The integral is evaluated using the transformation $t' \to (t - t')$. From Equations 3.127 and 3.148 the quantities of interest can be calculated. For example, for the case with N_0 chains per unit volume, the viscosity is given by

$$\eta_0 = \frac{\varsigma L N N_0 k_B T}{12kV}. \tag{8.69}$$

This shows that for the Rouse model, viscosity η scales linearly with MM M, since M scales linearly with L (at least for linear macromolecules). Similarly, the other material functions for the Rouse model in the case of dilute polymer solutions can be derived:

$$\lambda_0 = \left(\frac{\varsigma N_A K_\theta}{\pi^2 M_0 RT} \right) M^2 \propto M^2, \tag{8.70}$$

$$\eta_0 = \left(\frac{\varsigma_0 N_A K_\theta \rho}{6 M_0} \right) M \propto M, \tag{8.71}$$

$$J_e^0 = \left(\frac{2}{5\rho RT} \right) M \propto M. \tag{8.72}$$

where
ς_0 is the monomeric friction coefficient
M is the polymer MM
N_A is the Avogadro number
$K_\theta = \langle S^2 \rangle / M$ ($\langle S^2 \rangle$ is the mean-square radius of gyration (Figure 2.9))
$M_0 = M/N$ the repeat unit MM
R is the universal gas constant
T is the absolute temperature
ρ is the density

The Rouse theory is clearly not applicable to polymer melts of a MM greater than M_c for which entanglement plays an important role.

8.3.2 REPTATION MODEL

For polymer melts with $M > M_c$, an alternative theory is developed based on the conceptual idea of a macromolecule trapped in a network. Consider a macromolecule A entangled with other macromolecules. In three dimensions, such a system

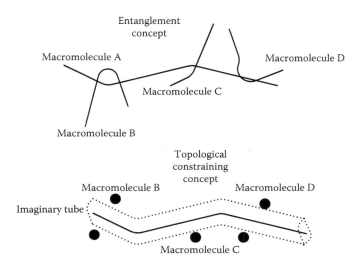

FIGURE 8.5 Schematic representation of a macromolecule A entangled in a network rep-resented by macromolecules B, C, and D. The network (upper representation) defines topo-logical constraints (black dots in lower representation) on the movement of macromolecule A. The overall spatial constraint can be thought of as defined by an imaginary tube through which the macromolecule needs to move when a force is exerted on the system.

can be thought of as a macromolecule A surrounded by a network of other macro-molecules that are obstacles (black dots) or topological constraints (Figure 8.5) for its relative motion.

The moving macromolecule A is not allowed to cross any of the obstacles. It can only move in a snake-like fashion in the axial direction within the confinement of the imaginary tube. This motion is given the name of reptation by its inventor Pierre Gilles de Gennes [42]. The original tube-model idea developed for describing the motion of a single free macromolecule in a network of cross-linked macromolecules is extended to uncross-linked polymer melts by Doi and Edwards [20]. Subsequently, it became convenient to think about the macromolecule as trapped in an imaginary tube as its motion can now be thought of as a form of 1D diffusion. The middle sections of the macromolecule must follow their neighbors along the tube contour while the poly-mer ends are free to explore the melt isotropically and create a new tube (Figure 8.6).

The connection between the reptation concept and the stress is as follows. In a polymer melt without stress, all macromolecules have attained their equilibrium

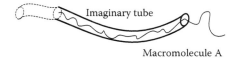

FIGURE 8.6 By reptation a macromolecule A diffuses through its original imaginary tube and creates a new tube segment. The "forgotten" portion of the original tube is indicated by the dotted line.

length and their configurations are completely random. If stress is induced, for example, by deforming the polymer melt, the configurations, that is, the tubes, will deviate from being random and the macromolecules may be stretched and become oriented. For small stresses, the randomness is dominant, while at high shear conditions, macromolecular stretch may contribute considerably. The induced stress will decrease over time. The macromolecules diffuse out of the nonrandom tubes induced by stress and establish new tubes that are randomly configured. The relaxation time of the stress is directly determined by the typical time of a diffusion process in one dimension. Since diffusion is a classical topic, much is known on how to model it.

First consider one diffusing macromolecule in a rigid grid. In reptation theory a macromolecule is represented by a Rouse chain, as modeled in Section 8.3.1. The segments of such a chain are labeled with index x, running from 0 to L. In the first instance it is assumed that the macromolecule has fixed length L, so the connecting springs are replaced by rigid rods. Initially, the corresponding tube has also length L, since the chain is assumed to have not enough room in the tube to curl. If a part of the macromolecule leaves the tube at one end, a similar part of the tube at the other end no longer contains a macromolecule making the effective tube shorter.

Since the macromolecule moves randomly, this happens at both ends making the tube shorter and shorter over time, with the part in the middle surviving on average the longest. This process is identical to a diffusion phenomenon like heat "leaking-away" in a rod via the endpoints, generating a temperature distribution that satisfies the standard diffusion equation. When, accordingly, a probability distribution function is considered

$$\psi(x, t) = \text{Prob}\{\text{segment } x \text{ is stil part of the tube at time } t\}. \tag{8.73}$$

As the function contains all information about the dynamics of the tube, the length σ of the surviving tube part follows on average from

$$\langle \sigma(t) \rangle = \int_0^L \psi(s, t)\, ds. \tag{8.74}$$

Here $\langle \ldots \rangle$ denotes averaging over all similar tube processes. The probability distribution $\psi(x, t)$ satisfies the diffusion equation

$$\frac{\partial \psi}{\partial t} = D \frac{\partial^2 \psi}{\partial x^2}, \tag{8.75}$$

with boundary conditions

$$\psi(x, 0) = \psi(L, 0) = 0 \tag{8.76}$$

and initial profile

$$\psi(x, t) = 1, \quad 0 < x < L. \tag{8.77}$$

This linear diffusion problem is easily solved. Just as for the Rouse model an expansion in basic functions can be used leading to

$$\psi(x, t) = \sum_{n=1}^{n=\infty} r_n(t)\phi_n(x). \tag{8.78}$$

Substituting this expansion into the diffusion equation and using the boundary and initial conditions gives the solution

$$\psi(x, t) = \sum_{n=1,3,5,\dots} \frac{4}{n\pi} \sin\left(\frac{n\pi x}{L}\right) e^{D\mu_n t} \tag{8.79}$$

with $\mu_n = -(n\pi/L)^2$. This demonstrates that this diffusion or tube renewal process has a dominant relaxation time

$$\lambda_0 = \frac{1}{\mu_1} = \frac{L^2}{\pi^2 D}. \tag{8.80}$$

The diffusion constant D is inversely proportional to the monomeric friction as already used in the Rouse model. It is also inversely proportional to the number of monomers N, since the longer the macromolecule, the more resistance it will experience when sliding through the tube. This leads to

$$D \propto \frac{1}{N\varsigma}. \tag{8.81}$$

Since both L and N scale linearly with MM M, it follows that λ_0 scales cubically with M:

$$\lambda_0 \propto \frac{L^2 N\varsigma}{\pi^2} \propto M^3. \tag{8.82}$$

For the Rouse model of a dilute polymer solution λ_0 scales with M^2, which leads to the conclusion that for large M, reptation is a much slower process.

The constitutive relation for the reptation model is found by starting from Equation 8.58 and using the dynamics of a Rouse model confined in a tube [4,20]. Eventually, this gives

$$\tau = G_0 M(t)\mathbf{Q}(E). \tag{8.83}$$

where
  G_0 is the plateau or relaxation modulus
  $\mathbf{Q}(E)$ is a tensor measuring the strain in the melt
  $M(t)$ is a memory function given by

$$M(t) = \sum_{n=1,3,5,\ldots} \frac{8}{n^2 \pi^2} e^{-2n^2 Dt/L^2}.$$ (8.84)

Note that this expression for the stress tensor shows separation between a time-dependent and a strain-dependent part, as is also the case for the K-BKZ model in Equation 8.41.

The reptation model relaxation modulus G_0 is independent of the MM for highly entangled macromolecules and the zero-shear viscosity scales as $\eta_0 \approx G_0 \lambda_0$, providing the scaling relations

$$G_0 \propto M^0$$ (8.85)

$$\eta_0 \propto M^3$$ (8.86)

$$J_e^0 \propto M^0$$ (8.87)

Experimental data, however, indicate that η_0 increases with $M^{3.4}$ for linear polymers. Graessley [43] and Lodge et al. [44] pointed out that in real polymer melts, a competition between reptation and other relaxation mechanisms exists. The viscosity values from the Doi–Edwards model, considering only reptation, should be taken as limit values. One of the competing mechanisms is contour length fluctuations. In the derivation above, the macromolecular representative chain is assumed to have a fixed length. In reality, this length fluctuates. Temporary shortening of the chain has the effect that the surviving part of the tube will decay faster than without contour length fluctuation.

Another effect to be accounted for is that the tube is not a fixed object over time. The tube consists of obstacles that stem from neighboring chains that themselves are also reptating, so these obstacles may disappear in the course of the time, simply because the associated chain has moved away. This effect is referred to as "constraint release (CR)." Both contour length fluctuation and constraint release cause the viscosity to scale in a way closer to the observed relation $\eta_0 \sim M^{3.4}$. Marrucci [45] realized that CR becomes more and more important if the shear rate increases. For these conditions, the fact that the chains have no fixed length but may stretch and shrink must be taken into account. For high shear rates, the effect of so-called convective constraint release (CCR) becomes an essential part of any description based on reptation theory.

Likhtman and Graham [46] developed such a model. To keep it numerically manageable, they introduced some reducing assumptions and coined the name Rolie-Poly model. Tchesnokov et al. [47] and Stepanyan et al. [48] extended the theory such that it includes all the mentioned effects, especially CCR and chain stretching. This leads to very good rheology simulations as long as the flow rate is moderate. At very high flow rates, however, the chains may stretch so extremely that they lead to violation of the assumptions underlying this theory. In fact, the tube picture of reptation theory seems to be no longer the most adequate description of polymer flow at very high flow rates, since effects such as CCR and stretching are no longer perturbations of the general scheme, but become dominant.

8.3.3　Branching

Branches restrict the relative number of the conformational modes of the macro-molecule. Depending on the detailed polymer architecture, they reduce the motion of the individual macromolecules. This is especially true when the branch length is larger than the entanglement MM M_e. Equivalently, branching restricts the reptation motion significantly leading to extremely high η_0. The increase of η_0 due to branching is easily explained in terms of the reptation picture. A good state-of-the-art review is given by McLeish [49], although the research in this field is still ongoing, for example, see Dubbeldam and Molenaar [50]. The branches on the longest continuous linear macromolecule, also called "backbone," are inside their own tubes. As long as the branches are entangled with the surroundings, they inhibit the backbone to reptate. Each of the branches has to retract itself from its own tube before the backbone may start to move in its tube. If the arms have fixed lengths, retraction would not take place, and no reptation would occur at all. However, the arms, acting as Rouse chains, may shrink to have zero length, but the chance for this to happen is small and becomes smaller with increasing arm length. The characteristic time of this slow process increases exponentially with the branch length or equivalently the number of entanglements per branch. For star-branched macromolecules, η_0 is given by

$$\eta_0 \propto A e^{\left(\alpha \frac{M_b}{M_e}\right)}, \tag{8.88}$$

where
M_b is the MM of the branch
M_e is the entanglement MM
A and α are constants

For comb type macromolecules, the η_0 enhancement depends on the branch length and the spacing between the branch points along the backbone [11].

The steady-state compliance J_e^0 is often found larger for branched macromolecules than for linear macromolecules, but it is seldom clear whether this is due to branching or to the accompanying broader MM distribution. According to the modified Rouse theory [6], J_e^0 is a linear function of M and for branched macromolecules it should be less than for linear macromolecules by a factor

$$g_2 = \frac{\langle S^4 \rangle}{\langle S^2 \rangle^2}, \tag{8.89}$$

with $\langle S^2 \rangle$ the mean-square radius of gyration (Figure 2.9). As a result, the J_e^0 of star-branched macromolecules is lower than for linear macromolecules at low MM, but larger for high MM. For comb and random-tree branched macromolecules, J_e^0 is always larger than for the corresponding linear macromolecules. For model comb branched macromolecules, the reduced steady-state compliance is given by

$$J_{eR} = J_e^0 \frac{c_m RT}{M_m} = 0.4 \left(\frac{c_m M_m}{\rho M_c'} \right)^b, \tag{8.90}$$

with

$$\begin{cases} b = 0 & \text{if } c_m M_m < \rho M_c', \\ b = -1 & \text{if } c_m M_m > \rho M_c', \end{cases} \tag{8.91}$$

where
  c_m is the backbone concentration
  ρ is the melt density
  M_c' is the critical MM

Equation 8.90 indicates that J_e^0 is exclusively due to the compliance of the backbone, and diluted with branches alone.

The shear-rate dependency of branched macromolecules is different from that of linear macromolecules. In general,

$$\eta_0 J_e^0 \dot{\gamma}_0 \approx 0.6, \tag{8.92}$$

with $\dot{\gamma}_0$ the shear rate at which the viscosity drops to 80% of η_0. Because J_e^0 is higher for branched compared to linear macromolecules, shear-dependent behavior is observed at lower shear rate. The nonlinear behavior of branched macromolecules has not been investigated systematically. However, after extensive shearing of branched polymer melts, the relaxation time for recovery can be of the order of hours.

8.3.4 Pom-Pom Model

Highly entangled polymers may show strikingly different behavior in different situations. An example is LDPE that has many, tree-like long branches. It shows strain hardening behavior in uniaxial extensional flow and strain softening behavior in shear flow. The phenomena of strain hardening and strain softening were already discussed in Section 3.6.2 in the context of shear flow and using the terms shear thickening and shear thinning. The present terms may refer to any type of flow. As for strain hardening, the behavior of LDPE is different from that of unbranched melts, but its strain softening behavior is very similar to that of unbranched polymers.

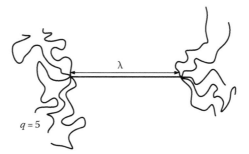

FIGURE 8.7 The Pom-Pom model describes the dynamical behavior of entangled macromolecules that have a stretchable backbone λ with at each end q dangling arms. Here, $q = 5$.

It turns out that the K-BKZ model, although being quite flexible, is not appropriate to predict both phenomena simultaneously. According to Samurkas et al. [51], the K-BKZ model fails to predict the observed degree of strain hardening in planar extension when the kernel functions are adjusted to fit the observed degree of strain softening in shear. This motivated McLeish and Larson [52] to develop a constitutive model that could describe the different types of LDPE behavior. The essential new element they wanted to incorporate is the presence of multiple branch points on one molecule. Any part of the backbone between two successive branch points has no free ends. The idea is that these parts may cause the material to harden in extensional flow. This led them to the so-called *Pom-Pom model*. The model is a macromolecular representation containing sufficient parameters to fit experimental data well, but has no resemblance to the real molecular structure of branched polymers. Its simplicity is inspired by the desire to keep the evaluation tractable in practice, which has favored its great popularity. In Figure 8.7, the chain structure that is used in the Pom-Pom model is sketched: a stretchable backbone with a number of q arms at both ends.

As is usual in reptation theory, the backbone and the arms are assumed to be confined by tubes. The backbone may not move unless all arms have retracted, but it may stretch and by that contribute to the stress. The arms will exert forces on the backbone and tend to stretch the backbone until it reaches a maximum tension. At higher stretch values, the arms will be drawn into the backbone tube. The essential features of this model are the presence of two distinct relaxation times, one for orientation and one for stretch, and the maximum on the stretching ability of the backbone. The combination of these two elements causes the model to show both strain hardening in extensional flow, uniaxial and planar, and extreme strain softening in shear flow.

The simplest version of the Pom-Pom model that still includes these aspects consists of the following set of equations. The stress tensor of a model with q arms is given by

$$\tau = \frac{15}{4} G_0 \varphi_b^2 \lambda^2(t) \mathbf{S}(t). \qquad (8.93)$$

Here, the orientation tensor $\mathbf{S}$ is introduced; its development in the time being governed by

$$\frac{\partial}{\partial t}\mathbf{A} = \mathbf{K}\cdot\mathbf{A} + \mathbf{A}\cdot\mathbf{K}^{\mathrm{T}} - \frac{1}{\tau_{\mathrm{b}}}\left(\mathbf{A} - \frac{1}{3}\mathbf{I}\right) \quad \text{with } \mathbf{S}(t) = \frac{\mathbf{A}(t)}{Tr\,\mathbf{A}(t)}. \tag{8.94}$$

The velocity gradient tensor $\mathbf{K}$ measures the deformation rate and the factor $\lambda(t)$ measures the stretch of the backbone. For $\lambda = 1$ there is no stretch. The maximum value for λ is q. As long as $\lambda < q$, its time evolution follows from

$$\frac{\partial}{\partial t}\lambda = \lambda(\mathbf{K}:\mathbf{S}) - \frac{1}{\tau_{\mathrm{s}}}(\lambda - 1). \tag{8.95}$$

Here the ":" symbol stands for full contraction of two matrices:

$$\mathbf{K}:\mathbf{S} = \sum_{i,j}K_{ij}\,S_{ji}. \tag{8.96}$$

The microscopic parameters of the model are the MM of the backbone M_{b}, the MM of the arms M_{a}, and the entanglement MM M_{e}, which measures the average MM between two successive entanglements in a tube. From these, the dimensionless numbers $s_{\mathrm{a}} = M_{\mathrm{a}}/M_{\mathrm{e}}$ and $s_{\mathrm{b}} = M_{\mathrm{b}}/M_{\mathrm{e}}$ are formed. The factor φ_{b} is defined as $\varphi_{\mathrm{b}} = s_{\mathrm{b}}/(2qs_{\mathrm{a}} + s_{\mathrm{b}})$ and the two relaxation times of the backbone and the arms, respectively, are given by

$$\tau_{\mathrm{b}} = \frac{4}{\pi^2}s_{\mathrm{b}}^2\varphi_{\mathrm{b}}\tau_{\mathrm{a}}(0)q, \quad \tau_{\mathrm{s}} = s_{\mathrm{b}}\tau_{\mathrm{a}}(0)q, \tag{8.97}$$

with $\tau_{\mathrm{a}}(0)$ the longest relaxation time of the arms, that is, the relaxation time associated with complete retraction of the arm.

8.4 MOLAR MASS DISTRIBUTION AND LINEAR VISCOELASTICITY

Empirical relations between structure and rheology were introduced in Section 8.1. Section 8.3 deals with microscopic theoretical concepts that explain how average MM and the presence of branching affect rheology. The present section explores similar tube-based theoretical considerations with regards to arbitrary ensembles of molecules and their rheology. Industrial polymers have a MMD, that is, they are non uniform. It is empirically known that MMD has large effects on shear thinning, which is strongly connected to the onset of melt fracture. This will be applied in Chapter 9. As an example, linear entangled polymers of arbitrary uniformity are discussed here.

Conventional reptation theory considers macromolecules in tubes that all have the same length. For non uniform systems, the tubes are made of macromolecules having a distribution of lengths. This changes the constraint release dynamics of the system. The double reptation (DR) concept of Tsenoglou [53] and Des Cloizeaux [54] accounts for this.

The essential features of MMD were introduced in Chapter 2. The log-based continuous function $w(M)$ $(dw/d \log M)$ is a practical descriptor for large ensembles of molecules. It is defined as

$$\int_{M=0}^{M \to \infty} w(M) d \log(M) = 1. \tag{8.98}$$

Its summarizing moments are

$$M_{\mathrm{w}} = \int_{M=0}^{M \to \infty} M \cdot w(M) \, d \log(M), \tag{8.99}$$

$$\frac{1}{M_{\mathrm{n}}} = \int_{M=0}^{M \to \infty} \frac{w(M)}{M} d \log(M), \tag{8.100}$$

$$M_{z+i} = \frac{\displaystyle\int_{M=0}^{M \to \infty} M^{2+i} w(M) \, d \log(M)}{\displaystyle\int_{M=0}^{M \to \infty} M^{1+i} w(M) \, d \log(M)}, \quad i \geq 0. \tag{8.101}$$

The DR equation with the so-called *single exponential* kernel relaxation function per mass is defined as

$$G(t) = G_{\mathrm{N}}^0 \left(\int_{-\infty}^{\infty} (e^{-t/kM^{3.4}})^{1/2} w(M) d \log(M) \right)^2. \tag{8.102}$$

In the limit for a uniform system, this equation reduces to

$$G(t) = G_{\mathrm{N}}^0 e^{-t/kM^{3.4}}, \tag{8.103}$$

which implies that the zero-shear viscosity (Equation 3.90) scales with $M^{3.4}$ and that steady-state compliance (Equation 3.148) is independent of M, as discussed in the previous sections. This follows from elementary calculus. Den Doelder [55] has numerically analyzed the effect of non uniformity on $G(t)$ via Equation 8.102 for a large number of MMD of strongly varying type. It was found that the effect η_0 was small, confirming empirical findings (see Section 8.1). In contrast, shear thinning and steady-state compliance increased significantly with increased non uniformity. A unified expression for J_e^0 was obtained in terms of a few characteristic moments, regardless of the shape of the distribution:

$$J_e^0 = \frac{1}{G_N^0} \left(\frac{M_z}{M_w} \right)^2 \left(\frac{M_{z+1}}{M_z} \right)^4. \tag{8.104}$$

This equation, based on microscopic theory, compares well with empirical relations based on limited types of MMD, as presented in Equation 8.7.

Predicting rheology of ensembles of long-chain branched, and linear chains of arbitrary uniformity is still a challenge. Empirical relations are only valid for limited subtypes, such as tubular high-pressure PE made in a specific reactor. Extensions of tube-based theories with hierarchical schemes of relaxation have been proposed by Larson, McLeish, and coworkers [56–58]. They offer the possibility for reverse-engineered polymers. Good results have been obtained for linear viscoelasticity. Further efforts focus on generalization to strong flows.

8.5 GENERAL OBSERVATION

Being able to accurately describe with mathematics the flow behavior of polymer melts and accordingly simulate the onset of polymer melt fracture is a valid target for research. The mathematical complexity is however quite challenging. It requires advanced mathematics to closely capture some of the experimental flow features. Several macroscopic and microscopic approaches have been formulated with varying degrees of successfully simulating some or nearly all rheologically important material functions. It is notable that irrespective of the continuum or molecular reasoning, the mathematical formalisms shape into similar equations. The main difference is the interpretation of parameters. Still, linking industrial polymer architecture measures to flow characteristic material functions remains mainly an empirical exercise.

REFERENCES

1. Bird, R. B., W. E. Stewart, and E. N. Lightfoot, *Transport Phenomena*. John Wiley & Sons Inc., New York, 1960.
2. Bird, R. B., R. C. Armstrong, and O. Hassager, *Dynamics of Polymeric Liquids*. Volume 1. *Fluid Mechanics*. John Wiley & Sons Inc., New York, 1988.
3. Larson, R. G., *Constitutive Equations for Polymer Melts and Solutions*. Butterworth Publishers, Boston, MA, 1988.
4. Macosko, C. W., *Rheology*. VCH Publishers, New York, 1994.
5. Tanner, R. I., *Engineering Rheology*. Clarendon Press, Oxford, U.K., 1985.
6. Ferry, J. D., *Viscoelastic Properties of Polymers*. John Wiley & Sons, Inc., New York, 1980.
7. Gedde, U. W., Molecular structure of crosslinked polyethylene as revealed by 13C nuclear magnetic resonance and infrared spectroscopy and gel permeation chromatography. *Polymer*, **27**:269–274 (1986).

8. Graessley, W. W., Viscoelasticity and flow in polymer melts. In *Physical Properties of Polymers* (J. E. Mark, Ed.). American Chemical Society, Washington, DC, 1984.

9. Bird, R. B., C. F. Curtiss, R. C. Armstrong, and O. Hassager, *Dynamics of Polymer Liquids*. Volume 2. *Kinetic Theory*. John Wiley & Sons Inc., New York, 1987.

10. Ottinger, H. C., *Stochastic Processes in Polymeric Fluids*. Springer Verlag, Berlin, Germany, 1996.

11. McLeish, T. C. B. and S. T. Milner, Entangled dynamics and melt flow of branched polymers. *Adv. Polym. Sci.*, **143**:195–256 (1999).

12. Mills, N. J., The rheological properties and molecular weight distribution of polydimethylsiloxane. *Eur. Polym. J.*, **5**:675–695 (1969).

13. Kurata, M., K. Osaki, Y. Einaga, and T. Sugie, Effect of molecular weight distribution on viscoelastic properties of polymers. *J. Polym. Sci.*, **12**:849–869 (1974).

14. Agarwal, P. K., A relationship between steady state shear compliance and molecular weight distribution. *Macromolecules*, **12**:342–344 (1979).

15. Lodge, A. S., *Elastic Liquids*. Academic Press, New York, 1964.

16. Lodge, A. S., *Body Tensor Fields in Continuum Mechanics*. Academic Press, New York, 1974.

17. Oldroyd, J. G., An approach to non-Newtonian fluid mechanics. *J. Non-Newtonian Fluid Mech.*, **14**:9–46 (1984).

18. Green, M. S. and A. V. Tobolsky, A new approach to the theory of relaxing polymeric media. *J. Chem. Phys.*, **14**:80–92 (1946).

19. Rouse, P. E., A theory of the linear viscoelastic properties of dilute solutions of coiling polymers. *J. Chem. Phys.*, **21**:1272–1280 (1953).

20. Doi, M. and S. F. Edwards, *The Theory of Polymer Dynamics*. Clarendon Press, Oxford, U.K., 1986.

21. Doi, M. and S. F. Edwards, Dynamics of rod-like macromolecules in concentrated solution. Part 1. *J. Chem. Soc. Faraday Trans. II*, **74**:1789–1802 (1978).

22. Doi, M. and S. F. Edwards, Dynamics of rod-like macromolecules in concentrated solution. Part 2. *J. Chem. Soc. Faraday Trans. II*, **74**:1802–1818 (1978).

23. Doi, M. and S. F. Edwards, Dynamics of rod-like macromolecules in concentrated solution. Part 3. *J. Chem. Soc. Faraday Trans. II*, **74**:1818, (1978).

24. Johnson, M. W. and D. Segalman, A model for viscoelastic fluid behavior which allows non-affine deformation. *J. Non-Newtonian Fluid Mech.*, **2**:255–270 (1977).

25. Johnson, M. W. and D. Segalman, An elastic porous molecule model for the molecular dynamics of polymer liquids. In *Mechanics Today*. Vol. 5 (S. Nemat-Nasse, Ed.). Pergamon Press, Oxford, U.K., 1980, pp. 129–137.

26. Johnson, M. W. and D. Segalman, Description of the non-affine motions of dilute polymer solutions by the porous molecule model. *J. Non-Newtonian Fluid Mech.*, **9**:33–56 (1987).

27. Gordon, R. J. and W. R. Schowalter, Anisotropic fluid theory: A different approach to the dumbbell theory of dilute polymer solutions. *Trans. Soc. Rheol.*, **16**:79–97 (1972).

28. Malkus, D. S., J. A. Nohel, and B. J. Plohr, Dynamics of shear flow of a non-Newtonian fluid. *J. Comp. Phys.*, **87**:464–487 (1990).

29. Malkus, D. S., J. A. Nohel, and B. Plohr, Analysis of new phenomena in shear flow of non-Newtonian fluids. *SIAM J. Appl. Math.*, **51**:899–929 (1991).

30. Malkus, D. S., Y. C. Tsai, and R. W. Kolkka, New transient algorithms for non-Newtonian flow. *Finite Elem. Fluids*, **8**:401–424 (1992).

31. Malkus, D. S., J. A. Nohel, and B. J. Plohr, Oscillations in piston driven shear flow of a non-Newtonian fluid. In *IUTAM Symposium on Numerical Simulation of Non-Isothermal Flow of Viscoelastic Liquids*. Dordrecht, the Netherlands, 1994, pp. 57–74.

32. Malkus, D. S., J. A. Nohel, and B. J. Plohr, Approximation of piston-driven flow of a non-Newtonian fluid. In *Control Theory, Dynamical Systems and Geometry of Dynamics* (K. D. Elsworthy, W. N. Everitt, and E. B. Lee, Eds.). Marcel Dekker, New York, 1993, pp. 173–192.

33. Aarts, A. C. T., Analysis of the flow instabilities in the extrusion of polymeric melts. PhD thesis, Eindhoven University of Technology, Eindhoven, the Netherlands, 1998.

34. Kaye, A., Non-Newtonian flow in incompressible fluids. Note No. 134, College of Aeronautics, Cranfield, U.K., 1962.

35. Kaye, A., An equation of state for non-Newtonian fluids. *Br. J. Appl. Phys.*, **17**:803–806 (1966)

36. Bernstein, B., E. A. Kearsly, and L. J. Zapas, A study of stress relaxation with finite strain. *Trans. Soc. Rheol.*, **7**:391–410 (1963).

37. Larson, R. G. and K. Monroe, The BKZ as an alternative to the Wagner model for fitting shear and elongational flow data of an LDPE melt. *Rheol. Acta.*, **23**:10–13 (1984).

38. Wagner, M. H., Analysis of time-dependent non-linear stress-growth data for shear and elongational flow of a low-density branched polyethylene melt. *Rheol. Acta.*, **15**:136–142 (1976).

39. Papanastasiou, A. C., L. E. Scriven, and C. W. Macosko, An integral constitutive equation for mixed flows: viscoelastic characterization. *J. Rheol.*, **27**:387–410 (1983).

40. Aarts, A. C. T. and A. A. F. van de Ven, Transient behavior and stability points of the Poiseuille flow of a K-BKZ fluid. *J. Eng. Math.*, **29**:371–392 (1995).

41. Flory, P. J., *Statistical Mechanics of Chain Molecules*. Wiley, New York, 1969.

42. De Gennes, P. G., Reptation of a polymer chain in the presence of fixed obstacles. *J. Chem. Phys.*, **55**:572–579 (1971).

43. Graessley, W. W., Entangled linear, branched and network polymer systems—Molecular theories. *Adv. Polym. Sci.*, **47**:67–117 (1982).

44. Lodge, T. P., N. A. Rotstein, and S. Prager, Dynamics of entangled polymer liquids: Do linear chains reptate? *Adv. Chem. Phys.*, **79**:1 (1990).

45. Marrucci, G., Dynamics of entanglements: A nonlinear model consistent with the Cox-Merz rule. *J. Non-Newtonian Fluid Mech.*, **62**:279–289 (2996).

46. Likhtman A. E. and R. S. Graham, Simple constitutive equation for linear polymer melts derived from molecular theory: Rolie-Poly. *J. Non-Newtonian Fluid Mech.*, **114**:1–12 (2003).

47. Tchesnokov, M. A., J. Molenaar, J. J. M. Slot, and R. Stepanyan, A constitutive model with moderate chain stretch for linear polymer melts. *J. Non-Newtonian Fluid Mech.*, **123**:185–199 (2004).

48. Stepanyan, R., J. J. M. Slot, and J. Molenaar, On the microscopic approach to the nonlinear dynamics of entangled polymer melts, *Europhys. Lett.*, **68(6)**:832–838 (2004).

49. McLeish, T. C. B., A tangled tale of topological fluids. *Phys. Today*, **61(8)**:40–45 (2008).

50. Dubbeldam, J. L. A. and J. Molenaar, Stress relaxation of star-shaped molecules in a polymer melt. *Macromolecules*, **42**:6784–6790 (2009).

51. Samurkas, T., R. G. Larson, and J. M. Dealy, Strong extensional and shearing flows of branched polyethylene. *J. Rheol.*, **33**:559–578 (1989).

52. McLeish, T. C. B. and R. G. Larson, Molecular constitutive equations for a class of branched polymers: The pom-pom polymer. *J. Rheol.* **42**:81–110 (1998).

53. Tsenoglou, C., Viscoelasticity of binary homopolymer blends. *ACS Polym. Prep.*, **28**:185–186 (1987).

54. Des Cloizeaux, J., Double reptation vs simple reptation in polymer melts. *Europhys. Lett.*, **5**:437–442 (1988).

55. Den Doelder, J., Viscosity and compliance from MM distributions using double reptation models. *Rheol. Acta*, **46(2)**:195–210 (2006).

56. Park, S. J., S. Shanbhag, and R. G. Larson, A hierarchical algorithm for predicting the linear viscoelastic properties of polymer melts with long-chain branching. *Rheol. Acta*, **44**:319–330 (2005).
57. Das, C., N. J. Inkson, D. J. Read, M. A. Kelmanson, and T. C. B. McLeish, Computational linear rheology of generally branch-on-branch polymers. *J. Rheol.*, **50**:207–234 (2006).
58. Das, C., M. Kapnistos, D. Auhl, I. Vittorias, J. den Doelder, T. C. B. McLeish, and D. J. Read, Computational framework for a priori prediction of rheological response of randomly branched polymers. In *Annual European Rheology Conference*, Cardiff, U.K., 2009.

9 Modeling Melt Fracture

Modeling is a scientific activity essential for obtaining a logical and quantitative description of physical phenomena. Typically, a model is a set of mathematical equations aimed at providing predictive answers to physical questions. The model-generated, that is, the simulated, numerical values, may be tested against experimental observational facts as summarized in Chapters 5 and 6. In practice, models contain both relations based on first principles and relations based on empirical practice. This chapter introduces such hybrid models related to polymer melt fracture. The first-principle relations are derived from the constitutive approach to rheology as presented in Chapters 3 and 8. In addition, different levels of empiricism are added to make the models practically useful. The level and the nature of empiricism depend on the selected constitutive equation and the desired connection to molecular information.

In continuum mechanical terms, the deformation of polymer melts under stress is described by the equations expressing conservation of mass, momentum, and energy. The material properties are expressed in terms of a viscoelastic constitutive equation. This modeling approach is very useful for describing and understanding many practical polymer melt flow phenomena. It makes these equations a logical choice to model polymer melt fracture. In addition, a good polymer melt fracture model should simplify the complex physical phenomena into clear and manageable equations without losing the essential physics of the issue at hand.

The modeling of polymer melt fracture still presents significant challenges. Critical is capturing, in a general fashion, the underpinning physics in a manageable mathematical framework. For practical purposes, a relatively simple mathematical model is introduced, that is, the relaxation oscillation (RO) model. In its basic format, the RO model requires substantial experimental input. The mathematics are however rich enough to allow for generalization and accurate prediction. The RO model can be refined to reduce the need for heuristic input making it predictive for the various melt fracture phenomena. In order to keep the mathematics manageable, a simplified analogue of an extruder, that is, a piston-driven slit die rheometer is used as a modeling reference frame. The modeling applies equally well to a flow channel with a capillary die (Figures 5.1 and 7.9). The modeling largely aims at determining the flow curve, that is, establishing a mathematical relation between the imposed constant flow rate and the measured total pressure in the reservoir. Additional elements of polymer architecture can be included that lead to linking flow characteristics and indirectly the extrudate appearance. The latter, however, remains an elusive target in terms of the exact extrudate topology.

9.1 THE RELAXATION-OSCILLATION MODEL

Melt fracture experimentation has led to the observation of either a discontinuous (non-monotonous) or continuous (monotonous) flow curve. The discontinuous flow curve consists of (at least) two steady-state branches separated by a transition region (see Chapter 5). The experimental fact of a discontinuity onset is a direct quantitative criterion for the occurrence of extrudate distortions. Accordingly, a first melt fracture model should be capable of describing the dynamics of the so-called spurt oscillations based on the two stable branches of the flow curve. In 1978, Weill [1–3] associated, for flow rates in the transition region, the evolution of the pressure oscillations in time with relaxation oscillations. Relaxation oscillations are cyclic phenomena during which potential energy is successively stored and relaxed. Van der Pol [4,5] studied relaxation oscillations for the first time in a triode circuit. One of the intriguing aspects of the phenomenon is that signal oscillations are maintained without periodic inputs. The energy source is steady. In the present system, the energy is provided by the steadily moving piston. The mathematical representation of relaxation oscillations requires a second-order differential equation or equivalently a system of two coupled first-order equations:

$$\frac{dy}{dt} = -x, \quad \varepsilon \frac{dx}{dt} = y - F(x),$$ (9.1)

where
   ε is a small parameter
   F is such that the system may show relaxation-oscillations (Figure 9.1c)

As presented by Molenaar and Koopmans [6], the flow in a capillary rheometer can be related to this mathematical formalism. For a constant-flow-rate experiment, the following parameters are involved: the piston velocity v_p; the reservoir volume $h(t)$ A, where $h(t)$ is the time-dependent distance between piston and capillary die and A the area of the piston; and the constant inlet flow rate Q_i, equal to v_pA. The total mass in the reservoir is given by $Ah\rho$, where $\rho(t)$ is the polymer melt density, which is assumed to be uniform over the reservoir. The mass flux leaving the reservoir and entering the capillary is equal to $\rho Q(t)$, where $Q(t)$ is the volume flux through the capillary die. The polymer in the die is assumed to be incompressible in view of the small volume of the die relative to the reservoir volume. The conservation of mass in the reservoir is represented by

$$A\frac{d}{dt}(h\rho) = -\rho Q.$$ (9.2)

The polymer density $\rho(t)$ is related to the pressure $P(t)$ in the reservoir by

$$\frac{1}{\rho}\frac{d\rho}{dt} = \chi\frac{dP}{dt},$$ (9.3)

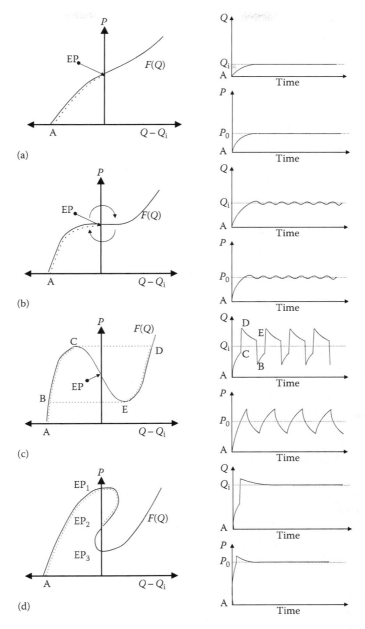

FIGURE 9.1 A schematic phase-plane representation of pressure P and flow rate difference $\Delta Q = Q - Q_i$ for four possible classes of F-functions and the associated solutions of the RO-model. Starting from A, initially the solution will move along $F(Q)$. The slope of $F(Q)$ determines the kind of steady state that will be reached. If the slope is positive (cases a and d) the steady state is constant in time. If the slope is vanishing (case b) or negative (case c) the steady state is periodic in time.

where χ is the (constant) melt compressibility. Using $dh/dt = -v_p$ and Equation 9.3, Equation 9.2 can be rewritten as

$$\frac{dP(t)}{dt} = -\frac{1}{A\chi h}\Delta Q(t),   \tag{9.4}$$

with the definition $\Delta Q = Q - Q_i$. Since $h(t)$ shows relatively little change with respect to the timescales of the "spurt" oscillations, it is considered as a constant for the present modeling purposes. Equation 9.4 shows that the reservoir pressure will increase (decrease) if $Q_i > (<) Q$. Equation 9.4 serves as the first RO-equation. From Equation 9.1 it follows that the second RO-equation should be of the form

$$\frac{d\Delta Q(t)}{dt} = \frac{dQ(t)}{dt} = \frac{1}{K}[P(t) - F(\Delta Q(t))].   \tag{9.5}$$

The F-function reflects the relationship between pressure and flow rate. It is identified as representative for the two stable branches of the flow curve. For modeling convenience, the two branches are connected by a straight line with negative slope as indicated by Durand et al. [7]. The resulting flow curve is determined by the combination of polymer characteristics and flow channel geometry. The proportionality factor $1/K$ measures the ability of the melt to adjust the pressure via the adjustment of the flow rate. For relaxation oscillations to occur, $1/K$ must be so large that P becomes almost instantaneously equal to F when Q changes.

For continuous flow curves, F is also continuous and the solution of the RO-model (Equations 9.4 and 9.5) always approaches the steady state exponentially in time. In the case of a flow curve that has two branches it can be shown [8] that the shape of the decreasing section connecting the two branches is not important, as long as it decreases in a monotonous fashion. The steady flow curve now leads to four classes of F-functions sketched in Figure 9.1.

The indicated trajectories in the phase plane are followed when starting from rest and suddenly imposing a constant piston speed. The F-function is translated horizontally when the imposed flow rate Q_i is changed, since it depends on $Q - Q_i$. Case (a) leads to a steady solution that is constant in time. Case (b) leads to small oscillations. Case (c) leads to relaxation oscillations. Case (d) leads to initial overshoots in $P(t)$ and $Q(t)$ and converges to a constant steady state. All these results correspond well with experimental results [7]. One of the merits of the RO-model is that it highlights the importance of compressibility in the reservoir. The RO-model correctly predicts the dependence of the oscillations on the imposed flow rate and the reservoir volume. It thereby leads to more insight into the physics of the system. However, the RO-model requires significant experimental input. The entire flow curve needs to be determined experimentally. This severely limits the predictive power of the model although it can successfully describe the extrusion of HDPE polymer melts [7,9], volcanoes [10], and chocolate [11].

9.2 COUPLING RO AND CONSTITUTIVE EQUATIONS

The RO-model provides a solid foundation for more sophisticated models that use less experimental input and are still able to describe the flow curve. In Chapter 7, where the understanding of melt fracture is highlighted, one of the topics relates to instabilities connected with the constitutive behavior of polymer melts [12–15]. The heuristically introduced F-function is replaced by a constitutive model (see Chapters 3 and 8) to reduce the experimental input for the RO-model. In the present case, the JSO model, introduced in Section 8.2.2, is selected as a typical example of such a constitutive model. Equation 9.4 for the mass balance over the flow channel is maintained. In line with the examples in Chapters 3 and 8, the slit die (Figure 3.2) is considered as flow channel geometry for further modeling purposes. Modification of the formulae to a capillary die is straightforward; no essential differences between the two geometries exist. Following the coordinate system shown in Figure 3.2, at the connection between the reservoir and die $x_1 = 0$ and at the die exit $x_1 = L$. The height of the die is $2H$. The flow in the die is assumed to be incompressible. The only nonvanishing component of the velocity vector is $v_1(t, x_2)$. Body forces and inertia can be neglected with respect to the stresses as follows from dimensional analysis [8,16]. As in Equation 3.82, the momentum balance equations then reduce to

$$\nabla \cdot \mathbf{S} = 0, \tag{9.6}$$

where $\mathbf{S}$ is given by Equation 8.34. At the centerline of the die the condition

$$\frac{\partial v_1}{\partial x_2}(t,0) = 0 \tag{9.7}$$

applies because of symmetry. The initially selected no-slip-boundary condition at the wall reads as

$$v_1(t, -H) = v_1(t, H) = 0. \tag{9.8}$$

The pressure in the die is a linear function of x_1 and independent from x_2 and x_3:

$$p(t, x_1) = P(t)\left(\frac{L - x_1}{L}\right), \tag{9.9}$$

where the atmospheric pressure is taken as the reference point, thus $p(t, L) = 0$. In a slit die, the stresses $\mathbf{D}$ and τ only depend on x_2. This has the consequence that only the x_1-component of Equation 9.6 is nontrivial. In integrated form, it reads as

$$\eta_0 \dot{\gamma} + \sigma_{12}(t, x_2) = -\frac{x_2}{L} P(t), \tag{9.10}$$

with σ_{12} defined satisfying Equation 8.35. Equation 9.10 expresses that the total shear stress in the die, given by the left-hand side, is linear in x_2 and attains its optimum at the die wall. In Example 8.2, the JS model has been applied to laminar flow. The results can be used here. The resulting formulae with $\boldsymbol{\tau}'$ replaced by $\boldsymbol{\sigma}$ are quoted. For convenience, new variables are introduced, namely

$$\sigma_+ = \frac{1}{2}((a-1)\sigma_{11} + (a+1)\sigma_{22}),$$

$$\sigma_- = \frac{1}{2}((a-1)\sigma_{11} - (a+1)\sigma_{22}). \tag{9.11}$$

In terms of these variables, the JSO equations read as

$$\lambda_0\dot{\sigma}_+ = -\sigma_+ + (a^2-1)\lambda_0\dot{\gamma}\sigma_{12}, \tag{9.12}$$

$$\lambda_0\dot{\sigma}_- = -\sigma_-, \tag{9.13}$$

$$\lambda_0\dot{\sigma}_{12} = -\sigma_{12} + (\lambda_0\sigma_+ + \mu)\dot{\gamma}. \tag{9.14}$$

From Equation 9.13, it is seen that σ_- can be neglected if time evolves.

The momentum balance equation (Equation 9.10) in the die needs to be coupled to the mass balance equation (Equation 9.4) in the reservoir. The latter equation relates P to Q, whereas the momentum balance equation (Equation 9.10) relates P to the local velocity v_1. The coupling is done by relating v_1 to Q:

$$Q(t) = W \int_{-H}^{H} v_1(t,x_2)\, dx_2 = -2W \int_{0}^{H} \dot{\gamma}(t,x_2)x_2\, dx_2, \tag{9.15}$$

where partial integration is used together with the conditions in Equation 9.8. As indicated in Figure 3.2, the height of the slit die is $2H$ and its width is W. The governing equations can be made dimensionless by scaling length variables by H, time by λ_0, and stress variables by the modulus μ, introduced in Equation 8.25. From now on, all variables in this section are dimensionless unless stated otherwise. No special notation is used. The mass balance equation (Equation 9.4) in the reservoir reads in dimensionless form as

$$\frac{dP}{dt} = \zeta\Delta Q(t) \tag{9.16}$$

with

$$\varsigma = \frac{H^3}{A\chi h\mu}. \tag{9.17}$$

If the dimensionless quantities σ_{12} and $\dot{\gamma}$ are replaced by the same quantities multiplied by $\sqrt{1-a^2}$, the parameter a is eliminated from Equations 9.12 and 9.14. The dimensionless versions of Equations 9.10, 9.12, and 9.14 are

$$\varepsilon\dot{\gamma} + \sigma_{12} = -x_2\beta P$$

$$\dot{\sigma}_+ = -\sigma_+ - \sigma_{12}\dot{\gamma} \qquad (9.18)$$

$$\dot{\sigma}_{12} = -\sigma_{12} + (\sigma_+ + 1)\dot{\gamma}$$

with the two dimensionless parameters

$$\varepsilon = \frac{\eta_0}{\mu\lambda_0}, \quad \beta = \frac{H}{L}. \qquad (9.19)$$

Equation 9.15 reads in dimensionless form as

$$Q(t) = -2\frac{W}{H}\int_0^1 \dot{\gamma}(t, x_2)x_2\, dx_2. \qquad (9.20)$$

The present model consists of Equations 9.16 through 9.20 together with the boundary conditions in Equation 9.8. The initial conditions are still to be chosen. It is instructive to analyze the steady state solutions first. To that end, the time derivatives in Equation 9.18 are set to zero. The stress σ_+ can be eliminated from Equation 9.18 and the steady shear stress σ_{12} can be expressed in terms of the steady shear strain rate $\dot{\gamma}$:

$$\sigma_{12}(\dot{\gamma}) = \frac{\dot{\gamma}}{1 + \dot{\gamma}^2}. \qquad (9.21)$$

Note that σ_{12} indirectly depends on the position x_2 via $\dot{\gamma}(t, x_2)$. Substitution of this steady state result into Equation 9.18 yields the relation

$$S_{\text{JSO}} \equiv \varepsilon\dot{\gamma} + \frac{\dot{\gamma}}{1 + \dot{\gamma}^2} = -x_2\beta P. \qquad (9.22)$$

For $\varepsilon > 1/8$, the total stress S_{JSO} is a continuously increasing function of $\dot{\gamma}$. For smaller ε, three regimes can be discerned. In Figure 9.2, the relation between S_{JSO} and $\dot{\gamma}$ is drawn for $\varepsilon = 0.01$. Branch I, with positive slope, ranges from the origin $(0, 0)$ to the local maximum at $(\dot{\gamma}_{\max}, S_{\max})$, approximately given by $(1 + 2\varepsilon, 1/2 + \varepsilon)$. The transition regime, with negative slope, ranges from the local maximum to the local minimum at $(\dot{\gamma}_{\min}, S_{\min})$, approximately given by $(\varepsilon^{-1/2} - 3/2\varepsilon^{1/2}, 2\varepsilon^{1/2} - \varepsilon^{3/2})$. Branch II, with positive slope, ranges from the local minimum to infinity. The local maximum is nearly independent of ε, while the local minimum shifts to the right when ε becomes smaller. For the polymer melts giving rise to "spurt," the steady stress–strain rate

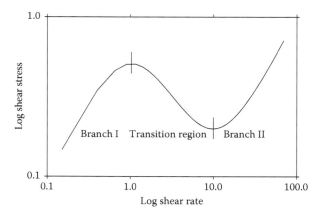

FIGURE 9.2   The local shear stress versus shear rate curve for the JSO model with $\varepsilon = 0.01$. For convenience, $x_2 < 0$ in Equation 9.22.

equation (Equation 9.22) is non-monotonous. Equation 9.22 is a local relation, valid for each position x_2 in the die. The transformation to the global flow curve of P versus Q, involving integration over the die cross-section via Equation 9.20, is governed by the time dependence of Equation 9.18 and is not unique. Therefore, it is not correct to directly associate the $S_{\mathrm{JSO}}(\dot{\gamma})$ relation with the F-function in Equation 9.5, which represents the experimental flow curve. With a non-monotonous steady stress–shear rate relation, solutions of Equation 9.22 can be constructed that lie on Branch I in the core region of the die and on Branch II in the outer layer, as shown in Figure 9.3.

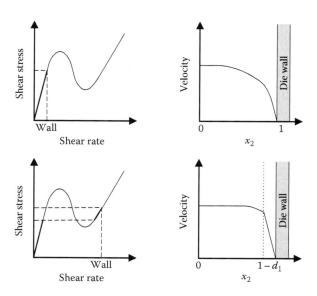

FIGURE 9.3   Schematic representation of possible steady-state velocity profiles that are solutions of the RO-model coupled to the JSO-model for $\varepsilon < 1/8$.

"Spurt" oscillations are expected to occur through oscillations in the thickness of this boundary (interfacial) layer.

The time-dependent model must be solved numerically. For simulations with controlled pressure, the mass balance equation (Equation 9.16) is not relevant. In this case, experiments have shown that no pressure oscillations can be measured. When increasing the pressure, a jump in the flow rate can be detected at a critical pressure. For experiments that start at very high pressures and gradually decrease the pressure, the flow rate remains high until a second, lower critical pressure is reached before it jumps back to low flow rates [17,18]. This is a hysteresis effect. Aarts and van de Ven [19] and Kolkka et al. [20] have shown that the JSO equations indeed give rise to such a hysteresis. For simulations in which the flow rate is controlled, Equation 9.16 must be taken into account. However, it was found [21] that the dynamics of the model equations do not give rise to the experimentally observed pressure oscillations. The boundary layer apparently has the freedom to adjust its thickness such that the integrated velocity profile is equal to the imposed flow rate. The model always converges to a constant steady solution.

In summary, although this extension of the RO-model has some attractive features, it does not seem to be in full agreement with the available experimental evidence.

9.3   SLIP-BOUNDARY CONDITIONS

From the experimental evidence and some theoretical considerations, it has become evident that the concept of wall slip is important for understanding melt fracture phenomena. Slip is a macroscopic concept that represents a finite velocity at the die wall. Slip at the wall becomes important if the slip velocity of the polymer melt is of the same order of magnitude as the mean flow velocity (Figure 7.3). A heuristic approach might be to relate the velocity at the wall v_s to the wall shear stress $\tau_w = |\tau_{12}(x_2 = H)|$ by way of Equation 7.2. In this equation, v_s and τ_w are nonnegative, absolute values. The stress-dependent coefficient of friction $\beta(\tau_w)$ may depend on wall material composition and polymer architecture. In practice, $\beta(\tau_w)$ can be so large that, effectively, the no-slip condition $v_s = 0$ holds. This is determined not only by the value of β, but also by the magnitude of the velocity gradient $\dot{\gamma}_w$ near the wall. The no-slip condition is valid if the increase in velocity over a small length scale is much larger than the velocity at the wall. The relative importance of slip is usually expressed in terms of the extrapolation length b, as defined in Equation 7.1. The extrapolation length b represents an effective widening of the flow channel, as illustrated in Figure 7.3. Just as the interfacial behavior of the system is characterized by the coefficient β, the bulk behavior is expressed in terms of the viscosity η, which relates the shear rate (or velocity gradient) to the shear stress:

$$\tau_{12} = \eta(\dot{\gamma})\dot{\gamma}. \tag{9.23}$$

In view of Equations 7.1, 7.2, and 9.23, b can be expressed in terms of the material parameters:

$$b(\tau_w) = \frac{\eta(\tau_w)}{\beta(\tau_w)}. \tag{9.24}$$

The no-slip assumption is reliable if b is much smaller than the die height H (for a slit die) or the die radius R (for a capillary die).

It is worth noting that not only the velocity at the wall but also the velocity profile in the bulk is determined by friction. The viscosity is a measure for the friction between molecules in the fluid phase. This bulk friction never leads to jumps in the velocity profile, because the molecules can move freely and are dragged along with each other. Therefore the bulk friction is characterized by the relation between velocity gradient and stress. On the other hand, the interfacial friction is characterized by the relation between velocity jump and stress at the die wall.

Following the same argument, it is also possible to rewrite the interfacial friction coefficient in terms of an effective interfacial viscosity η_i through a length scale $d \ll H$ (or R):

$$\beta = \frac{\eta_i}{d}. \tag{9.25}$$

The combination of this with Equation 9.24 leads to

$$b(\tau_w) = d \frac{\eta(\tau_w)}{\eta_i(\tau_w)}. \tag{9.26}$$

These continuum-level (macroscopic) slip relations may be derived from molecular (microscopic) models. The presence of a velocity jump at the wall can be caused by a number of different molecular mechanisms. Several mechanisms have been proposed for the stress dependence of the friction coefficient. Among others, Brochard and de Gennes [22], Migler et al. [23], Hatzikiriakos et al. [24,25], Stewart [26], Hill [27], and Drda and Wang [28] can be mentioned in this respect (see Chapter 7). In Figure 9.4, a qualitative molecular framework is shown that captures the basic ideas of these mechanisms and reflects the macroscopic flow phenomena of Figure 7.3.

Along the wall, a boundary layer exists that is populated by macromolecules attached to the wall and by bulk macromolecules that are entangled with the attached ones. The thickness of this layer is determined by the presence of attached macromolecules and may be time dependent. The layer ends at a distance from the wall where attached macromolecules are no longer present (on average). The macroscopic slip velocity v_s and the macroscopic wall shear stress τ_w introduced above should be associated with the (average) velocity and shear stress of the bulk macromolecules at the top of the boundary layer.

At low stress, many macromolecules have a strong physical bond with the die wall. They are considered *attached* to the wall. These molecules are also entangled with the nonattached flowing molecules (Figure 9.4a). The friction coefficient at low stress β_{ac} is found to be so large that the extrapolation length is in the order of $10\,\mathrm{nm}$. Using a typical value for the low-shear viscosity ($10^4\,\mathrm{Pa\,s}$) and Equation 9.24, one finds that $\beta_{ac} = 10^4\,\mathrm{Pa\,s}/10^{-8}\,\mathrm{m} = 10^{12}\,\mathrm{Pa\,s/m}$ for a highly entangled melt.

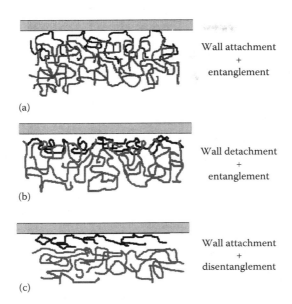

(a) Wall attachment + entanglement

(b) Wall detachment + entanglement

(c) Wall attachment + disentanglement

FIGURE 9.4 A schematic representation of the molecular mechanisms that can be associated with macroscopic slip: (a) no slip, (b) adhesive slip, and (c) cohesive slip.

There is a competition between the bonding force F_b and the force due to the flow of the entangled macromolecules F_f. At higher stresses when $F_f > F_b$, the bonded macromolecules are detached from the wall and a dramatic decrease in β from β_{ac} to β_{dc} can be observed (Figure 9.4b). Depending on the macromolecular-wall characteristics, this may happen in a narrow stress region around a critical value τ_{dc}, but the detachment may also increase gradually over a wide stress range. The former may be associated with strong slip and the latter with weak slip. This also depends on the bulk viscosity values at those stresses. Slip due to detachment is called *adhesive slip*. Alternatively when F_f is still smaller than F_b at elevated stress levels, a coil-stretch transition may occur in the bonded macromolecules at a critical stress τ_{as}. They disentangle from the nonattached macromolecules and effectively create a new interface characterized by β_{as} (Figure 9.4c). This so-called *cohesive slip* mechanism is often associated with strong slip. The polymer architecture can be imagined to determine the extent of the macromolecular entanglement as well as the physical bond between the macromolecules and the die-wall.

The magnitude of the transition or the change in β can be estimated. For both adhesive and cohesive slip, the effective interfacial friction can be considered as approximately equal to the wall friction of a flowing fluid of monomers. The friction coefficient may then be estimated from a typical monomeric viscosity and monomeric dimension, using Equation 9.25: $\beta_{dc} \approx \beta_{ac} \approx 10^{-2}\,Pa\,s/10^{-9}\,m \approx 10^{7}\,Pa\,s/m$. When the stress increases further after the transition, the value of β does not necessarily remain constant.

Example 9.1:   Relative Importance of Slip

The friction coefficient of an HDPE polymer at a stress of about 0.3 MPa and of water at any stress is found to be $\beta = 10^7$ Pa s/m. The viscosities of both materials are

$$\eta_{HDPE}(0.3\ MPa) = 10^3\ Pa\,s, \quad \eta_{water} = 10^{-3}\ Pa\,s. \tag{9.27}$$

By using Equation 9.24, it can be determined if the no-slip boundary condition may be applied in a slit die with $H = 0.001$ m. This leads to

$$b_{HDPE} = 10^{-4}\ m, \quad b_{water} = 10^{-10}\ m. \tag{9.28}$$

These results indicate that the no-slip condition is a very good approximation in the case of water, but in the case of HDPE the slip contribution to the velocity field is important and cannot be neglected.

At higher stresses, a bulk coil-stretch transition may occur in the bulk. Such a phenomenon is not fully interfacial and leads to a two-layered flow structure, where the outer layer has a much lower viscosity than the inner layer. This may be described by non-monotonous constitutive equations. When the flow is still laminar at high stresses, a second discontinuity in the flow curve may be initiated.

Within this framework, the effect of so-called *high* and *low energy die walls* can be explained. High-energy walls induce a strong polymer attachment, are often metallic (stainless-steel, aluminum) and are characterized by a high value for the work of adhesion W. These die walls are commonly used in polymer processing. During polymer melt flow, some detachment may take place (weak slip), but the coil-stretch transition occurs before the main attachment–detachment transition to the die wall takes effect. Low-energy walls with a low value of W can be created by coating the metallic die wall with fluoro-polymers or other additives that migrate to the die wall. For these walls, a coil-stretch transition does not take place since most, weakly attached molecules are detached at stresses lower than τ_{as}.

Based on the above considerations, slip models have been proposed that describe the onset point and slip magnitude as functions of shear stress [26,29,30] in a phenomenological way. A complete model should also be able to account for the hysteresis phenomenon that has often been observed with metallic walls.

Dubbeldam and Molenaar [31,32] studied the slip law within the framework of the reptation picture. They numerically simulated the dynamical behavior of attached macromolecules entangled with bulk macromolecules and indeed observed the slip phenomenon for a critical value of the bulk shear rate. This slip transition implies that if the bulk shear rate is gradually increased, the wall shear stress goes through a maximum, after which it suddenly drops down due to complete disentanglement of bulk and attached macromolecules.

The critical value of the bulk shear rate for which the slip velocity shows a sudden jump upward depends in an intricate way on a number of parameters. The values of these parameters determine how adhesive and cohesive slip interplay in a subtle fashion. This interplay is analyzed in depth by Tchesnokov et al. [33–35] and

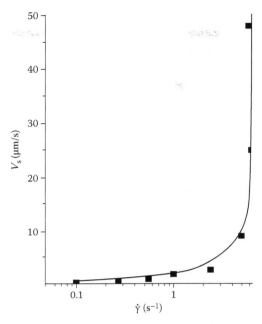

FIGURE 9.5 Slip velocity as a function of the bulk shear rate as measured in [37] (squares) and predicted by the slip model in [36] (solid line). In the model, the same parameters are used as were specified for the experimental system. The dramatic increase of the slip velocity at a critical shear rate is clearly shown.

Stepanyan et al. [36]. In their approach, they take into account all the aspects of the reptation theory and combine them with the effect of detachment. One essential outcome of this theory, necessary for any slip law, is the critical bulk shear rate for which the slip velocity suddenly starts to increase exponentially.

In Figure 9.5, an example for a particular configuration is given, where predicted data from [36] are compared with measured data from [37]. The critical bulk shear rate depends on the following important parameters: MM and topology, grafting density of attached macromolecules, adhesive energy needed to adsorb a macromolecule to the wall, wall material, temperature, and MMD. The effect of all these parameters has been studied and incorporated, so that the slip law can now be predicted for any system by evaluating explicit models. For polymer–polymer interfaces (PP/PS and PE/fluoropolymer) as studied by Lee at al. [38], the slip velocity was found to increase following a power-law dependence of shear stress.

An alternative slip model has been proposed by Adewale and Leonov [39] and is based on the work of the Russian school of Vinogradov et al. [40–42]. Their approach incorporates the hysteresis into the slip model. They determined flow curves for narrow MMD homologous series of PB and PI and found a correlation between the flow curve discontinuity onset point and the plateau modulus G_0, a bulk property. They also found that the transition region enlarges with increasing M_w and that this onset point is independent of M_w, MMD, and temperature. For their constant-pressure capillary experiments, only high-energy die walls were used. Their conclusion was that strong slip is due to a phase transition of the polymer near the wall. The fluid-phase polymer transforms under high shear into a high-elastic state and behaves more or less like a cross-linked (extremely high entangled) polymer. The critical stress for onset of slip can be derived from the Leonov model [43,44].

9.4   A RHEOLOGICAL MODEL INCLUDING WALL SLIP

The relevance of slip to melt fracture suggests the need for incorporating a slip model into the set of equations derived so far. The aim is to reach quantitative agreement between simulated results and capillary or slit die rheometer experimental flow curves and the observed extrudate distortions. For the mass balance in the reservoir, Equation 9.4 is used. For the momentum balance in the die, Equation 9.6 is applied. Contrary to the modeling approach in Section 9.2, it is assumed that the bulk flow is described by a shear thinning viscosity, capturing the essence of the die flow. The generalized Newtonian model (Equation 3.93) is applied. The total stress tensor $\mathbf{S}$ reads as

$$\mathbf{S} = -p\mathbf{I} + 2\eta\mathbf{D}, \tag{9.29}$$

where the viscosity η depends on the shear rate. For laminar flow in a slit die, all quantities in the model depend on t and x_2. Equation 9.10 is then replaced by

$$\eta(\dot{\gamma})\dot{\gamma} = -\frac{x_2}{L} P_{\text{die}}(t), \tag{9.30}$$

where, for later convenience, the pressure difference over the die is denoted by P_{die}. Elasticity is not considered to play an important role in the bulk flow within the die, but it is important for entry and exit effects and is incorporated in a pressure correction term:

$$P(t) = P_{\text{die}}(t) + P_{\text{loss}}(t). \tag{9.31}$$

The shear thinning relation $\eta = \eta(\dot{\gamma})$ of the polymer under consideration can be determined with standard rheometers (empirical) or derived from the MMD (first principle) as discussed in Section 8.4. A nonzero wall velocity v_w requires modification of Equation 9.15. In this case, the partial integration gives

$$Q(t) = W \int_{-H}^{H} v_1(t, x_2)\,dx_2 = 2WHv_s(t) - 2W \int_{0}^{H} \dot{\gamma}(t, x_2)x_2\,dx_2. \tag{9.32}$$

For the slip model, the phenomenological approach of Den Doelder et al. [21] is taken here, which is based on considerations by Greenberg and Demay [45]. This approach is satisfactory in demonstrating that any slip law of the kind used here will lead to relaxation oscillations as observed.

To incorporate a slip boundary condition, the slip velocity is assumed to be proportional to the power d of the absolute value of the wall shear stress:

$$v_s = G(t)\tau_w^d, \tag{9.33}$$

The parameter d is introduced for fitting purposes to experimental data, but is not essential for the present derivation. According to Equation 9.30, the shear stress at the wall is given by

$$\tau_w(t) = \frac{H}{L} P_{die}(t). \tag{9.34}$$

The essence of this slip model is the behavior of the proportionality coefficient G. It is assumed to vary between zero (stick) and G_{max} (total slip). The value of $G(t)$ at each instant depends on the value of τ at that instant. To that end, a so-called *switch-curve* $S(Q)$ is heuristically introduced (Figure 9.6).

The $S(Q)$ curve is defined by two experimental $F(Q)$ flow curve characteristics and separates the flow regimes where stick-boundary conditions apply from those where slip-boundary conditions hold. In contrast to the RO-model of Section 9.1, now only two characteristic flow curve points are needed—the endpoint of Branch I and the onset point of Branch II. It is important to realize that in principle, these points can be calculated for any specific extrusion system using the predictive models in Refs. [33–35], based on input parameters such as polymer mass and topology and adhesive energy of the wall–chain interaction. If one is only interested in fitting an observed oscillation, one can deduce these two points from the experimental flow curves. The definition of $G(t)$ is as follows: If, at time t, $\tau_w(t) < S(Q(t))$, with $Q(t)$ the flux at that time, then the coefficient $G(t)$ is assumed to exponentially converge to zero. If $\tau_w(t) > S(Q(t))$, $G(t)$ is assumed to exponentially converge to G_{max}. The speed of the increase and decrease is determined by a time constant t_{slip}. The value of this parameter has a great influence on the shape of the oscillations.

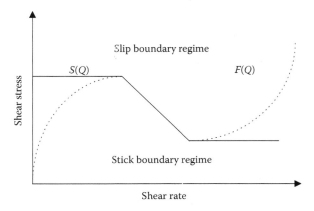

FIGURE 9.6 A schematic representation of a switch curve $S(Q)$ in relation to the flow curve $F(Q)$. The switch curve separates the stick-boundary regime from the slip-boundary regime.

To complete the model, the die entry pressure loss P_{loss} has to be specified. This is achieved experimentally by either the Bagley method or by using an orifice die ($L=0$) [18]. Both methods lead to approximately the same results for currently considered HDPE experiments, and show that P_{loss} explicitly depends on the flow rate entering the die:

$$P_{loss} = a_1 \dot\gamma_a^{a_2}. \tag{9.35}$$

The parameters a_1 and a_2 can be obtained from comparison of simulations with experimental data. The apparent wall shear rate $\dot\gamma_a$ is defined for a Newtonian flow in terms of the total mass flux Q. Its value is obtained from Equation 3.102 after the substitution of the value $n=1$. This leads to

$$\dot\gamma_a = \frac{3Q}{2H^2W}. \tag{9.36}$$

It is also possible to incorporate the pressure loss term as an increase in the die length L by an amount depending on Q. However, the method in Equation 9.35 is easier to implement. The entry pressure correction is assumed valid for both branches of the flow curve. This implies that the viscous resistance encountered in the converging flow near the die entry only depends on the integrated velocity field over the die cross section and not on the precise distribution of the velocities in the die (cf. plug flow versus Poiseuille flow).

The current model consists of Equation 9.4 in combination with Equations 9.29 through 9.36. As initial condition, the no-flow situation is chosen. The equations are numerically solved following a specific procedure. At a given time, the pressure P and the flow rate Q are known. As time increases, the reservoir length is reduced by the time step times the piston speed. The new P is calculated from Equation 9.4 by an Euler forward integration step [46,47]. To determine the new Q, Equations 9.29 through 9.36 have to be solved. They can be written in the form $Q=f(Q)$, so that a fixed-point method can be applied. The procedure starts by proposing a well-chosen initial value for Q. For a set Q, P_{loss} is calculated from Equation 9.35 and P_{die} from Equation 9.31. With Q and P_{die} known, G is calculated from the switch curve mechanism as described by Den Doelder et al. [21]. From P_{die}, the shear rate profile is calculated in combination with Equation 9.30 on a grid of discretized spatial positions along the x_2 axis, again using a fixed-point iterative procedure. This set is used to numerically calculate the integral in Equation 9.32. The integral is combined with the slip velocity calculated from Equation 9.33 to obtain Q from Equation 9.32. This new Q value is used as the new initial value for the numerical calculation routine. The procedure is repeated until the difference between the initial and the calculated Q becomes smaller than a certain value. Then a new time step is made. It is possible to modify Q_i, controlled by the piston speed and the reservoir height h at every time step. The entire calculation is carried out in dimensionless form [21]. Results of the model calculation can be compared with well-defined constant-rate capillary die experiments performed on HDPE by Durand and coworkers [7,18] (Figures 9.7 and 9.8).

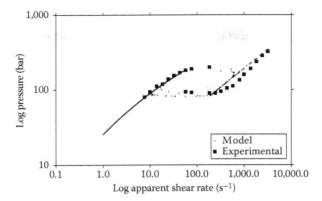

FIGURE 9.7 Comparison between a simulated (model) and experimental flow curve as obtained for an HDPE at 160°C using a constant-rate capillary rheometer with a die geometry of $L = 22.2$ mm, $2R = 1.39$ mm, and entry angle 180°. (From Durand, V., Ecoulements et instabilité des polyéthylènes haute densité, PhD thesis, Ecole des Mines de Paris (CEMEF), Sophia Antipolis, France, 1993.)

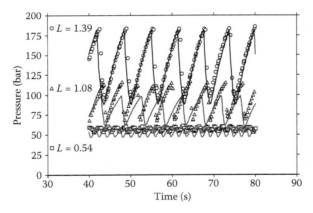

FIGURE 9.8 A comparison between simulated and experimental pressure oscillations obtained for an HDPE using a constant-rate capillary rheometer with different length dies, fixed die diameter $2R = 1.39$ mm, and entry angle 180° at 160°C.

The model accurately describes the characteristic aspects of the discontinuous flow curve such as the two steady-state Branches I and II. Furthermore, the die geometry dependence, time evolution, and dependence of the oscillation frequency and the shape of the measured pressure signal on reservoir volume and imposed piston speed are all described well.

In summary, the essential ingredients of the current model are (1) polymer compressibility χ in the reservoir, (2) the reservoir–die geometry, (3) the pressure loss terms due (to exit) and entry effects; (4) the shear-thinning viscous flow in the die; and (5) the empirical slip law.

9.5   BULK AND INTERFACIAL VISCOSITY BALANCE FOR DIFFERENT POLYMERS

Chapters 5 and 6 indicate that many polymers have been studied regarding their melt fracture behavior. However, the availability of a model that incorporates and unites all the melt fracture observations for various polymers would simplify the interpretation of the experimental observations. One of the most intriguing issues is why some polymers give rise to discontinuous flow curves while others only show one single branch without "spurt" and hysteresis features. The compressibility of the different polymers is of the same order of magnitude and the pressure correction magnitude does not seem to be crucial for any type of flow curve. Accordingly, the model of Section 9.4 suggests that the flow curve differences essentially arise from differences in the polymer bulk viscosity function and the interfacial behavior. The relation between bulk viscosity and polymer architecture is understood from microscopic constitutive models as presented in Chapter 8. Progress on the second polymer-differentiating element, the interfacial rheology, in relation to polymer architecture is being made (Section 9.3) [38]. However, much remains to be discovered there. This section illustrates how the models of Section 9.4 can be applied to identify the magnitude of slip needed to obtain continuous or discontinuous flow curves for polymers with known bulk viscosity.

To this end, two polymer types HDPE and LDPE were considered by Den Doelder [48]. He applied capillary rheometry to obtain flow curves, while the bulk viscosity was obtained from parallel plate oscillatory shear rheometry. Out of the many commercially available polymers in each class, two in each class were selected based on highest and lowest viscosity criteria. The four materials are referred to as HDPE-low, HDPE-high, LDPE-low, and LDPE-high, where high and low refer to the overall viscosity. For modeling purposes and simplicity (capturing the essential), the viscosity functions are described by the two-parameter power-law function (Equation 3.94) in the high shear rate and stress regime. For low flow rates and stresses, the error is substantial, but not relevant in view of the regime where extrudate distortions are observed. The power-law model relates viscosity η to shear stress τ and shear rate $\dot{\gamma}$ according to

$$\tau_{12}(\dot{\gamma}) = k\dot{\gamma}^n, \quad \eta(\dot{\gamma}) = k\left|\dot{\gamma}\right|^{n-1}, \quad \eta(\tau_{12}) = k^{1/n}\left|\tau_{12}\right|^{(n-1)/n}. \tag{9.37}$$

For the four polymers, the experimental shear rate and shear stress together with power-law viscosity fit are shown in Figure 9.9 at 190°C.

As before, for modeling purposes, the power-law fluid is considered to flow in a slit die type flow channel, and a slip equation is incorporated into the model. To eliminate the effect of die geometry, the flow curve is presented in terms of wall shear stress τ and apparent shear rate $\dot{\gamma}_a$. From Equation 9.32, it follows that the apparent shear rate has two contributions: one from the slip at the wall and one from the bulk:

$$\dot{\gamma}_a \equiv \dot{\gamma}_{a,slip} + \dot{\gamma}_{a,bulk} = \frac{3}{H} v_s(t) - \frac{3}{H^2} \int_0^H \dot{\gamma}(t, x_2) x_2 \, dx_2 \tag{9.38}$$

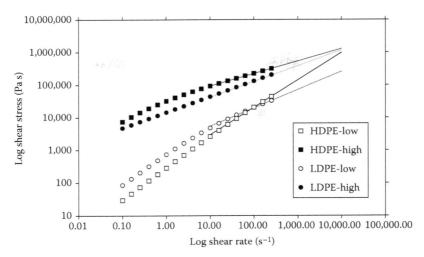

FIGURE 9.9 The power-law viscosity fit (line) and experimental shear rate–shear stress data (points) as obtained with a parallel plate rotational rheometer for two HDPE and two LDPE at 190°C. These commercial resins are selected for their widely different viscosity.

The combination of Equations 9.23, 9.30, and 9.34 shows that the stress is a linear function of x_2:

$$\tau_{12}(x_2) = \frac{-x_2}{H}\tau_w. \tag{9.39}$$

This can be used to write

$$\dot{\gamma}_{a,bulk} = \frac{3}{\tau_w^2}\int_0^{\tau_w}\dot{\gamma}(\tau_{12})\tau_{12}d\tau_{12}. \tag{9.40}$$

The substitution of Equation 9.37 yields for a power-law fluid:

$$\dot{\gamma}_{a,bulk} = \frac{3n}{2n+1}\left(\frac{\tau_w}{k}\right)^{1/n} = \frac{3n}{2n+1}\dot{\gamma}_w. \tag{9.41}$$

It is now possible to calculate the flow curve for the four model resins in the case of no slip (Figure 9.9).

After the combination of Equations 7.2, 9.38, and 9.41, the flow curve of a power-law fluid can be written in the form

$$\dot{\gamma}_a = \frac{3}{H\beta}\tau_w + \frac{3n}{2n+1}\left(\frac{\tau_w}{k}\right)^{1/n}. \tag{9.42}$$

To model and quantify the interfacial rheology for different polymer architecture and wall material, a mathematical formulation is made for the coefficient β in Equation 9.22, enabling a parameter variation study. The model to be used here reads as

$$\beta = \infty, \qquad \tau_w < \tau_c,$$

$$\beta = \beta_c \left(\frac{\tau_w}{\tau_c} \right)^{1-d}, \qquad \tau_w \geq \tau_c. \tag{9.43}$$

It contains the critical parameters τ_c, β_c, and the coefficient d already introduced in Equation 9.33. The effect of slip on the flow curve is shown in Figure 9.10. The

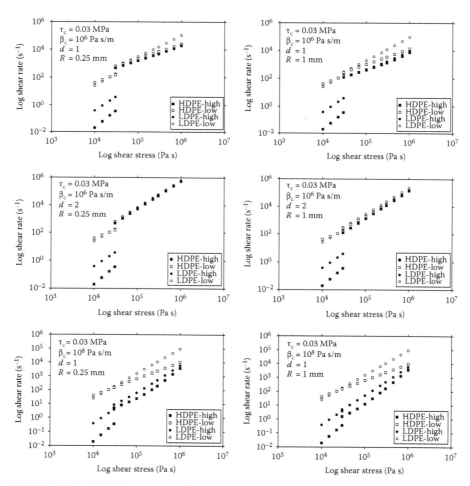

FIGURE 9.10 Model (power-law) flow curves assuming $\tau_c = 0.03\,\text{MPa}$, and considering the effect of a slip-boundary condition with $\beta_c = 10^6$ and $10^8\,\text{Pa s/m}$, for $d = 1$ and 2, and a radius $R = 0.25$ and 1 mm.

model calculations in the figure are valid for capillary dies, but similar results can be obtained for slit dies.

From Figure 9.10, it can be seen that the transition region in the discontinuous flow curve is smallest for the low viscosity polymers. The flow curves also show that if LDPE would have the same interfacial behavior as HDPE (same slip law), there would be a clear effect on the flow curve for small-enough die diameter. The selected low critical stress value for slip to occur at 0.03 MPa is associated with a low-energy (fluoro-polymer) wall. Lower stress values can hardly be measured with capillary rheometers. Experimentally, the transition regime is hardly ever found and the measured flow curves deal exclusively with Branch II. The same model can be applied to other polymers [48].

The combination of this model with experiments leads to the definition of the governing interfacial relations. From the literature, it is known that HDPE polymers display the largest transition region, LLDPE polymers have a smaller transition region and LDPE and PP polymers have very small or none for high-energy walls. For these polymers, Den Doelder [48] observed that no slip is found for the low viscosity polymers up to stresses of 0.4 MPa, even for small diameter capillary dies ($2R = 0.5$ mm). A high slip velocity is required in order to have an impact on the flow curve. Accordingly, when the slip velocities that are found for the high viscosity polymers would also be valid for the low viscosity polymers, then a noticeable impact on the flow curve should be found experimentally. Therefore, this analysis leads to the conclusion that the interfacial behavior of polymers is a function of the nature of the interface and the polymer architecture. The quantification of this function is possible with the present modeling approach.

9.6 FLOW CURVE AND MELT FRACTURE RELATION

The models presented in the previous sections have identified the various elements that play a role in determining the flow curve. However, it is still required to relate the flow curve with the observable extrudate distortions. Of particular practical importance is the definition of the onset flow conditions for the observation of melt fracture. Additionally, an accurate prediction of the shape of the extrudate would be a definite scientific breakthrough. To date, these two challenges still require significant research. This implies that only qualitative order of magnitude considerations can be presented regarding modeling polymer melt fracture onset and extrudate distortions' type and shape.

A model that relates the flow curve to melt fracture should distinguish at least three main types of extrudate distortions, each originating in a specific region of the flow channel. Volume distortions are initiated in the die entry region. "Spurt" distortions are initiated in the die land region. Surface distortions are initiated in the die exit region. Upon increasing the piston speed, the encountered extrudate distortion sequence depends on the polymer architecture, which in turn affects the local stress (Chapter 8).

In a flow channel (Figure 9.11), these regions are characterized by a distinctive polymer melt velocity profile and deformation. The region close to the piston is characterized by a slow laminar flow. The die entry region encloses the sudden

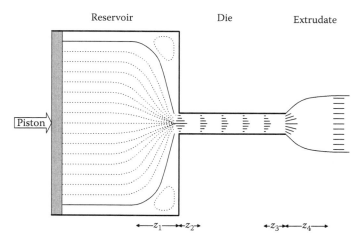

FIGURE 9.11 Model flow channel that indicates the various regions characterized by a distinctive polymer melt deformation as schematically visualized in terms of the velocity profile.

geometrical contraction and is characterized by a strong extensional flow along the center flow line and, in particular, close to the entry corners (Figure 7.12). The steady state laminar flow in the die region begins some distance downstream from the die entry and is governed by the viscous flow that determines the bulk flow curve. The die exit region encloses the transition from a constraining wall to a free boundary. It is also characterized by a strong extensional flow and a cooling of the extrudate. The final region (the extrudate) is characterized by a stress-free plug flow.

9.6.1 "Spurt" Distortions

The models developed in Sections 9.1 through 9.5 are directly coupled to the "spurt" distortions. "Spurt" is a die land phenomenon and the die land flow curve modeling provides the insights to predict for which conditions the "spurt" oscillations take place. The extrudate appearance during the various phases in a "spurt" oscillation cycle depends on the die entry and exit effects.

9.6.2 Surface Distortions

Surface distortions are related to the polymer deformation near the singularity at the die exit region. The exit region is considered to have a limited effect on the flow curve (some authors mention a change of slope, see Section 7.1.6). Pressure fluctuations in the reservoir or in the die appear to be too small to exceed the noise levels of the pressure detectors in experimental setups (with transducers placed before the die entry and inside the die). For a laminar flow, the velocity profile at a given flow rate in the die exit region changes from a quasi-parabolic profile at a distance z_3 upstream of the exit to a flat profile at a distance z_4 downstream of the exit. The onset of surface distortions can be modeled to occur when the (elongational) stress and strain rate locally surpass a critical value. The elongational strain rate $\dot{\varepsilon}_E$ can be identified with the change in velocity in the outermost layer over the longitudinal distance $z_3 + z_4$. For the case when the extrudate diameter is about the same as the die diameter ($2R$), this change in velocity Δv_{wall} equals $Q_{bulk}/\pi R^2$. The average strain rate equals $\Delta v_{wall}/(z_3 + z_4)$. However, the maximum strain rate is larger. When assuming that the onset of surface distortions occur when

$$\frac{2\Delta v_{wall}}{z_3 + z_4} > \dot{\varepsilon}_{E,c}, \tag{9.44}$$

the critical condition becomes

$$Q_{bulk} > \frac{\pi R^2 (z_3 + z_4)\dot{\varepsilon}_{E,c}}{2} \quad \text{or} \quad \dot{\gamma} > \frac{2(z_3 + z_4)\dot{\varepsilon}_{E,c}}{R}. \tag{9.45}$$

This illustrates that the appropriate flow curve variable, which is related to the onset of surface distortions is the apparent wall shear rate times the die diameter. The conversion to the corresponding wall shear stress leads to a smaller dependence on the die diameter due to the slope of the flow curve. This explains why the stress is often found to be the appropriate dimensionless parameter for defining the onset of melt fracture. Experimentally, the critical stress may seem to be diameter independent due to problems with determining the surface distortion onset point accurately. The above reasoning allows for an order of magnitude calculation. Assume that $z_3 + z_4 = 5$ mm. For LLDPE, the strain rate can be related to the elongational stress τ_E [49]:

$$\tau_E(\text{Pa}) = 10^5 \dot{\varepsilon}_E(\text{s}^{-1}). \tag{9.46}$$

From Göttfert Rheotens® (Göttfert Werkstoff-Prüfmaschine GmbH, Buchen, Germany) experiments, a general criterion for rupture [50,51] is $\tau_E > 10^6$ Pa. The combination of these equations allows the calculation of critical apparent elongational rates in relation to various die diameters. Migler et al. [52] have fine-tuned this approach experimentally, proposing the derived elongational property reconfiguration rate to unite observations for high- and low-energy walls.

Thus, the connection between surface distortions and the flow curve is made via a conversion of die land main flow to die exit stretching flow, combined with a rupture criterion based on local stress and deformation. Allal et al. [53] have modeled the rupture criterion in terms of molecular variables. The Griffith elasticity theory is used to clarify the role of plateau modulus, weight-average MM M_w, and MM between entanglements M_e. A coupling is made between shear stress inducing die land wall slip, and elongational stress leading to skin rupture. Dispersity effects are taken into account via weighted integration over MM. The reptation theory tube renewal time decreases strongly with increasing MMD, leading to the postponed onset of surface distortions, in agreement with the experimental observations presented in Chapter 6.

9.6.3 Volume Distortions

The connection between volume distortions and the flow curve is made via a conversion of die land main flow to die entry stretching flow, combined with flow instability criteria related to contraction flow phenomena such as vortex formation, and the loss of symmetry. Most models in this area are empirical, based on the systematic variation of entrance flow conditions for different geometries and polymers, as discussed in Chapters 5 through 7. Unification is sought via macroscopic rheological descriptions, capable of distinguishing polymer response based on differences in shear and, particularly, elongational flow. The experiments by Hurlimann and Knappe [54] are useful reference points. Rate-dependent strain hardening is a key feature here. The connection to the microscopic scale is possible via the models presented in Chapter 8. The procedure is to use microscopically based constitutive equations to connect MMD and topology to these key rheological functions, and apply them in the specific contraction flow situation of interest. Combined with macroscopic stress/deformation-based criteria for the onset of flow instabilities, and with additional models on the relaxation of these instabilities while moving through the die to the exit, the volume-distorted extrudate appearance can be predicted.

9.7 GENERAL OBSERVATION

It is clear that "spurt" melt fracture in terms of a discontinuous flow curve is understood and modeled in great detail on both a macroscopic and microscopic level. However, to the extent that they exist, the surface and volume distortion models are rather phenomenological at present. In Chapter 7, it is argued that the three different types of extrudate distortions can be considered as being due to a similar microscopic mechanism. This is associated with the entanglement/disentanglement of macromolecules due to high stresses, and occurring at different scales and locations. Mainly, extensional or stretching flow regimes are critical. Such an insight opens a view for the future modeling of all distortions in a single mathematical model. A unified quantitative model could be built upon the concept of relaxation oscillations.

The relaxation oscillations of "spurt" distortions are due to the cycling of macromolecules between entangled and disentangled states. It happens along the entire die wall and in combination with density fluctuations in the reservoir.

Surface distortions can be modeled as local relaxation oscillations near the die exit. They occur when the overall die land stress is still rather low. Mathematical rigor can be obtained when the stress profile along the entire die wall, including the exit, is known. An example of a criterion for the onset of surface distortion oscillations could be that the exit stress should exceed the critical stress for disentanglement. The length of this section could depend on the polymer architecture under consideration. In such a model, the same slip law used for "spurt" may be used for surface distortions. Explicitly associating macroscopically extensional flow and microscopically molecular stretching may, in combination with local boundary conditions, offer a path forward.

Volume distortions could also be modeled as relaxation oscillations. The entanglement dynamics for this type of melt fracture are to be considered at the die entry and as an interfacial relaxation oscillation between highly extended polymer layers.

The extension difference between microscopic polymer layers can be viewed as inducing oscillating density differences that cause a buckling of layers. Eventually, this leads to a distortion of the laminar flow pattern into a 3D vortex formation near the corners of the reservoir. The resulting melt flow instabilities that are observed as a volume-distorted extrudate are thus induced by the large acceleration of the polymer in going from the reservoir into the die.

Numerical accuracy and deeper knowledge of polymer properties are required to determine the precise stress levels and the criteria for entanglement/disentanglement, combined with their influence on the formation of vortices and other non-laminar flow phenomena. Such a unified model might offer a satisfactory combination of being quantitative (mathematical rigor), having easy-access model input parameters (from polymer architecture or from basic polymer characterization), and describing the important features of extrudate distortions with predictive capability.

REFERENCES

1. Weill, A., Oscillations de relaxation du polyéthylène de haute densité et défauts d'extrusion. PhD thesis, Université Louis Pasteur, Strasbourg, France, 1978.
2. Weill, A., About the origin of sharkskin. *Rheol. Acta*, **19(5)**:623–632 (1980).
3. Weill, A., Capillary flow of linear polyethylene melt: Sudden increase of flow rate. *J. Non-Newtonian Fluid Mech.*, **7(4)**:303–314 (1980).
4. Van der Pol, B., On relaxation oscillation. *Philos. Mag.*, **2(11)**:978–994 (1926).
5. Van der Pol, B., Forced oscillations in a circuit with non-linear resistance. *Philos. Mag.*, **3(13)**:65–80 (1927).
6. Molenaar, J. and R. Koopmans, Modeling polymer melt flow instabilities. *J. Rheol.*, **38(1)**:99–109 (1994).
7. Durand, V., B. Vergnes, J. F. Agassant, E. Benoit, and R. J. Koopmans, Experimental study and modeling of oscillating flow of high density polyethylenes. *J. Rheol.*, **40(3)**:383–394 (1996).
8. Grob, M. J. H. B., Flow instabilities in a polymer extrusion process. MSc thesis, Eindhoven University of Technology, Eindhoven, the Netherlands, 1994.
9. Ranganathan, M., M. R. Mackley, and P. H. J. Spitteler, The application of the multipass rheometer to time-dependent capillary flow measurements of a polyethylene melt. *J. Rheol.*, **43(2)**:443–451 (1999).
10. Denlinger, R. P. and R. P. Hoblitt, Cyclic eruptive behavior of silicic volcanoes. *Geology*, **27(5)**:459–462 (1999).
11. Ovaici, H., M. R. Mackley, G. H. McKinley, and S. J. Crook, The experimental observation and modelling of an "Ovaici" necklace and stick-"spurt" instability arising during the cold extrusion of chocolate. *J. Rheol.*, **42(1)**:125–157 (1998).
12. Georgiou, G. C. and M. J. Crochet, Time-dependent compressible extrudate-swell problem with slip at the wall. *J. Rheol.*, **38(6)**:1745–1755 (1994).
13. Georgiou, G. C. and M. J. Crochet, Compressible viscous flow in slits with slip at the wall. *J. Rheol.*, **38**:639–654 (1994).
14. Georgiou, G. C., Extrusion of a compressible Newtonian fluid with periodic inflow and slip at the wall. *Rheol. Acta*, **35**:531–544 (1996).
15. Georgiou, G. C. and D. Vlassopoulos, On the stability of the simple shear flow of a Johnson-Segelman fluid. *J. Non-Newtonian Fluid Mech.*, **75**:77–97 (1998).
16. Den Doelder, C. F. J., Sharkskin and "spurt" in JSO modelled polymer extrusion. PhD thesis, Eindhoven University of Technology, Eindhoven, the Netherlands, 1996.

17. Bagley, E. B., I. M. Cabott, and D. C. West, Discontinuity in the flow curve of polyethylene. *J. Appl. Phys.*, **29**:109–110 (1958).
18. Durand, V., Ecoulements et instabilité des polyéthylènes haute densité. PhD thesis, Ecole des Mines de Paris (CEMEF), Sophia Antipolis, France, 1993.
19. Aarts, A. C. T. and A. A. F. van de Ven, Transient behavior and stability points of the Poiseuille flow of a K-BKZ fluid. *J. Eng. Math.*, **29**:371–392 (1995).
20. Kolkka, R. W., D. S. Malkus, M. G. Hansen, G. R. Ierley, and R. A. Worthing, "Spurt" phenomena of Johnson-Segalman fluid and related models. *J. Non-Newtonian Fluid Mech.*, **29**:303–335 (1988).
21. Den Doelder, C. F. J., R. J. Koopmans, J. Molenaar, and A.A.F. van de Ven, Comparing the wall slip and the constitutive approach for modelling "spurt" instabilities in polymer melt flows. *J. Non-Newtonian Fluid Mech.*, **75**:25–41 (1998).
22. Brochard, F. and P. G. de Gennes, Shear dependent slippage at polymer/solid interface. *Langmuir*, **8**:3033–3037 (1992).
23. Migler, K. B., H. Hervet, and L. Leger, Slip transition of a polymer melt under shear stress. *Phys. Rev. Lett.*, **70(3)**:287–290 (1993).
24. Hatzikiriakos, S. G., I. B. Kazatchkov, and D. Vlassopoulos, Interfacial phenomena in the capillary extrusion of metallocene polyethylenes. *J. Rheol.*, **41(6)**:1299–1316 (1997).
25. Anastasiadis, S. H. and S. G. Hatzikiriakos, The work of adhesion of polymer/wall interfaces and its association with the onset of wall slip. *J. Rheol.*, **42(4)**:795–812 (1998).
26. Stewart, C. W., Wall slip in the extrusion of linear polyolefins. *J. Rheol.*, **37(3)**:499–513 (1993).
27. Hill, D. A., Wall slip in polymer melts: A pseudo-chemical model. *J. Rheol.*, **42(3)**:581–549 (1998).
28. Drda, P. P. and S. Q. Wang, Stick-slip transition at polymer melt/solid interfaces. *Phys. Rev. Lett.*, **75(14)**:2698–2701 (1995).
29. Hatzikiriakos, S. G., Wall slip of linear polyethylenes and its role in melt fracture. PhD thesis, McGill University, Montreal, Canada, 1991.
30. Hatzikiriakos, S. and J. M. Dealy, The effect of interface conditions on wall slip and melt fracture of high-density polyethylene. In *SPE ANTEC*, Montreal, Canada, 1991, 49, pp. 2311–2314.
31. Dubbeldam, J. L. A. and J. Molenaar, Self-consistent dynamics of wall slip. *Phys. Rev. E*, **67**:11803–11814 (2003).
32. Dubbeldam, J. L. A. and J. Molenaar, Dynamics of the "spurt" instability in polymer extrusion. *J. Non-Newtonian Fluid Mech.*, **112**:217–235 (2003).
33. Tchesnokov, M. A., J. Molenaar, and J. J. M. Slot, Dynamics of molecules adsorbed on a die wall during polymer melt extrusion. *J. Non-Newtonian Fluid Mech.*, **126**:71–82 (2005).
34. Tchesnokov, M. A., J. Molenaar, J. J. M. Slot, and R. Stepanyan, A molecular model for cohesive slip at polymer melt/solid interfaces. *J. Chem. Phys.*, **122**:214711–214722 (2005).
35. Tchesnokov, M. A., Modeling of polymer flow near solid walls. PhD thesis, University of Twente, Enschede, the Netherlands, 2005, ISBN 9036521866.
36. Stepanyan, R., J. J. M. Slot, J. Molenaar, and M. A. Tchesnokov, A simple constitutive model for a polymer flow near a polymer-grafted wall. *J. Rheol.*, **49**:1129–1151 (2005).
37. Leger, L., E. Raphael, and H. Hervet, Surface anchored polymer chains: Their role in adhesion and friction. *Adv. Polym. Sci.*, **138**:185–225 (1999).
38. Lee P. C., H. E. Park, D. C. Morse, and C. W. Macosko, Polymer-polymer interfacial slip in multilayered films. *J. Rheol.*, **53(4)**:893–915 (2009).
39. Adewale, K. P. and A. I. Leonov, Modeling "spurt" and stress oscillations in flows of molten polymers. *Rheol. Acta*, **36**:110–127 (1997).

40. Borisenkova, E. K., V. E. Dreval, G. V. Vinogradov, M. K. Kurbanaliev, V. V. Moiseyev, and V. G. Shalganova, Transition of polymers from the fluid to the forced high-elastic and leathery states at temperatures above the glass transition temperature. *Polymer*, **23(1)**:91–99 (1982).

41. Vinogradov, G. V., V. P. Protasov, and V. E. Dreval, The rheological behavior of flexible chain polymers in the region of high shear rates and stresses, the critical process of "spurt"ing, and supercritical conditions of their movement at T>Tg. *Rheol. Acta*, **23**:46–61 (1984).

42. Vinogradov, G. V., A. Y. Malkin., Y. G. Yanovskii, E. K. Borisenkova, B. V. Yarlykov, and G. V. Berezhnaya, Viscoelastic properties and flow of narrow distribution polybutadienes and polyisoprenes. *J. Polym. Sci., A2*, **10**:1061–1084 (1972).

43. Leonov, A. I. and A. N. Prohunin, *Nonlinear Phenomena in the Flow of Viscoelastic Polymer Fluids*. Chapman & Hall, London, U.K., 1994.

44. Larson, R. G., *Constitutive Equations for Polymer Melts and Solutions*. Butterworth Publishers, Boston, MA, 1988.

45. Greenberg, J. M. and Y. Demay, A simple model of the melt fracture instability. *Eur. J. Appl. Math.*, **5**:337–358 (1994).

46. Gear, C. W., *Numerical Initial Value Problems in Ordinary Differential Equations*. Prentice-Hall, Englewood Cliffs, NJ, 1971.

47. Press, W. T. H., B. P. Flannery, A. A. Teukolsky, and W. T. Vetterling, *Numerical Recipes*. Cambridge University Press, Cambridge, U.K., 1989.

48. Den Doelder, J., Design and implementation of polymer melt fracture models. PhD thesis, University of Eindhoven, Eindhoven, the Netherlands, 1999, ISBN 90-386-0701-6.

49. Constantin, D., Linear-low-density polyethylene melt rheology: Extensibility and extrusion defects. *Polym. Eng. Sci.*, **24(4)**:268–274 (1984).

50. Rutgers, R. P. G., An experimental and numerical study of extrusion surface instabilities for polyethylene melts. PhD thesis, Department of Chemical Engineering, University of Cambridge, Cambridge, U.K., 1998.

51. Wagner, M. H., V. Schulze, and A. Göttfert, Rheotens mastercurves and drawability of polymer melts. *Polym. Eng. Sci.*, **36**:925–935 (1996).

52. Migler, K. B., Y. Son, F. Qiao, and K. Flynn, Extensional deformation, cohesive failure, and boundary conditions during sharkskin melt fracture. *J. Rheol.*, **46(2)**:383–400 (2002).

53. Allal, A., A. Lavernhe, B. Vergnes, and G. Marin, Relationships between molecular structure and sharkskin defect for linear polymers. *J. Non-Newtonian Fluid Mech.*, **134**:127–135 (2006).

54. Hürlimann, H. P. and W. Knappe, Relation between the extensional stress of polymer melts in the die inlet and in melt fracture. *Rheol. Acta*, **11(3–4)**:292–301 (1972).

10 Preventing Melt Fracture

Limitations in the understanding of melt fracture have not barred people from finding solutions that alleviate the observable extrudate distortions. Typical approaches include the addition of some (preferably cheap) "stuff," polymer blending, or the modification of processing equipment. The use of "stuff" is not always related to melt fracture. In fact, most of the time, "stuff" is targeted on achieving very different effects, for example, thermal stabilization, reducing cost, and coloring. Sometimes, it may have (unwittingly) unexpected benefits. In general, "stuff" is applied to expand the applicability of polymers into very diverse markets—from packaging to more durable applications. This implies that nearly all commercially available polymers contain a certain level of low-MM organic and/or inorganic additives—from catalyst residues to antioxidants to all sorts of agents that improve the polymer functionality. These plastics are often applied in blends with other plastics for the same reasons as mentioned above. The influence of many of these additives or plastic blends on improving processing, and reducing or preventing melt fracture is mostly an art. Some of the more systematic studies have unfortunately limited extrapolation value.

The advantage of the additive or blend approach is its ease to implement. The modification of processing equipment is often more difficult and costly. It requires some insight into which variables affect melt fracture. Furthermore, such modifications often need applicability for a wide range of commercially available polymers in order to be economically viable to implement. In this chapter, some procedures are reviewed that can be practically implemented by plastic converters to prevent melt fracture.

10.1 ADDITIVES

The effect of additives on the melt fracture behavior of polymers is complex and often not very well understood. Little information is available that explains their influence on melt fracture. Additives are an integral part of commercial polymers; they turn them into plastics. Typically, a range of additives are present in commercial polymers in low concentration (<1% wt). Higher concentrations of such additives often have limited usefulness either as no additional benefits can be observed or the added cost impacts the economics. The influence of each one or all may be difficult to determine. Researchers should be aware of their presence. Mostly, additives (Table 10.1) are organic molecules of relatively low MM and are referred to as antioxidants, pigments, dyes, antistatics, UV stabilizers, thermal stabilizers, plasticizers, or slip agents [1–48]. The last class includes lubricants, fluoro-elastomers, and polysiloxanes. Not all additives have been studied in relation to melt fracture. Only a few are identified as having an effect or are commercially viable to prevent melt fracture

TABLE 10.1

A Selection of Common Types of Additives Used in Commercial Polymers, i.e., Plastics

Additive	Function
Anti-blocking agents	Substances that prevent plastics films from sticking together, and are used to facilitate handling
Anti-fogging agents	Substances that improve film clarity by preventing any water from condensing as droplets on the film surface
Antioxidants	Substances that prevent polymer oxidation during processing and service life
Anti-static agents	Substances that reduce or eliminate the static electric charges accumulating during processing on the produced objects
Compatibilizers	Substances that are added to mixtures of dissimilar polymers to enable them to become more homogeneously mixed
Coupling agents	Substances bonded to the surfaces of one component (usually inorganic materials) to improve compatibility with polymers and increase the long-term stability of the mixture
Diluents	Solvents or any organic substance reducing the viscosity of the polymer
Fillers	Substances used in larger concentrations (>1% wt) and typically of an inorganic nature that minimize polymer use and/or improve application performance
Flame retardants	Substances reducing or preventing combustion
Heat stabilizers	Substances preventing the thermal degradation of the polymer
Impact modifiers	Substances improving the mechanical properties of polymers for impact
Light stabilizers	Substances preventing ultraviolet degradation of the polymer
Lubricants	Substances that reduce viscosity and prevent polymers from sticking to the mold wall
Nucleating agents	Substances that promote or control the crystal formation in semicrystalline polymers
Optical brighteners	Fluorescent organic substances that correct discoloration or enhance whiteness
Peroxides	Radical forming oxygen substances that cross-link and/or degrade polymers
Pigments and dyes	Substances that change the color of the polymer
Plasticizers	Substances that space out the polymers to facilitate their movement leading to enhanced flexibility and ductility
Processing aids	Substances that migrate to the die surface during processing to prevent surface distortion

both for reasons of relative performance, that is, the additive provides sufficient differentiation and the lowest cost.

Most additives do not interact chemically with the polymer, that is, they do not change the polymer architecture. In fact, most additives aim at preserving the polymer architecture under the often harsh thermo-mechanical polymer processing conditions. However, for a number of applications such as wire and cable coatings, and pipes and profiles, additives are intended to react and modify the polymer architecture. This happens during the processing and changes the rheological character of the melt, possibly inducing a melt fracture feature not observed with the unmodified polymer. The additive either grafts onto the polymer, induces cross-linking, or degrades the polymer to a lower MM. The latter is done for isotactic homo-PP using peroxides and is referred to as "vis-breaking," that is, reducing the viscosity. This kind of reactive additive is just one ingredient out of many as used in a formulation. The combination of polymers (often different classes), reactive, nonreactive organic additives and inorganic fillers can induce complex fluid flow behavior. In such cases, it becomes difficult to identify the polymer origin of melt fracture although the macroscopic fluid-flow criteria for melt fracture initiation remain valid.

Fillers are additives used in much higher concentration (>1% wt) and typically aim at replacing the polymer with a much cheaper inorganic substance, for example, calcium carbonate. In other cases, fillers target specific performance criteria such as an application's weight increase (e.g., barium salts), haptics (e.g., calcium carbonate), color (e.g., titanium dioxide), and barrier (e.g., mica or talc). The effects on melt fracture vary depending on the overall characteristics of the fluid flow. Of all fillers, carbon black is probably the most controversial in terms of its effect on melt fracture. Not all carbon black is the same and different types originating from different production methods can either induce or prevent melt fracture compared to the natural polymer [49–53].

10.1.1 SLIP AGENTS

Slip agents are typically low-MM organic molecules. For PVC and rubber, they are often known as lubricants and are applied to improve overall processing [18,40,41]. They act by generating a coating onto the metal surface that prevents the sticking of the polymer melt to the processing equipment. In polyolefin polymers, these lubricants are long chain fatty acids, alcohols, or amides with a low compatibility to the nonpolar polymer. Their concentration in the polymer granules is usually low (<2000 ppm) [42]. During processing, the lubricant coating induces slip and reduces surface distortions. Chauffoureaux et al. [43] show that these lubricants increase the velocity close to the die wall, which suggest their potential to reduce extrudate surface distortion. However, in practice, this type of lubricant may start to accumulate and chemically decompose over time at the die exit and on chill rolls (plate-out). Subsequently, die lines may be induced on the extrudate.

The most widely studied slip agents are the fluoro-polymers. Their commercial success, in particular, in combination with linear polyolefin polymers, is closely associated with their ability to reduce surface distortions. They are most often referred to as polymer processing aids (PPAs). The combination of LLDPE with fluoro-polymers is most often studied [1,4,5,7,9,11,13,14,16,17,19–22,24,26,28–31,33–35,38,39,45]. A range of fluoro-polymers exists but all are used in a similar fashion [37]. The fluoro-polymer is the die construction material or a die coating. The coating is applied either before processing takes place using solutions or sprays or via gradual deposition during a continuous extrusion. In the latter case, a blend of a fluoro-polymer concentrate (masterbatch) and the polymer is extruded. The fluoro-polymer slowly migrates (typically 15–30 min) to the die wall during extrusion [45]. A continuous feed is required as the coating wears off. In the presence of other additives such as anti-blocking agents or pigments, the beneficial effect on melt fracture may be limited [12]. An anti-blocking agent is composed of microscopic hard particles, for example, silica, that create microscopic imperfections into the extrudate surface to reduce the blocking (i.e., sticking together of, e.g., films). The type, shape, and size of the particles seem to be critical. The fluoro-polymer concentration in the polymer for optimum effect seems to be between 500 and 1000 ppm [5]. The action of polysiloxanes is comparable to that of the fluoro-polymers.

The mechanism of how fluoro-polymers and related slip agents or PPAs work can be explained in terms of the stick-slip mechanism [54–63]. The slip agents attach readily to the die wall either via direct coating or via a slow migration in a continuous extrusion process, and create a slippery, low surface tension surface. The nonpolar polyolefin polymer is incompatible with the slip-agent. A slippery surface is created that prevents the stick-slip process, which causes observed stress oscillations. It reduces the extensional stress induced by the interfacial tension between die wall and polymer melt and thus postpones the observable surface distortions. In principle, higher flow rates in the presence of slip agents or PPAs still can induce polymer melt fracture either as a surface- or volume-distorted extrudate if the extensional stress along the die land near the exit or at the die entrance becomes too high. For example, this will be the case when spurt occurs, as demonstrated by Migler et al. [64,65].

Other additives acting as PPAs and affecting melt fracture by postponing the appearance of surface distortions are boron nitride [66,67], nanoclay [39,68], and diatomite/PEG [69]. Kulikov et al. [46] use silanols cured by boron oxide as a lubricant to LLDPE in low concentration (<0.1 wt%) to delay melt fracture. They also indicated that adding further plate-like particles (kaolin, mica, boron nitride) to such a lubricant helped delay the onset flow rate for surface distortions to appear. Hong et al. [47] explored the use of hyperbranched polymers. They are similar to dendrimers but do not have precise end-group multiplicity and functionality resulting in a far more irregular branching structure. Hexadecanoate-terminated polyester with generation 3–6 [70] is immiscible with LLDPE, reduces the viscosity of the blend and migrates to the surface inducing a delay in the flow rate onset of surface distortions. Wang et al. [48] investigate the effect of an incompatible hyperbranched PE and a compatible linear highly branched PE [71,72] in a metallocene LLDPE on the onset of melt fracture. At concentrations of 3–5 wt%, the hyperbranched PE had an effect while the other chain topology proved to be ineffective.

10.1.2 POLYMER BLENDS

The blending of two or more polymers (actually plastics) to reduce the viscosity at the desired processing rate is a common procedure to prevent melt fracture. Blending LDPE with LLDPE is a standard industry practice that typically reduces or delays surface distortions during processing [73]. The blending effectively reduces the viscosity of the LLDPE at high processing shear rates and increases the resistance to extensional stretch-related failure. The amount of LDPE blending is determined by the performance requirements of the extruded film. Herranen and Savolainen [74] show the influence of blending LLDPE and LDPE on the onset of melt fracture for online extrusion. The flow curve measurements are combined with ultra-sound velocity detection. For the LLDPE, base resin melt fracture is observed. Blending 10% LDPE with LLDPE gives surface distortions at lower shear rates and shear stresses than the base LLDPE while volumetric distortion appears at the same values. Increasing the amount of LDPE in the blend causes the surface distortions to disappear and only volume distortions to remain. It should be noted that a delayed melt fracture onset strongly depends on the homogeneity of the blend and the classes of LDPE/LLDPE used. Overall, the lower concentrations of LDPE can have a beneficial effect by delaying the onset flow rate for the surface distortions of LLDPE.

Blending is also practiced in plastics other than PE. In these cases, the blending becomes more of a formulation of multiple components. This may induce phase behavior making the interpretation of the origin of melt fracture more complex. Additional mechanisms such as interfacial tension associated with particle size or lubrication and plastication become dominating. For preventing melt fracture, it is critical to understand the precise composition of the formulation and the molecular makeup of the individual components. Typically, the highest MM or lowest modulus polymer component needs modification. Perez et al. [75] studied blends of mLLDPE with LDPE. Surface distortions are postponed to higher rate with an increasing level of LDPE above 10 wt%. Delgadillo-Velazquez and Hatzikiriakos [76] studied blends of an LLDPE with different LDPEs. At low levels of LDPE (<20%), there is limited effect of the onset stress and rate of both sharkskin and spurt. The pure LDPEs show a critical stress and rate that strongly vary between the materials, depending on their structure. Blends with more than 20% LDPE do not show stick-slip oscillations. Den Doelder et al. [77] showed that the critical stress for the onset of the volumetric distortions of various LDPEs decreased with increasing M_w, at similar M_w/M_n and LCB levels, with the onset stress varying from 0.07 to 0.2 MPa.

10.1.3 WAX

Low-molar mass hydrocarbon waxes can be used to improve the surface quality of the extrudate. However, as is the case with many additives, only low concentrations can be used (<10% wt). Higher concentrations tend to negatively influence the final performance properties of the extruded article. Although a lubricating effect is anticipated, waxes mainly reduce viscosity of the polymer. Laun showed that adding 12.5% wt wax to PP has no influence on the slip velocity and only lowers the bulk viscosity (Figure 10.1).

Blyler and Hart [79] added 10% wt linear, low-MM ($M_w = 1600$ g/mol) wax to a linear PE ($M_w = 200,000$ g/mol). A significant drop in the viscosity is observed shifting Branch I of the discontinuous flow curve to higher shear rates. The magnitude of the discontinuity diminished and Branch II shifted to lower shear rates. For 20% wt wax, the discontinuity was barely visible and disappeared completely at 30% wt. The onset of the extrudate distortion still occurred at the same stress regardless of the blend composition.

10.1.4 FILLERS

Few papers are available presenting a systematic study on the influence of fillers and reenforcing agents on melt fracture [53,80–82]. Kleinecke [81,82] observed little influence on the shape of the flow curve by adding 10% 5 μm glass spheres to

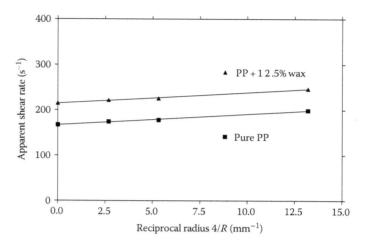

FIGURE 10.1 Adding wax to PP reduces the viscosity of the polymer and does not affect the slip velocity as measured using the Laun method [78]. The slope of the curve indicating the slip velocity is identical. (Courtesy of M. Laun from The Dow Chemical Company.)

HDPE. Ma et al. [53] studied flow patterns of LDPE with added calcium carbonate, titanium dioxide, or carbon black. A 5% by volume concentration suppressed the entrance vortices. For HDPE, Dospisil et al. [80] observed only a minor effect on the flow curve discontinuity of HDPE containing 5%–10% wt calcium carbonate or carbon black. 10% wt carbon fibers reduced the pressure oscillations. Pressure oscillations disappeared completely when changing to a higher melt index HDPE (0.1–0.3 g/10 min), using a conical die and adding 7% fumed silica or carbon black. Talc did not show this effect.

Rosenbaum et al. [38] and Hatzikiriakos and Migler [39] reported on the effects of small amounts of boron nitride (~0.1% wt) to eliminate surface distortions and substantially postpone the onset of the flow curve discontinuity. Boron nitride is an inorganic compound widely used as a nucleating agent. It is available in the form of platelets, thermally stable and chemically inert. The effect is observed for capillary experiments with metallocene PE and fluoro-polymers.

10.2 EXTRUDER AND PROCESSING CONDITIONS

Any method that reduces the local (elongational) stress during flow can delay the onset of melt fracture. The most obvious method (but not an economic option) is to reduce the flow rate. In particular for delaying surface distortions, an alternative is increasing the die gap or radius at constant throughput. The resulting lower shear rate reduces the die exit stress. For extrusion blown film, increasing the die gap may however aversely affect the optical appearance and the mechanical property balance in machine and cross direction. However, a very small widening of the die gap (a few μm or less) may have a significant improvement on the extrudate surface quality and the optical appearance of the film. This also implies that clean and smooth die surfaces, in particular, at the die lips, can prevent surface distortions. Damaged dies or dies that have substantial die-lip deposits induce die lines or melt fracture lines or bands with more intense low-amplitude high-frequency surface distortions [83,84].

Increasing the melt temperature is often tried to reduce surface distortions, but this may not always have the desired effect [62,85,86]. The wanted overall reduction of viscosity requires significant temperature changes (>30°C) in view of the relatively low activation energy of flow for most polyolefins (HDPE, LLDPE, metallocene-based PEs and PP). In contrast, LDPE and most of the other thermoplastics tend to give volume distortions where small temperature changes are ineffective. Depending on the polymer, even low temperature extrusion may show improved extrudate surface quality [85]. Perez-Gonzalez et al. [87] showed beneficial distortion-postponing effects for a mLLDPE at low T, which they related to flow-induced crystallization. In fact, crystallization increases the polymer's resistance against locally induced extensional stretching. Santamaria et al. [88] obtained similar results for a series of mPEs. Miller and Rothstein [89] and Miller et al. [90] used programmed die exit cooling and heating to influence surface distortions via temperature gradients. Introducing a temperature difference between the die and the melt can help but may be difficult to implement in practice. This effect can now be understood in terms of the stick-slip mechanism or, more precisely, the local extensional stress as explained in Chapter 7.

In those cases where the polymer only shows volume distortions, the above methods are of little or no consequence. Typically, the melt fracture originates somewhere in the feed block or die as a consequence of vortex formation induced by the mechanical constraints or strong elongational flows. The combination of flow simulation techniques and rheology variations may locate the potential cause and suggest mechanical modifications of the flow channel. Before engaging into a significant feed block and die design effort, it is important to ensure that the observable volume distortion is not caused by other extrusion phenomena. Alternate explanations may be an inhomogeneous feeding of the die, polymer surging associated with screw design, or important temperature differences along the extruder.

Another approach for improving the extrudate surface quality is changing the construction material of the die [91,92]. Some of the suggested metals or other materials may however wear very quickly or are very expensive to implement. Dao and Archer [93] used aluminum and stainless steel with PB and observed stick-slip transitions and conjectured it to relate to the flow-induced disentanglement of surface-adsorbed and bulk polymer molecules. Experiments with polished R-brass substrates led to much lower steady-state shear stresses without any stick-slip transition. The oxidation of the R-brass substrates at elevated temperatures dramatically increased steady-state shear stresses and the corresponding transient stresses showed a weak stick-slip process. Kulikov and Hornung [94] proposed silicon rubber or fluorinated silicon rubber coatings to eliminate LLDPE surface distortions.

In the literature, several other options have been explored, with mixed success. As a consequence, they are not widely adopted in the industry. Mechanically inducing a slip layer along the die using air or an inert gas is very difficult to maintain under stable conditions [95]. Introducing an oscillating die and/or mandrel is again

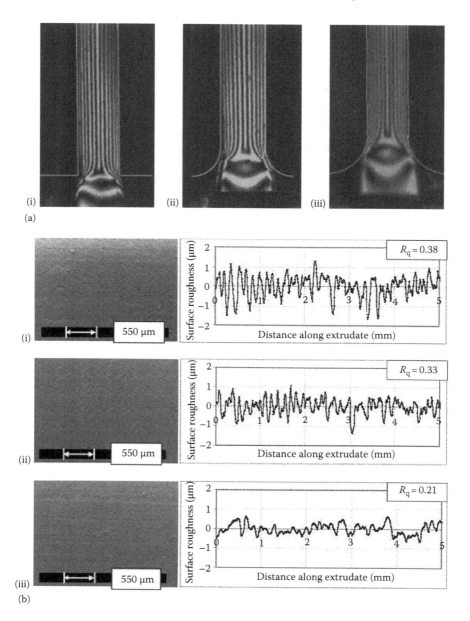

FIGURE 10.2 Modifying the die exit through tapering may prevent the appearance of surface distortions at equal flow rates. Arda and Mackley [98] showed the effect of die exit curvature for LLDPE in a slit die through flow birefringence (a) and measuring the extrudate roughness (b). Flow birefringence was measured at 0.05 rpm at 190°C. The surface roughness of tapes resulting from a 0.6 rpm extrusion at 190°C was measured using a profilometer and visualized with SEM. (From Arda, D. A. and Mackley, M. R., *J. Non-Newtonian Fluid Mech.*, 126, 47, 2005. With permission from Elsevier.)

mechanically possible but difficult to practice and often too costly to implement [96]. Ultrasound approaches suffer from similar shortcomings as the previous method [97]. The aim is to relax the stress build up through a mechanically induced vibration. One other possibility in particular for delaying the onset of surface distortions is the reshaping of the die exit edges by tapering (Figure 10.2). Lee and Park [50] observed an increase of the melt fracture onset shear rate for LDPE and SBR by increasing the die exit angle up to 45°. Higher angles gave identical effects as the reference die. Arda and Mackley [98] showed that this approach moves the detachment point of the melt from the die exit back into the flow channel. Depending on the tapering the die exit stress is reduced, that is, a more gradual die exit stress is built up, partly offset by a slightly enlarging exit die gap. This is a viable option in a number of processing technologies. In other cases, it may have an adverse effect on the dimensional control of the extrudate (in view of extrudate swell). Combeaud et al. [99] studied PS volume distortions with designs with half-entry angles of 30°, 45°, and 90°, as well as a smooth convergent (trumpet shape). They find no effect on the flow curve, but a strong effect on the onset rate of the distortions, especially going from smooth convergent via 30°–45°. From 45° to 90° the change was very limited. Pol et al. [100] substantiated similar findings with numerical flow simulations.

Arda and Mackley [98] studied the effect of surface roughness in a slit die for a 1 MI LLDPE. The critical stress for onset of surface distortions was found to increase from 0.16 to 0.19 MPa when the die insert surface roughness decreased from 1.5 to 0.05 μm. Further validation beyond the academic findings is required for considering its general industrial feasibility.

10.3   DEALING WITH MELT FRACTURE

It is clear that melt fracture presents itself in many different shapes and forms. Given the nature of polymers, melt fracture cannot be made to disappear; at best it can be delayed to higher extrusion flow rates. The fact that melt fracture is not observed in combination with certain processing conditions does not imply that the flowing polymer melt cannot generate melt fracture. Once it is observed there are several approaches on how to deal with it. However, there are no single remedies. The paradigm in the plastics industry is that for a certain application a specific processing machine is needed. For plastic converters, these machines are very expensive capital investments. These fixed assets are not to be changed or modified easily and without good reason. The appearance of melt fracture then becomes the burden of the polymer producers. They need to present polymer architecture options that meet the performance requirements of the application and at least the performance provided by polymers on offer by other polymer producers. The design of a new polymer architecture is often a long process. It may even be restricted by regulations stipulated in norms and laws. Moreover new polymer architecture does not provide converters with the immediate solution needed for the issue at hand. For those situations, some general practical approaches and preventive measures can be considered. What follows are guidelines and recommendations based on the present state-of-the-art understanding of melt fracture. There is no guarantee that when these guidelines or recommendations are followed, a solution for the problem will be found. Moreover, nearly all

TABLE 10.2

Checklist of Polymer Melt Fracture Remedies

Surface Distortions	Die Exit Phenomenon	Comments
General remedy	Reduce polymer melt (elongational) stress at die exit	
Ways to do this		
	Increase die gap at constant flow rate	Most effective, a few micron may suffice
	Lower flow rate	Impacts operating cost
	Add slip agents	Concentration depends on severity and adds cost
	Add viscosity-reducing agent	Low viscosity polymers or oligomers may affect performance and add cost
	Blend polymers to reduce viscosity	May affect performance depending on concentration
	Increase melt temperature	Not very effective
	Increase the die temperature relative to melt (if possible)	May have an adverse effect—lowering die temperature is then a solution
	Decrease melt temperature	May have an adverse effect—depends on polymer
	Clean dies	Remove accumulated debris
	Smooth dies	Repair any damaged die
	Taper die and/or mandrel	Requires detailed die design and melt flow simulation
	Change die and/or mandrel material of construction	Expensive and prone to wear
Volume Distortions	**Die Entry Phenomenon**	
General remedy	Reduce elongational stress and vortex forming capacity	
Ways to do this		
	Lower flow rate	Impacts operating cost
	Add viscosity-reducing agent	Low viscosity polymers or oligomers may affect performance and add cost
	Blend polymers to reduce viscosity	May affect performance
	Modify feedblock	Requires detailed die design and melt flow simulation

processed polymers are in fact formulations, that is, polymer blends including several additives and possibly reactive components. This requires in depth knowledge of the formulation composition and the polymers architecture to begin formulating a hypothesis leading to a solution for melt fracture. Typically, each formulation is very specific to the desired application. Extrapolations of a solution need to be critically evaluated before the implementation of a different formulation is considered.

10.3.1 MELT FRACTURE CHECKLIST

First, it needs establishing that the observable extrudate distortion is in fact melt fracture. Extruder surging, residual water, inhomogeneous melts, impurities, and partly cross-linked polymers may all cause the appearance of a melt fracture–like feature.

Second, a distinction needs to be made between surface and volume distortions. It defines the nature and originating location of melt fracture, that is, at the die exit or upstream in the die and feedblock.

Third, finding solutions requires articulating a hypothesis on possible causes and testing the validity of that assumption. A systematic theoretical or experimental test of the problem's hypothesis then leads to a potential solution and knowledge supported by data.

It should be noted that most extruders wear over time. Screw geometries and melt flow channels may be dimensionally affected thus changing the processing conditions and flow rates.

If all the Table 10.2 recommendations fail, polymer selection criteria should be discussed with the polymer manufacturers. Equally, as for the measures converters can take, most polymer architecture remedies that apply to surface distortions will not work for volume distortions.

Balancing product performance with processing is a complex matter governed mainly by rheology. Lower MM polymers, branched polymers, or polymers with a very broad MMD are potential options that lower the overall processing viscosity and associated stresses. The matter becomes quickly more intricate for optimized converter processes. The "hand-waving" recommendations need to be quantified and tailored to the processing and performance requirements. As a consequence, polymer architectures may become highly differentiated plastics only to be processed melt fracture free within certain operating windows. A close collaboration between converter, processing equipment producer, and polymer manufacturer is desired but often constrained because of confidentiality issues.

For those desiring the functionality or aesthetics of a distorted extrudate, all the above methods should be ignored and in fact reversed.

10.4 GENERAL OBSERVATION

From the previous chapters, methodologies can be developed to rationalize and understand the Edisonian efforts practiced for decades in industry. All workable methods that prevent or minimize the severity of melt fracture tend to reduce the (shear and extensional) stress exerted on the polymer during flow. Essentially, viscosity is lowered, slip is induced, vortices are avoided, or die gaps are widened.

An important distinction needs to be made between surface distortions and volume distortions. The originating mechanism and onset location is different. The additionally observed "spurt" phenomenon is a periodic cycling through surface and volume distortions associated with melt compressibility and die-land slip, and as such not an additional "new" melt fracture phenomenon. Furthermore, it is observed only on the odd occasion in industrial practice. Still missing to complete the understanding of polymer melt fracture, however, is the quantification of the preventive methods in terms of formulation makeup and polymer architecture. The additional variables of polymer-processing conditions and constraints complicate the effort. The development of a mathematical frame that captures existing knowledge may help to indicate avenues for further research and point to possible, more effective methods to prevent melt fracture. Many of the present solutions are mechanical design compromises. For converters, such a framework must have a continuum mechanical nature, that is, a simulation tool in terms of macroscopic variables: flow, temperature, pressure, and geometrical constraints. For the polymer manufacturers, it requires a molecular level understanding where the polymer architecture variables are defined in terms of rheological material functions. Ultimately, connecting the two would be the perfect match. Practically, this is challenging academically but also industrially as competition does not encourage the exchange of knowledge.

In conclusion and referring again to Noam Chomsky, the problem of polymer melt fracture has become one that has its solution, with plenty of insight, based on increased knowledge, and thus we have more than just a notion of what we have to look for.

REFERENCES

1. Dwight, D. W., Surface analysis and adhesive bonding. I. Fluoropolymers. *J. Colloid Interface Sci.*, **59**:447–455 (1977).
2. Rudin, A., A. T. Worm, and J. E. Blacklock, Fluorocarbon elastomer processing aid in film extrusion of linear low density polyethylenes. *J. Plast. Film Sheeting*, 1:189–204 (1985).
3. Foster, G. N. and R. B. Metzler. Olefin polymer compositions containing silicone additives and their use in the production of film material. US Patent 4535113 (1985).
4. Heckel Jr., T. A. and F. J. Rizzo, Melt extrudable composition of perfluorocarbon polymers. US Patent 4617351 (1986).
5. Rudin, A., A. T. Worm, and J. E. Blacklock, Fluorocarbon elastomer aids polyolefin extrusion. *Plast. Eng.*, **42**:63–66 (1986).
6. Nam, S., Mechanism of fluoroelastomer processing aid in extrusion of LLDPE. *Int. Polym. Process.*, **1**:98–101 (1987).
7. Chu, S. C., Extrudable blends of polyethylene with siloxanes and films thereof. European Patent Application EP 260004 A2 (1988).
8. Johnson, B. V. and J. M. Kunde, Polyolefin processing aid versus additive package? *Plast. Eng.*, **44**:43–46 (1988).
9. Duchesne, D. and B. V. Johnson, Extrudable thermoplastic hydrocarbon polymer composition and process aids. US Patent 4855360 (1989).
10. Leung, E., D. Goddard, and F. H. Ancker, Process for processing thermoplastic polymers. US Patent 4857593 (1989).
11. Sano, A., M. Yoshizumi, H. Hirata, K. Matsuura, and H. Suyama, Polyethylene-fluorinated polyolefin blends with good melt fluidity. European Patent Application EP 333478 A2 (1989).
12. Blong, T. J. and D. Duchesne, Antiblock/processing-aid interferences. *Plast. Compd.*, **13**:50–52, 56–57 (1990).
13. Logothetis, A. L. and C. W. Stewart, Melt processable tetrafluoroethylene copolymer compositions. European Patent Application EP 395895 A1 (1990).
14. Radosta, J. A., Talc anti-blocks for maximized LLDPE blown film performance. *J. Plast. Film Sheeting*, 7:181–189 (1991).

15. Chapman Jr., G. R. and D. E. Priester, Polymer compositions containing extrusion processing aids comprising a fluorocarbon elastomers and vinylidene fluoride polymers. PCT International Application WO 9105009 A1 (1991).

16. Chapman Jr., G. R., D. E. Priester, C. W. Stewart, and R. E. Tarney, Fluoropolymer process aids containing functional groups for polymers with poor melt processability. PCT International Application WO 9105013 A1 (1991).

17. Stewart, C. W., R. S. McMinn, and K. M. Stika, A model for predicting slip velocity during extrusion with fluoropolymer processing additives. In *SPE ANTEC*, Detroit, MI, 50, 1992, pp. 1411–1414.

18. Parker, H. Y. and D. L. Dunkelberger, Processing aids for poly(vinyl chloride). *J. Vinyl Technol.*, **15**:62–68 (1993).

19. Stewart, C. W., A model for predicting slip velocity during extrusion with fluoropolymer processing additives. E.I du Pont de Nemours, Private communication (1993).

20. Stewart, C. W., Wall slip in the extrusion of linear polyolefins. *J. Rheol.*, **37**:499–513 (1993).

21. Hatzikiriakos, S. G. and J. M. Dealy, Effects of interfacial conditions on wall slip and sharkskin melt fracture of HDPE. *Int. Polym. Process.*, **8**:36–43 (1993).

22. Hatzikiriakos, S. G., The onset of wall slip and sharkskin melt fracture in capillary flow. *Polym. Eng. Sci.*, **34**:1441–1449 (1994).

23. Amos, S. E., Effects of processing aids on new generation Ziegler-Natta catalyzed LLDPE blown film extrusion. In *SPE ANTEC*, San Francisco, CA, 52, 1994, pp. 2789–2793.

24. Rosenbaum, E. E., S. G. Hatzikiriakos, and C. W. Stewart, Flow implications in the processing of tetrafluoroethylene/hexafluoropropylene copolymers. *Int. Polym. Process.*, **10**:204–212 (1995).

25. Hatzikiriakos, S. G., P. Hong, W. Ho, and C. W. Stewart, The effect of Teflon coatings in polyethylene capillary extrusion. *J. Appl. Polym. Sci.*, **55**:595–603 (1995).

26. Driver, F., A. P. Plochocki, and S. I. Kogan, Extrusion of fluoropolymers: Effect of the polymer/metal interface in the die on onset of the unstable melt flow. *SPE ANTEC*, Boston, MA, 53, 1995, pp. 47–50.

27. Cardinal, J.-C. and D. Priester, Broad spectrum processing aids based on multi-functional segments. In *AddCon 95: Worldwide Additives and Polymer Modifiers Conference*, Le Grand-Saconnex, Switzerland, 1995, Paper 21.

28. Kazatchkov, I. B., S. G. Hatzikiriakos, and C. W. Stewart, Extrudate distortion in the capillary/slit extrusion of a molten polypropylene. *Polym. Eng. Sci.*, **35**:1864–1871 (1995).

29. Rosenbaum, E. E., S. G. Hatzikiriakos, and C. W. Stewart, The melt fracture behavior of Teflon resins in capillary extrusion. In *SPE ANTEC*, Boston, MA, 53, 1995, pp. 1111–1115.

30. Taylo, J. W. R., S. K. Goyal, N. D. J. Aubee, and N. K. K. Bohnet, Polyethylene with reduced melt fracture. US Patent 5459187 A (1995).

31. Chiu, R., J. W. Taylor, D. L. Cooke, S. K. Goyal, and R. E. Oswin, Extrudable olefin polymer composition with few melt defects and its use for blown film and coated wire and cable. US Patent 5550193 (1996).

32. Caronia, P. J. and S. F. Shurott, A process for extrusion. European Patent Application EP 743161 A2 (1996).

33. Xing, K. C. and H. P. Schreiber, Fluoropolymers and their effect on processing linear low density polyethylene. *Polym. Eng. Sci.*, **36**:387–393 (1996).

34. Priester, D. E., Increasing pelletizing extruder output with fluoropolymer alloys. In *SPE ANTEC*, Toronto, Ontario, Canada, 55, 1997, pp. 58–62.

35. Aten, R. M., C. W. Jones, and A. H. Olson, Tetrafluoroethylene copolymer. PCT International Application WO 9707147 A1 (1997).

36. Xing, K. C., W. Wang, and H. P. Schreiber, Extrusion of non-olefinic polymers with fluoroelastomer processing aid. In *SPE ANTEC*, Toronto, Ontario, Canada, 55, 1997, pp. 53–58.
37. Hauenstein, D. E., D. J. Cimbalik, and P. G. Pape, Evaluation of process aids for controlling surface roughness of extruded LLDPE. In *SPE ANTEC*, Toronto, Ontario, Canada, 55, 1997, pp. 3002–3010.
38. Rosenbaum, E. E., S. K. Randa, S. G. Hatzikiriakos, C. W. Stewart, D. L. Henry, and M. D. Buckmaster, A new processing additive eliminating surface and gross melt fracture in extrusion of polyolefins and fluoropolymers. In *SPE ANTEC*, Atlanta, GA, 56, 1998, p. 676.
39. Hatzikiriakos, S. G. and K. B. Migler (Eds.), *Polymer Processing Instabilities*. Marcel Dekker, New York, 2005.
40. Parker, H.-Y. and J. L. Allison, *Processing Aids*. John Wiley & Sons, New York, 1990.
41. Mascia, L., *The Role of Additives in Plastics*. Edward Arnold Co., London, U.K., 1974.
42. Kalyon, D. M. and M. Khemis, Effects of slip additives on the rheology of low density polyethylene resins and ultimate properties of blown films. In *SPE ANTEC*, New Orleans, LA, 42, 1984, pp. 136–140.
43. Chauffoureaux, J. C., C. Dehennau, and J. Van Rijckevorsel, Flow and thermal stability of PVC. *J. Rheol.* **23**(1):1–24 (1979).
44. Isignaro, C. B., Polymer processing aids eliminate melt fracture in LLDPE blown film. *Plast. Addit. Compd.*, **4**(1): 20–22 (2002).
45. Bigio, D., M. G. Meillon, S. B. Kharchenko, D. Morgan, H. Zhou, S. R. Oriani, C. W. Macosko, and K. B. Migler, Coating kinetics of fluoropolymer processing aids for sharkskin elimination—The role of droplet size. *J. Non-Newtonian Fluid Mech.*, **131**(1–3):22–31 (2005).
46. Kulikov, O., K. Hornung, and M. Wagner, Silanols cured by borates as lubricants in extrusion LLDPE. *Rheol. Acta*, **46**(5):741–754 (2007).
47. Hong, Y., J. J. Cooper-White, M. E. Mackay, C. J. Hawker, E. Malmström, and N. Rehnberg, A novel processing aid for polymer extrusion: Rheology and processing of polyethylene and hyperbranched polymer blends. *J. Rheol.*, **43**(3):781–793 (1999).
48. Wang, J., M. Kontopoulou, Z. Ye, R. Subramanian, and S. Zhu, Chain-topology-controlled hyperbranched polyethylene as effective polymer processing aid (PPA) for extrusion of metallocene linear-low-density-polyethylene. *J. Rheol.*, **52**(1):243–260 (2008).
49. Sinha, D., S. Kole, S. Banerjee, and C. K. Das, Dependence of flow behaviour on carbon black distribution in polyblend systems. *Rheol. Acta*, **25**:507–512 (1986).
50. Lee, Y. S. and O. O. Park, The effect of asymmetries of die exit geometry on extrudate swell and melt fracture. *Korean J. Chem. Eng.*, **11**:1 (1994).
51. Kim, S., A criterion for gross melt fracture and its relation with molecular structure. PhD thesis, McGill University, Montreal, Canada, 2000.
52. Mongruel, A. and M. Cartault, Non-linear rheology of styrene butadiene rubber filled with carbon black or silica particles. *J. Rheol.*, **50**(2):115–135 (2006).
53. Ma, C.-Y., J. L. White, F. C. Weissert, and K. Min, Flow patterns in carbon black filled polyethylene at the entrance to the die. *J. Non-Newtonian Fluid Mech.*, **17**:275–287 (1985).
54. Wang, S. Q., P. A. Drda, and A. Baugher, Molecular mechanisms for polymer extrusion instabilities: Interfacial origins. In *SPE ANTEC*, Toronto, Ontario, Canada, 55, 1997, p. 1006.
55. Wang, S. Q. and P. Drda, Molecular instabilities in capillary flow of polymer melts. Interfacial stick-slip transition, wall slip, and extrudate distortion. *Macromol. Chem. Phys.*, **198**:673–701 (1997).

56. Wang, X., B. Clement, P. J. Carreau, and P. G. Lafleur, HDPE and LLDPE extrudate roughness study by screening design. In *SPE ANTEC*, Detroit, MI, 50, 1992, pp. 1256–1258.

57. Wang, S. Q. and P. A. Drda, Superfluid-like stick-slip transition in capillary flow of linear polyethylene melts. 1. General features. *Macromolecules*, **29**:2627–2632 (1996).

58. Wang, S. Q., P. A. Drda, and Y. W. Inn, Exploring molecular origins of sharkskin, partial slip, and slope change in flow curves of linear low density polyethylene. *J. Rheol.*, **40**:875–898 (1996).

59. Wang, S. Q. and P. A. Drda, Stick-slip transition in capillary flow of polyethylene. 2. Molecular weight dependence and low temperature anomaly. *Macromolecules*, **29**:4115–4119 (1996).

60. Wang, S. Q. and P. A. Drda, Stick-slip transition in capillary flow of linear polyethylene: 3. Surface conditions. *Rheol. Acta.*, **36**:128–134 (1997).

61. Drda, P. P. and S.-Q. Wang, Stick-slip transition at polymer melt/solid interfaces. *Phys. Rev. Lett.*, **75**:2698–2701 (1995).

62. Barone, J. R., N. Plucktaveesak, and S. Q. Wang, Interfacial molecular instability mechanism for sharkskin phenomenon in capillary extrusion of linear polyethylenes. *J. Rheol.*, **42**:813–832 (1998).

63. Yang, X., H. Ishida, and S. Q. Wang, Wall slip and absence of interfacial flow instabilities in capillary flow of various polymer melts. *J. Rheol.*, **42**:63–80 (1998).

64. Migler, K. B., C. Lavallée, M. P. Dillon, S. S. Woods, and C. L. Gettinger, Visualizing the elimination of sharkskin through fluoropolymer additives: Coating and polymer–polymer slippage. *J. Rheol.*, **45**(2):565–581 (2001).

65. Migler, K. B., Y. Son, F. Qiao, and K. Flynn, Extensional deformation, cohesive failure, and boundary conditions during sharkskin melt fracture. *J. Rheol.*, **46**(2):383–400 (2002).

66. Kazatchkov, I. B., F. Yip, and S. G. Hatzikiriakos, The effect of boron nitride on the rheology and processing of polyolefins. *Rheol. Acta*, **39**:583–594 (2000).

67. Lee, S. M., G. J. Nam, and J. W. Lee, The effect of boron nitride particles and hot-pressed boron nitride die on the capillary melt flow processing of polyethylene. *Adv. Polym. Technol.*, **22**(4):343–354 (2003).

68. Hatzikiriakos, S. G., N. Rathod, and E. B. Muliawan, The effect of nanoclays on the processability of polyolefins. *Polym. Eng. Sci.*, **45**(8):1098–1107 (2005).

69. Liu, X. and H. Li, Effect of diatomite/polyethylene glycol binary processing aid on the melt fracture and the rheology of polyethylenes. *Polym. Eng. Sci.*, **45**(7):898–903 (2005).

70. Malmström, E., M. Johansson, and A. Hult, The effect of terminal alkyl chains on hyperbranched polyesters based on 2,2-bis~hydroxymethyl-propionic acid. *Macromol. Chem. Phys.*, **197**:3199–3207 (1996).

71. Ye, Z. and S. Zhu, Newtonian flow behavior of hyperbranched high-molecular-weight polyethylenes produced with a Pd-diimine catalyst and its dependence on chain topology. *Macromolecules*, **36**:2194–2197 (2003).

72. Ye, Z., F. Alobaidi, and S. Zhu, Melt rheological properties of branched polyethylenes produced with Pd- and Ni-diimine catalysts. *Macromol. Chem. Phys.*, **205**:897–906 (2004).

73. Kurtz, S. and H. G. Apgar Jr, Linear low density ethylene hydrocarbon copolymer containing composition for extrusion coating. US Patent 4339507, Union Carbide Corporation, 1982.

74. Herranen, M. and A. Savolainen, Correlation between melt fracture and ultrasonic velocity. *Rheol. Acta.*, **23**:461–464 (1984).

75. Perez, R., E. Rojo, M. Fernandez, V. Leal, P. Lafuente, and A. Santamaría, Basic and applied rheology of m-LLDPE/LDPE blends: Miscibility and processing features. *Polymer*, **46**:8045–8053 (2005).

76. Delgadillo-Velazquez, O. and S. G. Hatzikiriakos, Processability of LLDPE/LDPE blends: Capillary extrusion studies. *Polym. Eng. Sci.*, **47**(9):1317–1326 (2007).
77. Den Doelder, J., R. Koopmans, M. Dees, and M. Mangnus, Pressure oscillations and periodic extrudate distortions of long-chain branched polyolefins. *J. Rheol.*, **49**(1):113–126 (2005).
78. Laun, H. M., Squeezing flow rheometry to determine viscosity, wall slip and yield stresses of polymer melts. In *Polymer Processing Society, PPS12*, Sorrento, Italy, 1996, pp. 31–33.
79. Blyler, L. L. J. and A. C. J. Hart, Capillary flow instability of ethylene polymer melts. *Polym. Eng. Sci.*, **10**:193–203 (1970).
80. Dospisil, D., J. Kubat, P. Saha, J. Trlica, and J. Becker, Melt flow instabilities of filled HDPE. *Int. Polym. Proc.*, **13**:91–98 (1998).
81. Kleinecke, K.-D., Zum Einfluss von Füllstoffen auf das rheologische Verhalten von hochmolekularen Polyethylensmeltzen. I Das Fliesverhalten in Sher- und Dehnströmung. *Rheol. Acta*, **27**:150–161 (1988).
82. Kleinecke, K.-D., Zum Einfluss von Füllstoffen auf das rheologische Verhalten von hochmolekularen Polyethylensmeltzen. I Untersuchungen in der Einlaufströmung. *Rheol. Acta*, **27**:162–171 (1988).
83. Gander, J. D. and A. J. Giacomin, Review of die lip buildup in plastics extrusion. *Polym. Eng. Sci.*, **37**:1113–1126 (1997).
84. Ding, F. and A. J. Giacomin, Die lines in plastics extrusion: Film blowing experiments and numerical simulations. *Polym. Eng. Sci.*, **44**(10):1811–1827 (2004).
85. Venet, C., Propriétés d'écoulement et défauts de surface de résins polyéthylènes. PhD thesis, Ecole des Mines de Paris, Sophia Antipolis, France, 1996.
86. Cogswell, F. N., Stretching flow instabilities at the exits of extrusion dies. *J. Non-Newtonian Fluid Mech.*, **2**:37–47 (1977).
87. Perez-Gonzalez, J., L. de Vargas, V. Pavlinek, B. Hausnerova, and P. Saha, Temperature-dependent instabilities in the capillary flow of a metallocene linear low-density polyethylene melt. *J. Rheol.*, **44**(3):441–451 (2000).
88. Santamaria, A., M. Fernandez, E. Sanz, P. Lafuente, and A. Munoz-Escalona, Postponing sharkskin of metallocene polyethylenes at low temperatures: The effect of molecular parameters. *Polymer*, **44**:2473–2480 (2003).
89. Miller, E. and J. P. Rothstein, Control of the sharkskin instability in the extrusion of polymer melts using induces temperature gradients. *Rheol. Acta*, **44**:160–173 (2004).
90. Miller, E., S. J. Lee, and J. P. Rothstein, The effect of temperature gradients on sharkskin surface instability in polymer extrusion through a slit die. *Rheol. Acta*, **45**(6):943–950 (2006).
91. Ramamurthy, A. V., Wall slip in viscous fluids and influence of materials of construction. *J. Rheol.*, **30**:337–357 (1986).
92. Shaw, M. T., Detection of multiple flow regimes in capillary flow at low shear rates. *J. Rheol.*, **51**(6):1303–1318 (2004).
93. Dao, T. T. and L. A. Archer, Stick-slip dynamics of entangled polymer liquids. *Langmuir*, **18**:2616–2624 (2002).
94. Kulikov, O. and K. Hornung, A simple way to suppress surface defects in the processing of polyethylene. *J. Non-Newtonian Fluid Mech.*, **124**:103–114 (2004).
95. Arda, D. R. and M. R. Mackley, Sharkskin instabilities and the effect of slip from gas assisted extrusion. *Rheol. Acta*, **44**(4):352–359 (2005).
96. Brauns, Ya. A., Strength of viscoelastic liquid in quasi static three-dimensional tension, *Mech. Compos. Mater.*, **14**(3):434–439 (1978).
97. Chen, J., Y. Chen, H. Li, S.-Y. Lai, and J. Jow, Physical and chemical effects of ultrasound vibration on polymer melt in extrusion. *Ultrason. Sonochem.*, **17**(1):66–71 (2009).

98. Arda, D. R. and M. R. Mackley, The effect of die exit curvature, die surface roughness and a fluoropolymer additive on sharkskin extrusion instabilities in polyethylene extrusion. *J. Non-Newtonian Fluid Mech.*, **126**(1):47–61 (2005).

99. Combeaud, C., Y. Demay, and B. Vergnes, Experimental study of the volume defects in polystyrene extrusion. *J. Non-Newtonian Fluid Mech.*, **121**(2–3):175–185 (2004).

100. Pol, H. V., Y. M. Joshi, P. S. Tapadia, A. K. Lele, and R. A. Mashelkar, A geometrical solution to the sharkskin instability. *Ind. Eng. Chem. Res.*, **46**(10):3048–3056 (2007).

Index